sweet stuff

sweet stuff

An American History of Sweeteners
from Sugar to Sucralose

Deborah Jean Warner

Published in Cooperation with
ROWMAN & LITTLEFIELD PUBLISHERS, INC.

Smithsonian Institution
Scholarly Press

WASHINGTON, D.C.
2011

Published by SMITHSONIAN INSTITUTION SCHOLARLY PRESS

P.O. Box 37012, MRC 957
Washington, D.C. 20013-7012
www.scholarlypress.si.edu

In cooperation with
ROWMAN & LITTLEFIELD PUBLISHERS, INC.

Published in the United States of America
by Rowman & Littlefield Publishers, Inc.
A wholly owned subsidiary of The Rowman & Littlefield Publishing Group, Inc.
4501 Forbes Boulevard, Suite 200, Lanham, Maryland 20706
www.rowmanlittlefield.com

Estover Road
Plymouth PL6 7PY
United Kingdom

British Library Cataloguing in Publication Information Available

Library of Congress Cataloging-in-Publication Data:
Warner, Deborah Jean.
    Sweet stuff : an American history of sweeteners from sugar to sucralose /
Deborah Jean Warner.
        p. cm. — (A Smithsonian contribution to knowledge)
    Includes bibliographical references and index.
    ISBN 978-1-935623-05-2 (cloth : alk. paper)
    1. Sweeteners—United States—History. I. Title.
    TP421.W37 2011
    664'.10973—dc22   2010049492

Printed in the United States of America

⊗™ The paper used in this publication meets the minimum requirements of
American National Standard for Information Sciences—Permanence of Paper for
Printed Library Materials, ANSI/NISO Z39.48-1992.

CONTENTS

Acknowledgments		vii
Introduction		1
1.	Sugar Refining in New York City	5
2.	Molasses	31
3.	Cane Sugar in Louisiana	47
4.	Cane Sugar in Florida	67
5.	Beet Sugar: Profitable and Patriotic	85
6.	Corn, Chemistry, and Capitalism	109
7.	Cane Syrup and Corn Syrup	123
8.	Dextrose, High-Fructose Corn Syrup, and Specialty Sugars	133
9.	The Sorghum Rage of the Gilded Age	145
10.	Maple Sugar and Syrup	159
11.	Honey	169
12.	Saccharin	181
13.	Cyclamates	195
14.	Aspartame and Sucralose	209
Notes		215
Selected Bibliography		275
Index		277
About the Author		291

ACKNOWLEDGMENTS

I begin by thanking the many librarians and archivists who have made this research pleasant as well as possible. Within the Smithsonian, these include Chris Cottrell, Jim Roan, Lou Rossignol, and Stephanie Thomas at the National Museum of American History Library; Michael Hardy at Smithsonian Libraries Research Annex; Kirsten Van der Veen and Lilla Vekerdy at the Dibner Library; Leslie Overstreet and Daria Wingreen-Mason at the Cullman Library; Erin Rushing in the Digital Services Division; and those who facilitated interlibrary loans. The National Agricultural Library in Beltsville, Maryland, is a welcoming and wonderful place to work, as are the American Philosophical Society Library in Philadelphia and the Library of Congress. I also thank Eugene Morris and his colleagues at the National Archives in College Park, Maryland; Sonya Rooney, who helped me work through the Monsanto papers at the Washington University Archives in Clayton, Missouri; Irene Coffey, who facilitated access to the Ware papers at the Franklin Institute in Philadelphia; Gabe Harrell of Special Collections at Louisiana State University Libraries in Baton Rouge; the archivists who made available the Charles F. Chandler papers at the Rare Book and Manuscript Library at Columbia University, New York; those who look after the Ira Remsen papers in the Special Collections, Milton S. Eisenhower Library, at Johns Hopkins University, Baltimore; and Beth Sherwood and Jennae Biddiscombe at the Louisiana State Museum in New Orleans. Thanks also to the Smithsonian Institution for its ambiance, support, and research-opportunity funds.

Among the many friends and colleagues at the National Museum of American History who have helped me understand the science and technology, as well as the social and cultural histories, of sweet stuff, I would especially acknowledge Dwight Bowers, Richard Doty, Paul Forman, Rayna Green, Jim Hughes, Paula Johnson, Kimberly Kelly, Peggy Kidwell, Stacey

Kluck, Bonnie Lilienfeld, Harry Rand, Harry Rubenstein, Fath Davis Ruffins, Ann Seeger, Roger Sherman, Kathy Sklar, Jeffrey Stine, Sherry Stout, Steve Velasquez, and Diane Wendt. Included here are such Smithsonian fellows and research associates as Benjamin Cohen, Marcel LaFollette, and Emily Pawly. Informants in the field include Marc Aronson; Luis Avila at Columbia University; Bruce Boynton of the National Honey Board; Shirley Cherkasky and other members of the Culinary Historians of Washington; Stephen Clarke at Florida Crystals; Rebecca Roddenbery Cline in Cairo, Georgia; Richard Feltoe at the Redpath Sugar Museum; James Johnson at the U.S. Beet Sugar Association; Suzanne Junod at the Food and Drug Administration History Office; Royce Lowery of Carson Ann Syrup; Gwendolyn Mayer at the Hudson Library and Historical Society in Hudson, Ohio; Erin McLeary; Ilene Miller of Miller's Honey Company; Quincy Montz in Reserve, Louisiana; Sandra Oliver of Food History News; William Outlaw at Florida State University at Tallahassee; William Patout of M. A. Patout & Son Ltd.; I. Barton Smith at the Bee Research Laboratory; Charlie Steen of Steen's Syrup; and Virginia Whitfield of Whitfield Foods Inc.

Finally, special thanks to Robert Friedel for urging me to focus on people, to Michael Schiffer for his critical read of each chapter, to Tom Foley for his advice on science and style, and to Jack Warner for his consistent support throughout this project and so much more.

INTRODUCTION

B efore the introduction of sugar, sweet stuff was so scarce that most people ate none at all, except perhaps in an occasional piece of fruit or spoonful of honey. Sugar came to Europe at the time of the Crusades, but, as it was expensive, only the most privileged members of society enjoyed it. Although European plantations in Brazil and the West Indies increased the availability of sugar, Robert Boyle, a wealthy and well-informed Anglo-Irish man of science, could note in the early 1660s that sugar was "(at least in these Western Regions) an almost recent discovery," adding that it preserved meats and rendered other foods "exceeding grateful to the Taste."[1] England was soon the largest sugar-consuming country in the world, with annual per-capita consumption reaching four pounds in 1704, eighteen pounds in 1800, and ninety pounds in 1900. Consumption in the United States lagged, but not by much. The average American today consumes some 150 pounds of sugars each year, plus substantial amounts of artificial sweeteners. *Sweet Stuff* tells the story of how this change occurred and how sweeteners have affected key aspects of the American experience—including diet and economy, science and technology, labor and capital, and politics and popular culture.

Honey and maple sugar were the primary sweeteners in the early English colonies in North America, but by developing a robust trade with the West Indies, the northern colonists gained access to sugar and its related products, molasses and rum. England's imposition of heavy taxes on this sweet stuff, especially its insistence on collecting these taxes, led the colonists to consider the advantages of independence. Following the American Revolution, Congress recognized that a sugar tariff was the most efficient, and perhaps the most equitable, way to raise the monies needed to run the government. Thus, until the establishment of the income tax in the early

1

twentieth century, the sugar tariff constituted the single largest source of federal funds.

The Louisiana Purchase brought sugar-cane lands into the nation, as did the subsequent acquisitions of Florida, Texas, Hawaii, and Puerto Rico. By the 1850s, the *New York Times* was describing sugar as one of the "first necessities" of the American people, second only to salt. As production and consumption increased, sugar would be seen as a good test of the country's "civilization and cultured taste" and a "chief test of the nation's prosperity." By the turn of the century, as the United States was taking in about one-fourth the world's production, a senator from Louisiana boasted that Americans had become "the greatest sugar consumers on earth."[2]

Gross statistics mask the ways and extent to which sugar consumption correlated with class. Wealthy families in colonial America tended to favor refined white sugar, while most Americans made do with molasses and raw sugar (often known as muscovado). While people of means could afford well-rounded and nutritious diets, molasses was an essential part of the diet of many Americans—a situation that remains as true today as it was in the colonial period. As anthropologist Sidney Mintz has observed, as factories expanded, sweet stuff provided the inexpensive calories that enabled workers to remain away from home and on the job for many hours each day—and with these calories came such problems as pellagra, diabetes, and tooth decay.[3]

Like their counterparts abroad, colonial Americans used sugar to make palatable such newly available and bitter beverages as coffee, tea, and cocoa. As sugars became more readily available, jams, pastries, ice creams, candies, and other such treats became common. Commercially processed and packaged foods, widely available since the late nineteenth century, rely on sweet stuff to improve their color, texture, and shelf life. Thus, most of the sweet stuff in our diet today comes from foods that we might not regard as sweet.

As their sweet tooth lengthened, Americans began promoting sugar independence. A host of public-private partnerships brought sorghum from China and Africa to the United States in the 1850s, improved the plant, and developed more efficient means of expressing the juice and turning it into syrup. With help from the U.S. Department of Agriculture and local agricultural extension agents, beet sugar became a thriving industry in the latter years of the nineteenth century. We find similar patterns in the histories of cane sugar, cane syrup, maple, and honey.

Sugar production and refining were among the first important industries to benefit from scientific research and analysis. Scientists worked out the chemical composition of such sugars as sucrose, glucose, and fructose. With their polariscopes, they determined the saccharine strength (purity) of cane and beets grown under different conditions, as well as that of sugars on the market and in the customs house. Following the lead of European scientists who produced glucose from starch, American scientists succeeded in making it from corn, a plant that grows plentifully in North America. As the new industry began to flourish, scientists analyzed such products as syrup and candy to determine the extent to which they were adulterated with inexpensive glucose. Scientists would later learn how to turn ordinary corn syrup into high-fructose corn syrup. Sugar scientists also formed the nucleus of the community that coalesced into the American Chemical Society. Inventors and engineers were also important in this story, developing ever more efficient tools and machines for cutting cane, transporting molasses, producing sugar, and performing numerous other tasks.

Scientists also discovered the sweet taste of substances that were not sugars at all but would serve as artificial sweeteners. The first of these was saccharin, found by accident in 1878, which became commercially available in the 1890s. Because of its cost advantage, saccharin was widely used in soda pop and other prepared foods. Because it was nonnutritive, it was attractive to those seeking sweetness without the attendant calories. So, too, were cyclamate, aspartame, and sucralose, the major artificial sweeteners that came on the market in the late twentieth century.

Ferocious debates, both fair and foul, over the existence and importance of side effects attended the proliferation of new sweets. Some concerns were raised by dispassionate observers, but some reflected raw commercial interests. Clinical studies designed to resolve the issues convinced no one but true believers. Laboratory experiments tended to be equally inconclusive, except when the test animals were given massive doses of the suspect substances. The tariff provoked equally ferocious debates over the relative merits and political clout of farmers who grew the sweet plants, factory owners who produced sugars, sugar refiners, and consumers. They also raised concerns about foreign lands that produced sugar and purchased America's goods or otherwise contributed to America's national security.

From the point of view of labor, sugar presents an ugly picture. Cultivating cane and producing sugar have always been arduous and dangerous occupations, initially performed in the West Indies and Louisiana by slaves brought from Africa who labored under the lash of plantation owners and overseers who tended to be particularly vicious. Indeed, it was well known that sugar slaves had worse and shorter lives than slaves who worked with such crops as cotton, indigo, rice, and tobacco. After emancipation, many former slaves found that they could not survive except by working in cane fields and sugarhouses. Many progressive Americans saw beet sugar as a means of avoiding goods produced by slaves, but they tended not to notice the extent to which beet growers relied on immigrants who could not obtain safer or more lucrative occupations and often made ends meet by enlisting the work of their women and children. Refinery work was better than work in the fields, though fires were frequent, and here, too, workers tended to be immigrants recently arrived in the country.

From the point of view of capital, however, sweet stuff was a great investment. This is clearly evidenced by the great houses that planters built in the West Indies and along the rivers and bayous of Louisiana, the mansions that refiners built along the avenues of New York and other major cities, and the many great art collections now residing in public museums.

chapter 1

SUGAR REFINING IN
NEW YORK CITY

The cultivation of sugar cane, a tall grass known scientifically as *Saccharum officinarum*, began in New Guinea about nine thousand years ago and eventually spread to China and India. Persians were producing granulated sugar from the juice by the fifth century AD, Islam spread cane culture throughout the Mediterranean, and Columbus brought cane to the Caribbean on his second voyage in 1493. Europeans then established sugar plantations in Brazil and on islands in the West Indies, and as they came to appreciate the amount of labor involved in this enterprise, they brought African slaves to the Americas to do the work.

Since the sugar in cane degrades quickly after harvesting, cane cannot be shipped long distances. Planters therefore established mills on or near their plantations, where the juice was extracted from the cane and heated in a series of pans until it formed raw sugar and molasses (a thick, sweet liquid that would not crystallize). Then, because the tropical regions where cane grows well tend not to have much fuel, the raw sugar was sent to European cities to be refined.

Raw sugar was suitable for many purposes, but for delicate dishes it had to be clarified, or purged of its extraneous matter. Some cookbooks offered instructions for this task. The seventeenth-century *Booke of Sweetmeats*, of which Martha Washington had a manuscript copy, told the cook to add a beaten egg white to a pound of sugar dissolved in a pint of water. When the mixture came to a boil, the egg white would coagulate, capture some of the impurities in the sugar, and rise to the top. The resultant dross could then be removed with a spoon or by straining through a cloth.[1]

Commercial refiners (sometimes known as sugar bakers) followed a similar path but on a larger scale, and they sometimes used ox blood as the coagulant rather than egg whites. When the sugar was ready to crystallize, they put it into inverted conical molds with a hole at the bottom through which the molasses could drip out. To make "clayed" sugar, they covered the tops of the molds with wet clay, designed so that water dripping through the molds would carry the remaining impurities to the bottom. When sufficiently dry, the sugar was removed from the molds and wrapped in inexpensive blue paper, the color of which masked the sugar's yellow tint. Customers used iron nippers to break off small pieces as needed and tongs to lift lumps into their cups of tea, coffee, or chocolate (Figure 1).

We know from maps that several refineries operated in North America in the late seventeenth century, and newspaper advertisements indicate that several more were soon to follow. A prominent merchant named Nicholas Bayard announced in the *New York Gazette* for August 17, 1730, that he had erected a "House for Refining all sorts of Sugar and Sugar Candy, and [had] procured from Europe an experienced artist in that Mystery. At which Refining House all persons in City and Country may be supplied by Wholesale and Re-tail with both double and single Refined Loaf Sugar, as also Powder and Shop-Sugars, and Sugar Candy at Reasonable Rates." As the name suggests, double-refined sugar was cleaner than that which had been refined only once. Some refiners also offered brown and muscovado

Figure 1: Ad for George Webster, grocer, at the Sign of the Three Sugar Loaves, from *New York Journal; or, The General Advertiser* (November 19, 1772), 786. Courtesy Library of Congress.

(another name for raw) sugar. Powdered sugar was refined sugar that had been pounded in a mortar or crushed on a board.[2] A few years later, when American refiners were cutting into the profits of refiners in the mother country, there was said to be interest in bringing to Parliament a bill to prohibit the distilling of rum and the refining of sugar in the British colonies of North America.[3]

By 1856, when ten refineries were operating in New York with several others under construction, the *New York Times* reported that the city ranked foremost among the sugar markets of the world and that the importation of foreign refined sugars had "almost entirely ceased." The *Journal of Commerce* reported that the city was "nearly encircled by enormous refining establishments, easily recognized by their lofty walls and chimneys." Some cost $800,000 to build, and a few had machinery that could purify crude sugars "in the most expeditious and effectual manner." About $3 million were invested in the business, and the industry employed some eighty-eight hundred men. Sugar refining was soon the largest and most lucrative manufacturing business in New York, and New York was the dominant center of sugar refining in the country. Several factors contributed to this situation. Along with extensive transportation facilities, wrote one historian, there were New York's "long pre-eminence as the nation's leading port, its numerous credit institutions for financing large importations of raw sugar, the availability of inexpensive anthracite coal for powering its refineries," and its large pool of "relatively cheap unskilled labor."[4]

Havemeyers & Elder

The Roosevelts and other leading New York families were involved with sugar refining in the colonial period. After the American Revolution, as old money sought new ventures, immigrants took over the business. Among the first were William and Frederick Havemeyer, two brothers from Hanover who had worked in London before establishing a sugar bakery in lower Manhattan in 1807. Over the course of the ensuing hundred years, a succession of sons, sons-in-law, and grandsons joined the business and—by means of smart management, shrewd investments, new technologies, and naked aggression—came to control much of the sugar refining in the country.

The first refinery in Brooklyn, far from the crowded streets of lower Manhattan, was that of Havemeyers & Elder. Begun in 1856 by Frederick C. Havemeyer Jr. and George Elder, it flourished under the leadership of Havemeyer's sons. Henry Osborne Havemeyer worked in the office while still a boy and was soon known as "the shrewdest man in his line." He saw to the tasks of minimizing the cost of refining, maximizing the return on investment, and besting the competition. His more technically adept brother, Theodore Augustus Havemeyer, had visited English and German refineries and incorporated the best practices at the family's firm.[5] By 1869, the refinery was processing one hundred tons of sugar each day. By 1876, the refining time had been reduced to just twenty-four hours, and more than one thousand men on the payroll produced over one hundred million pounds of sugar annually.[6]

Havemeyers & Elder faced the waterfront and so could have proprietary docks onto which the raw sugar was easily unloaded. At the rear of the property, trains brought coal for the boilers and staves for barrels and took the refined sugar to customers in other cities, eventually all across the country. Indeed, Havemeyers & Elder convinced the Erie Railroad to establish a freight depot in Brooklyn so that, as the *New York Times* explained, this railroad would be able to "keep the immense sugar business of the country to itself" and New York refiners would have an increasingly strong position vis-à-vis refiners in Boston, Philadelphia, and Baltimore (Figure 2).[7]

Havemeyers & Elder adopted a host of sophisticated techniques. Steam powered the machinery and heated the vacuum pans that removed water from the sugar. The vacuum pan, patented in 1813 by English chemist Edward Charles Howard, operated on the principle that decreased pressure lowered the temperature at which water boils. It thus offered substantial savings on fuel and diminished the amount of sugar that was burned or split (*inverted* was the technical term) into glucose and fructose.[8]

Robert and Alexander Stuart, sons of a Scottish immigrant who settled in New York, patented a process for manufacturing candy using syrup or sugar "purified by steam" in 1833. Judges at a fair sponsored by the American Institute saw the Stuarts' sugar as "without exception the most excellent and finest . . . that ever came under our notice." The Stuarts' sugar was also said to have been prepared "without the use of blood, or any other disgusting material." That is, the Stuarts used bone black (or animal charcoal) filters to decolorize their sugar and give it a "very beautiful appearance."[9]

Havemeyers & Elder used these techniques as well. Theodore Havemeyer imported from Cuba a new process for revivifying bone black, the most expensive element of the business. He also devised a carriage to move heavy sugar molds from one place to another and obtained the rights to a

Figure 2: Lithograph of Havemeyers & Elder sugar refinery. Atlantic Publishing & Engraving Co., ca. 1870. Warshaw Collection of Business Americana—Sugar. Courtesy Archives Center, National Museum of American History, Smithsonian Institution.

patent for making sugar molds from paper or pasteboard rather than from heavy iron (Figure 3).[10]

By 1866, Havemeyers & Elder had thirty thousand sugar molds as well as seventeen centrifugal machines that dried the sugar by spinning it around, in essentially the same way that a modern clothes dryer does. A French inventor had devised a centrifugal machine for cloth in 1837, and patents for centrifugals suitable for sugar were issued in England, France, and the United States soon thereafter. Peter Möller, a young Hanoverian who worked in yet another New York sugarhouse, would later recall that the proprietors, Edward and George Woolsey, "spent a great deal of time and money" before getting a centrifugal machine that actually worked. Eventually, however, their machine reduced labor demands by about 30 percent and worked an "evolution in the trade."[11]

Owners of tropical mills who invested in these new technologies were soon producing a range of sugars suitable for direct consumption but that contained enough molasses to provide color and taste. These were offered under such names as "turbinado" (from the Spanish word for centrifugal) and "demerara" (this tended to be darker than turbinado, not having been washed with steam), "plantation white," and "plantation yellow." Although

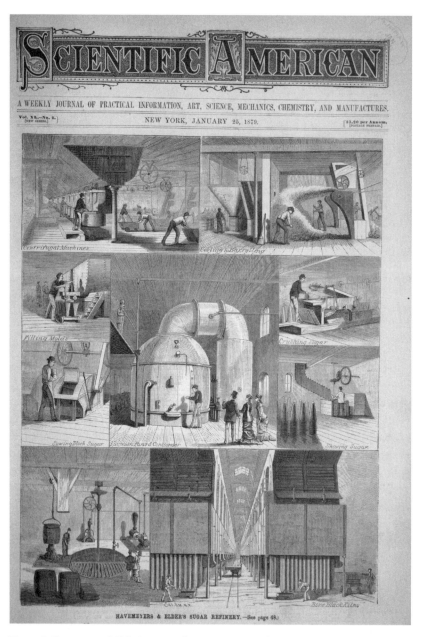

Figure 3: Havemeyers & Elder's sugar refinery. From *Scientific American* 40 (January 25, 1879), front cover. Courtesy Smithsonian Institution Libraries.

originally less expensive than the best refined white sugar, these direct-consumption sugars are sold today at premium prices.

Nineteenth-century advertisements mention a wide range of other sugars as well—including double-refined (this was available as loaf, crushed or ground), Circle A double-refined crushed, refined white, refined yellow, refined yellow-C, steam-refined white, and New Orleans sugar—but do not explain the meanings of these terms. And then there was the granulated sugar that we now take for granted. This became widely available in the latter half of the nineteenth century, thanks to the introduction of devices that dried and jiggled the "soft" sugar that came out of centrifugal machines. Charles Hersey was the most successful of the several inventors who tackled this problem, and the Hersey Manufacturing Company in South Boston produced many of the "granulators" used in sugar factories and refineries around the world.[12]

The Sugar Trust

Economic historians describe the American economy at midcentury as one of perfect competition, with many firms operating on a somewhat level playing field and no one firm controlling the market. As production began to exceed demand, however, comity gave way to cutthroat competition. Those with sufficient capital were able to reap the rewards of new technologies and increased economies of scale and to use expanding transportation and communications networks to reach customers across the nation and around the world. They could also form trade agreements with friends and force competitors from the field.[13]

Sugar refining fit this pattern to a tee. The refineries built in the 1850s and 1860s were soon producing more sugar than they could sell at a profit; thus, some were forced to close. By 1878, only about twenty-five refineries remained in the United States; fifteen were in New York.[14] Two years later, having recognized the extent to which unregulated competition was cutting into their profits, prominent refiners in New York, Boston, and Philadelphia agreed to curtail production until demand improved.[15] Yet, when fire demolished the Havemeyers & Elder refinery in 1882, it was quickly rebuilt and made into the largest and most efficient sugarhouse in the world.[16] In 1887, having realized that informal agreements would not suffice, the Havemeyers led the push to form the Sugar Refineries Company. With an initial capitalization of $50 million, this new firm acquired the assets of eighteen large refineries, closed some of them, and consolidated the work of others.

Having adopted the trust mechanism that John D. Rockefeller used so successfully to build his petroleum empire, the Sugar Refineries Company became known as the Sugar Trust. There were other trusts at the time, but the Sugar Trust was widely recognized as being among "the most successful from the point of view of growth, control, and profit to its members."[17]

In 1887, after New York courts decided that the Sugar Refineries Company was a combination of corporations and thus violated the state's corporate laws, the constituent firms transferred their allegiance to the American Sugar Refining Company (ASRC), a firm incorporated in New Jersey, which had recently passed a law allowing corporations to hold stock in other corporations. The Havemeyer brothers led the ASRC until their deaths, turning it into what is generally deemed one of the first large, publicly held U.S. industrial corporations.[18]

The ASRC kept improving its facilities, claiming each time to have the most modern refinery in the world. A refinery in Chalmette, Louisiana, across the river from New Orleans, opened in 1909. New refineries in Brooklyn and Baltimore would come on board after World War I, each project costing $8 to $10 million. The latter covered some fifteen acres, faced the waterfront, and could handle four oceangoing vessels at once. By means of the Pennsylvania and Baltimore & Ohio railroads, it reached customers throughout the middle Atlantic and central states, becoming the principal distribution point on the eastern seaboard for sugar for the Midwest.[19]

The Sugar Tariff

Despite widespread smuggling, taxes on sugar and molasses coming into the British colonies of North America provided a whopping 97 percent of the revenue collected by British agents in the years between the Sugar Act of 1764 and the outbreak of the Revolution. It is thus not surprising that Americans saw the tariff as the best means of raising the monies needed to retire the debt incurred during the war and run the new federal government. Following British practice, Americans taxed sugar on the basis of quality. The first tariff, enacted on July 4, 1789, assessed raw sugar at one cent per pound, refined sugar at three cents per pound, and grades in between at one and a half cents per pound. Given the market prices at that time, these few pennies amounted to a very large percentage ad valorem. Syrup was a nonnumerated item and so paid a mere 15 percent ad valorem. The specific rates changed from time to time, but not the principle ·of differentiation so favorable to those who imported and refined raw sugar.

All tariffs invite fraud, and this one was no exception. Problems arose as dealers and importers sought ways to lessen their tax burden. The first came to light in 1831 when the governor of Louisiana informed Secretary of the Treasury Louis McLane about a scam that was diminishing the federal revenue and harming the prosperity of Louisiana: imported barrels labeled molasses or syrup were often filled with cane juice that had been boiled almost to the point of granulation or with granulated sugar dissolved in warm water. Congress then sought a full accounting of frauds pertaining to the introduction of sugar in the form of syrup. When convinced that a problem really existed, it amended the tariff so that syrup paid the same rate as sugar of the same degree of refinement.[20]

McLane turned to Benjamin Silliman, professor of chemistry and natural philosophy at Yale College. Although Silliman was forced to report that science had not yet provided a solution to the problem at hand, this was one of the first important instances in which the federal government sought scientific advice.[21]

The Tariff of 1870, which taxed sugar at one to four cents per pound depending on grade, was applauded by those who processed raw sugar in their increasingly modern facilities but decried by those who imported relatively pure sugar from more progressive mills. Some even argued that the tariff discriminated unjustly in favor of low-grade sugars and that since its adoption, no "good sugars" had been imported into the United States. Moreover, low-grade sugars had been bought in the West Indies "at a higher price than higher grades, because the duties on the lower grade were less."[22]

By the 1880s, the tariff was seen as burdensome to consumers, especially to the poor people who relied on sugar for much of their caloric intake. With the federal budget running a surplus, there were calls for reform. The McKinley Tariff of 1890, although the most protectionist ever, eliminated the tax on sugar, both raw and refined, thereby removing a key advantage for American refiners. When Congress set out to reverse the major provisions of this tariff, the American Sugar Refining Company made substantial contributions to the campaign coffers of both political parties. Resisting these blandishments, the House of Representatives voted to keep sugar on the free list in 1894. The Senate then added an amendment favoring the refiners, and the final Wilson-Gorman bill stated that raw sugar would pay an ad valorem tax of 40 percent based on the purchase price at the country of origin, refined sugar would pay that amount plus one-eighth of a cent per pound, and refined sugar from countries such as Germany that gave an export bounty to their refiners would pay one-eighth of a cent per pound more. Grover Cleveland, who had campaigned against provisions of this sort, let the bill become law without his signature.[23]

The twentieth century would see numerous lengthy debates over tariffs and price supports. While the details of the debates are beyond the scope of this book, it is important to recognize how important these policies were for government and industry. In 1920, for instance, the American Sugar Refining Company claimed that it was the largest duty-paying corporation in the country, paying over $24 million in tariffs each year.[24]

Challenges to the Sugar Trust

The Arbuckle brothers of New York held the patents for machines that measured a pound of ground coffee, poured it into a paper container, and sealed the package. When packaged coffee proved profitable, the Arbuckles began thinking about packaging sugar. When the American Sugar Refining Company would not consider a discount on large quantities, the Arbuckles built their own refinery. John Arbuckle gained market share by selling directly to retailers, offering substantial discounts to wholesalers, and even lowering the price of sugar below the cost of production. Henry Havemeyer matched him step for step. Eventually recognizing the futility of the struggle, these two headstrong men agreed to live and let live. Fearing prosecution under the recently enacted Sherman Antitrust Act, however, they kept their agreement informal.[25]

In 1901, when Americans were producing beet sugar for between three and four cents per pound, the ASRC decided it was time to gain control over this nascent enterprise. Accordingly, it purchased substantial shares in large beet-sugar firms ranging from Michigan to California. One holdout was the American Beet Sugar Company, a mammoth firm that was dumping its excess sugar in communities along the Missouri River. The ASRC responded by cutting its prices in this area. The two sides eventually reached an agreement according to which the ASRC would acquire half of the common stock of the American Beet Sugar Company and receive three cents for every pound of beet sugar it sold.[26]

The ASRC and the Law

In 1892, after the ASRC arranged to purchase a controlling interest in four refineries in Philadelphia, the Department of Justice ordered the district attorney in Philadelphia to bring suit, charging that the ASRC and associated firms had "monopolized the manufacture of refined sugar, and also the interstate commerce therein, within the United States." The case, *United States v. E. C. Knight*, was quickly dismissed; the court found that the defendants may have monopolized sugar refining but not interstate commerce, and so they were not guilty under the provisions of the Sherman Antitrust Act. That is, the ASRC might be tried in state court, but the federal government had no standing in the case. The government lost again in the circuit court of appeals and then in the U.S. Supreme Court, which decided that Congress had a right to prohibit monopolies in commerce but not in manufacture.[27]

Theodore Roosevelt, who assumed the presidency in 1901, was determined to rein in the rogue acts of corporations. His chief agent in this campaign was Henry L. Stimson, a corporate lawyer whom he tapped to become a U.S. attorney for the Southern District of New York. One tactic Stimson would use was the Elkins Act, which criminalized railroad rebating. His first case involved 1.84 million pounds of sugar that the ASRC had sent by the New York Central Railroad to a wholesale grocer in Detroit, Michigan. After Stimson showed that the rebate enabled cane sugar refined in the East to compete with beet sugar in the Midwest, the judge found the railroad guilty on all counts and imposed a fine of just $18,000. A second case, this one filed against the ASRC, ended in the same way. Subsequent cases were dropped after the ASRC agreed to plead guilty to Stimson's remaining indictments and pay a relatively modest fine of $150,000.[28]

Another issue concerned the places where federal agents inspected samples of sugar coming into the country. Refiners in lower Manhattan had to send their samples to the Customs House located nearby, but refiners in Brooklyn, who had docks of their own, could have their sugar "landed and 'inspected' wholly in private"; thus, they had only to secure the friendship of a single inspector to get their sugar graded "as they please[d]." This problem received little attention until 1903, when Richard Parr, a sampler at the New York Custom House, shared his concerns with William Loeb, a former classmate then working in the White House. Loeb raised the matter with President Roosevelt and suggested that a treasury agent might get the evidence needed to prove the case. In 1907, now serving as a T-man in New York, Parr visited the Havemeyers & Elder docks in Brooklyn. There, as he would later tell a federal grand jury, he found seventeen large platform scales rigged so as to register false weights in favor of the company and was offered bribes to forget what he saw. Stimson and his assistant, the young Felix Frankfurter, easily obtained an indictment, and the ASRC agreed to repay the $134,000 it owed the government for the case at hand plus $2 million for duties unpaid during the previous dozen years.[29]

Stimson then brought a criminal case against the dock superintendent of the refinery and several company weighers. After a short but well-publicized trial, the men were found guilty and sentenced to a year in jail. Convinced that the men had not acted on their own, Stimson's successor sought indictments against the secretary of the ASRC, the former cashier, and the former superintendent of the Williamsburg refinery and several others in positions of authority. Faced with compelling evidence, often splashed on the front pages of newspapers, the jury decided to convict. Roosevelt later described this matter as "probably the most colossal fraud ever perpetrated in the Customs Service," and Parr received a reward of $100,000.[30]

William Loeb, who served as collector of the Port of New York during the William Howard Taft administration, found fraud to be much more rampant than he had imagined. Sugar samples at the ASRC docks in Brooklyn and Jersey City were routinely taken from the bottom of sacks where the molasses had settled, and low-grade samples were routinely substituted for higher-grade ones. Moreover, ASRC men doctored sugar so that it appeared to be of lower grade and bribed federal employees to get favorable tests. With this evidence in hand, Loeb set about cleaning house to an extent theretofore not seen in the federal civil service.[31]

Congressional action began when Thomas Hardwick, a Democrat from Georgia, urged the House of Representatives to investigate the ASRC in light of the Sherman Antitrust Act. After extensive hearings in Washington, DC, and New York, the Hardwick Committee concluded that, "from its organization, in 1891, to 1907 almost every step in the industrial development of the American Sugar Refining Company was part of a carefully laid campaign to obtain and maintain a monopoly in inter-State trade in sugar, to obtain complete control, and to forestall, destroy, and weaken competition." The committee went on to say that much of the success of the ASRC was due to the "genius" of Henry O. Havemeyer, and that since his death in 1907, the firm had begun to mend its ways.[32]

While the Hardwick Committee was conducting its investigation, the Justice Department was preparing a case against the American Sugar Refining Company and others. The pretrial hearings began in April 1912 and were ongoing when Woodrow Wilson was sworn in as president the following year. Wilson was not eager to pursue the matter, and Earl Babst, the president of the ASRC, helped the nation by alleviating sugar shortages during World War I. In 1922, when enthusiasm for trust busting was at low ebb and sugar was again readily available, the ASRC admitted that it had at one time violated the Sherman Antitrust Act but did so no longer. The several parties then signed a consent decree according to which the defendants were enjoined from "combining and conspiring among themselves to restrain interstate and foreign trade." There were no other penalties.[33]

Refinery Work

In 1872, the *New York Times* reported that sugar refineries demanded but a small amount of skilled labor, and "probably not ten per cent of all hands employed require any special knowledge." The situation was much the same in 1896 when the *Brooklyn Eagle* reported that few "Americans" sought work in the local sugar refineries, "perhaps" because the wages were not high and the work was hot, exhausting, and dangerous. Steam boilers and sugar dust tended to explode. Fires were frequent. Iron hooks and vats of boiling sugar were hazardous, and not only for the laborers. The young George Havemeyer was caught in the machinery and killed in his family's refinery in 1861.[34]

The sugar bakers of New York demanded $1.25 for a day's pay in 1853, but to no avail. By 1900, after a spate of generally futile strikes, the average rate had risen to $1.56.[35] Since refineries tended to run day and night, this was for a twelve-hour shift, a situation still common in the 1920s. When refineries were shut for cleaning or repairs, the workers were summarily dismissed. Thus, the coal shortages that forced Havemeyers & Elder to shut down in 1887 were said to have caused much suffering among their employees.[36]

Immigrants filled most of the unskilled refinery jobs: Irishmen and Germans predominated in the nineteenth century and Poles and Lithuanians in the early twentieth. Women joined the workforce as packers when refineries began offering sugar in one-, two-, and five-pound boxes and bags (Figure 4).

In conjunction with each refinery, a cooperage produced the barrels needed to transport the sugar, and to supply these, the ASRC owned vast amounts of timberland, several lumber camps and mills, and hundreds of miles of railroad.[37] The coopers were among Havemeyers & Elder's most highly skilled workers—at least, that was the case before 1873, when the firm fired several hundred unionized coopers and replaced them with barrel-making machines that could be tended by "nonsociety" men.[38] When a fire swept through a Brooklyn cooperage in 1900, the *New York Times* reported

Figure 4: "Filling and Sewing Bags of Granulated Sugar, New York." Probably at the ASRC refinery in Brooklyn. Keystone stereograph, 1916. Courtesy Prints & Photographs Division, Library of Congress.

that the Italian workmen in that shop "were known only by numbers, the same as the laborers in the Havemeyer sugar refinery."[39]

In 1922, when ASRC opened a refinery in Baltimore, it employed one thousand people. By 2001 the number had fallen to around five hundred. In the 1930s, the ASRC was producing sixteen million pounds of sugar a day, or between a quarter and a third of the country's total. But as the firm's five refineries had been largely mechanized, a mere five thousand men were on the payroll. This move toward automation would culminate in 1960 when the ASRC opened a new refinery in the Charlestown section of Boston. This facility would produce as much sugar as the old refinery in South Boston but employ half as many workers.[40]

A Sugar Refinery Workers Union was organized in the 1930s and affiliated with the International Longshoremen's Association. Striking in Philadelphia in 1938, the union asked that wages for unskilled labor be raised from sixty-three to sixty-five cents an hour. Management resisted, saying that southern sugar mills paying fifty cents an hour could undersell the northern refineries.[41] When a similar strike was called against refiners in Brooklyn and Yonkers in 1941, the ASRC offered to raise the hourly wage from seventy to seventy-five cents. A spokesman for the firm said this would be thirty-five to forty-five cents an hour above the wage paid in the South.[42]

Refinery workers shared in the general prosperity of the post–World War II period. That would change in 1988 when Tate & Lyle acquired the sugar departments of Amstar (see "The End of an Era" below), closed the refinery in Charlestown, and laid off more than two hundred people, most of them described as union workers who had been with the company for years and earned between $12 and $15 per hour with good benefits. A decade later, having found that it was no longer profitable to process raw sugar in Brooklyn, Tate & Lyle repurposed this venerable facility. Henceforth, the Brooklyn workers would simply take semiprocessed liquid sugar off a barge from the plant in Baltimore and crystallize it into granules. At the same time, Tate & Lyle announced it would eliminate sick days and most seniority rights, cut one hundred jobs, and contract out as many jobs as it pleased, all in violation of the union contract. Some 284 workers then took to the picket lines, only to cave in February 2001. The refinery closed in January 2004.[43]

Science and Sugar Refining

Sugar was among the first industries to rely on scientific expertise for quality control. The Stuarts, mentioned above, were said to have "expended time and money freely, employing experts of the highest rank, such as Professor [John] Torrey, a famous chemist in his day."[44] As refining became more competitive, reliance on science increased. Paul Casamajor, a Cuban-born Harvard graduate who had studied medicine and chemistry in Paris, landed a job with Havemeyers & Elder in 1867. At his death in 1887, he was said to be the leading sugar expert in the world, earning the phenomenal annual salary of $20,000.[45]

Charles F. Chandler, professor of chemistry and academic dean of the Columbia School of Mines, became a consultant to the New York Steam Sugar Refinery in 1868, spending two hours a day analyzing sugar and earning $1,500 a year.[46] As he came to appreciate the extent to which the sugar industry depended on science, Chandler established a program in sugar analysis at the School of Mines. Graduates found ready employment in refineries from New York to New Orleans. Some set up private laboratories for sugar analysis. Some wrote books on the subject.[47] Several were active members of the American Chemical Society—indeed, most of the founding members of the American Chemical Society were sugar chemists, and the early issues of the *Journal of the American Chemical Society* are replete with articles pertaining to sugar analysis. The Havemeyers provided funds for Havemeyer Hall, the chemistry building erected on Columbia's new Morningside Heights campus in 1897.[48]

The sugar beetle, a nasty creature that lived in raw sugar, came to attention at midcentury when Arthur Hill Hassall, a London physician with a microscope, found beetlelike "animalcules" in nineteen samples of raw sugar. Hassall's report was published by Parliament's Select Committee on Adulteration of Food, Drinks and Drugs and widely reported in the United States. So, too, was that of a Scotsman named Robert Nichol, who said that raw sugar contained immense numbers of disgusting-looking insects—up to one hundred thousand in every pound—and so should never be used for dietetic or domestic purposes. Citing Professor Cameron, a public analyst in Dublin, Nichol went on to say that the *Acarus sacchari* was "formidably organized, exceedingly lively, and decidedly ugly" and that it caused the disease popularly known as "grocer's itch." He published a lithograph of a much-enlarged acarus that had been found in Mauritius sugar and drawn from life by the eminent London microscopists Smith, Beck & Beck. As late as 1891, *Willett & Gray's Weekly Statistical Sugar Trade Journal* was saying week after week that the *Acari sacchari* "DO NOT occur in Refined Sugar of Any quality, because they cannot pass through the charcoal filters of the refinery, and because Refined Sugar does not contain any nitrogenous substances upon which they could feed" (Figure 5).[49]

But even if sugar was clean, it could still be adulterated. When some sugar on the market was said to be tainted with glucose (that is, corn sugar), Havemeyers & Elder denounced the allegation and provided funds so the New York Board of Health could investigate the matter. Pierre de Peyster Ricketts, a chemist who had earned his PhD working under Chandler's supervision, did the work and found the sugars in question to be "absolutely unadulterated."[50]

Finally, there was the problem of determining quality for commercial purposes. To handle this, wholesale buyers and sellers agreed in 1907 to form the New York Sugar Trade Laboratory, and because the stakes were so high, they took pains to attract a chemist who would be above reproach. Charles A. Browne, the laboratory's first director, had earned a PhD from the University of Göttingen and served as chief of the sugar laboratory at the U.S. Department of Agriculture (USDA). His *Handbook of Sugar Analysis* (1912) offered practical advice for use in research, technical, and control laboratories. In 1923, when Browne was named chief chemist at the USDA, the New York laboratory turned to Frederick W. Zerban (he had changed his name from Fritz around the time of World War I), a German-born chemist with a PhD from the University of Munich, who had spent many years analyzing sugar in Louisiana and South America.[51]

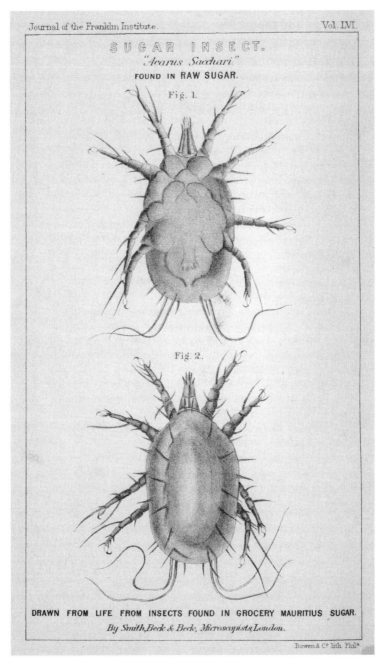

Figure 5: The sugar mite. From *Journal of the Franklin Institute* 56 (1868), plate 56. Courtesy Smithsonian Institution Libraries.

Sugar Analysis and the Tariff

The early American tariffs assumed that customs agents could easily determine the difference between one grade of sugar and another. But as modern machinery enabled cane-sugar makers in the tropics and beet-sugar makers in Europe to produce a greater range of sugars, the problem became much more difficult. Congress took note of this situation in 1861, writing a tariff that defined the distinctions among various grades of sugar in terms of the Dutch Color Standard. This was of a set of clear glass jars, each containing a different sample of sugar ranging from dark brown (No. 6) to almost white (No. 23), filled and sealed by an association of sugar brokers in Amsterdam and sanctioned by the Dutch government.

The Dutch Color Standard may have been inexpensive and easy to use, but it could not determine the saccharine content of a sugar sample. For that purpose one relied on a technique that began with Jean-Baptiste Biot, a French savant who found that when polarized light was sent through various substances, some of them would rotate the plane of polarization. Biot also found that solutions of cane sugar and beet sugar affected polarized light in the same way and that solutions of higher concentration rotated the light to a greater degree. Recognizing the practical import of his discovery, Biot issued a set of instructions for the medical, scientific, and industrial uses of the polariscope.[52] Jean Baptist François Soleil, a highly accomplished instrument maker in Paris, then developed the saccharimeter—that is, a polariscope intended specifically for sugar analysis.[53]

Early accounts of polariscopes and saccharimeters—the words were often used interchangeably—were wonderfully effusive. An 1848 report noting that Biot's beautiful discovery was being used in sugar refineries in Paris went on to say that the Philadelphia refinery of J. S. Lovering & Company had "also availed themselves of this process; and in their establishment may be seen practically illustrated, on a large scale, the advantages to be attained by the union of science and art." In 1852, a reporter saw a saccharimeter in the window of Benjamin Pike Jr., proprietor of the leading scientific instrument shop in New York, and explained that Soleil had transformed Biot's ideas, "so far rather barren in practical value," into "the most beautiful and certain means of determining the amount and richness" of sugar. In 1874, noting that a chemist who analyzed sugars in New York had acquired a new polariscope made to order in Germany that cost nearly $500, the *American Grocer* proclaimed, "O, glorious Science, beautiful handmaiden of Commerce and all Arts, we can never sufficiently honor thee for thy invaluable services." William Havemeyer would later give two "valuable polariscopes" to the chemistry department of New York University.[54]

In 1876, concerned that high-quality sugar was coming into the country without payment of the requisite tax, the secretary of the Treasury sent a sample of questionable sugar to the National Academy of Sciences and was informed that it was dark in color but of high saccharine content. John Sherman, who succeeded to the Treasury when Rutherford B. Hayes became president later that year, authorized customs agents to impound any sugar that appeared to have been colored and not identified as such. The test came when agents seized 712 bags of sugar that arrived at the Port of Baltimore from the Demerara region of British Guiana. The ensuing trial provided compelling evidence that color did not necessarily correlate with quality. Indeed, the Demerara planters argued that whatever color was in the sugar belonged to it. Moreover, the color was not taken out "as it would be in the making of sugar for the English market, or for the market of any country where there was a rational tariff. So long as the United States Government choose to maintain a sliding scale of sugar duties, and adhere to the preposterous system of a colour test, by which the darker the colour of the sugar, no matter how rich in saccharine matter it may be, the lower the duty—so long will Demerara sugar planters be entitled to take advantage of their absurd regulations." Subsequent investigations showed that the Demerara planters were telling the truth. By using bichromate of lime rather than animal charcoal, they could make sugar of middling color but exceptionally high purity, and they did so "under the positive instructions" of the American importers.

As the extent of this problem became apparent, Secretary Sherman informed customs agents that they should assess taxes on the basis of crystallizable sugar rather than color. That is, while the tariff act specified the Dutch Color Standard, agents should rely on polariscopic analysis. Accordingly, the Treasury purchased expensive Soleil-Scheibler polariscopes made in Berlin for the customs house laboratories in New York, Boston, Philadelphia, Baltimore, New Orleans, St. Louis, and San Francisco.[55] The men who imported low-grade sugar were pleased with this decision. Those who imported more refined sugar were not. Some of them sued the collector of customs in New York, alleging that polariscopic analysis had caused them to be overtaxed on a cargo of sugar brought from Cuba. The circuit court for the Southern District of New York found that the Treasury was at fault because Congress had not authorized chemical or optical tests. When the Supreme Court upheld this decision, the Treasury was forced to refund some $2 million in sugar duties and instruct the customs houses to "collect duties on apparent color of sugar as imported." The Ways and Means Committee of the House of Representatives had been considering polariscopes for several years, but this Court decision forced its hand. Thus, the tariff that went into effect in May 1883 stated clearly that the polariscopic test should be the basis for assessing customs duties on all sugars below No. 13 Dutch in color, and the Dutch Color Standard should be used for all higher-grade sugars. The revision of 1894 raised the bar between raw and refined to No. 16 Dutch. Neither of these decisions, however, stopped the complaints about the reliability of the polariscopes and the methods of testing. William McMurtrie, president of the American Chemical Society, told Congress in 1900 that if the instruments used in the customs houses and industry were carefully calibrated and tested and the methods standardized, "all of these disputes will be removed and a great deal of time and worry and expense would be obviated." Congress established the National Bureau of Standards the following year and charged the new agency with attending to this problem. In short order, through standardized instruments and test procedures, as well as routine comparisons of results obtained by different customs houses, the Bureau of Standards settled the issue.[56]

Sugar Brands

When the question of adulteration came up in the early 1880s, Havemeyers & Elder put a label on every package stating that the sugar inside was absolutely pure. But as the typical package at that time was a barrel or half-barrel from which grocers would scoop out the amount requested, the typical consumer probably paid little attention to the brand in question. That situation would change around the turn of the century as grocers and manufacturers alike began emphasizing the modernity of their operations. The Franklin Sugar Refinery of Philadelphia, a constituent member of the American Sugar Refining Company, offered granulated sugar in family-sized boxes. Advertisements announced, "The Barrel Man's the Only Loser. Ninety-nine out of one-hundred women would rather buy goods in packages than goods in bulk. Mainly because they think they're cleaner, and so they are."[57]

　　Havemeyers & Elder introduced two new brands around 1898: Eagle Brand granulated sugar and Crystal Domino Sugar, so named because it came in lumps shaped like dominoes. Sugar cubes were certainly not new. William Möller had patented a machine for cutting block sugar as early as 1861, and Havemeyers & Elder had been offering cubes at least since

1879. What was new was the name. So, too, was the national campaign—carried out in newspapers and the new mass-circulation magazines—that presented Domino Sugar as "A Triumph in Sugar Making!" "Every piece alike—and every piece sparkles like a cluster of diamonds, the result of its perfect crystallization." It came in five-pound boxes "just like fancy candy." Because these boxes were "packed at the refinery and opened in the household," there was no intermediate handling, hence "no dirt, no waste, no possible adulteration." The ASRC went a step further in 1911, boasting that Professor L. B. Allyn, a noted member of the pure-food crusade, had set his seal of approval on Crystal Domino Sugar and other products of the firm.[58]

Promotion of Domino sugars was a prime concern of Earl Babst, a lawyer who served the ASRC for forty years, first as president and later as chairman of the board. When he joined the firm in 1915, Babst was known as "the most successful merchandiser among the food product industrial leaders of the United States" and a "master of the art of creating a demand for packaged food products."[59] In 1919, when the ASRC was offering individual sugar cubes wrapped in paper for high-class restaurants, the *Wall Street Journal* reported that sales were slight but the idea helped the public associate the Domino brand with "articles of superior merit." By 1920, one could purchase Domino granulated sugar, tablet sugar, powdered sugar, confectioners' sugar, brown sugar, and golden syrup. Domino Superfine Table Sugar, so fine it dissolved completely in iced tea or frosted coffee, would follow in 1929. Domino Brownulated Sugar—brown sugar that "Pours and Stores like Granulated"—made its appearance in 1964.[60]

The Sugar Institute and the Sugar Research Foundation

Sugar production plummeted during World War I but recovered quickly, and soon the market was glutted. Obviously concerned about this matter, the executives of the leading American sugar companies came together in December 1927 to consider ways of stabilizing the industry. Under the leadership of Earl Babst of the ASRC, these men formed an organization that would compile statistics, examine tariffs, and collect other information of interest to the sugar industry. By January 1928, the Sugar Institute had obtained approval from the federal government and organized under the laws of the state of New York.[61]

The first order of business was expanding consumption. Americans were then eating 119 pounds of sugar each year—as well as substantial amounts of other sweeteners—and if each would ingest an additional twenty pounds, consumption might catch up with production. To make this happen, the Sugar Institute flooded newspapers and magazines with promotional "articles." Some claimed that sugar improved the taste of nutritious foods and so made them more palatable to picky eaters. Some quoted doctors who said that sugar was "okay" for teeth, and sweet snacks would prevent afternoon fatigue. Some quoted an unnamed international authority to the effect that "rounded slimness" was preferable to the "flat, skinny figure" of the flapper era. As the Great Depression deepened, Americans were told that sugar could be used to can fruits and vegetables and improve the taste of cheap cuts of meat. Seeking new industrial uses for sugar, the Sugar Institute gave money to the Pittsburgh-based Mellon Institute for Industrial Research to see if sugar could be used in wood preservatives, textile finishing, and the manufacture of adhesives.[62]

Despite the Sugar Institute's repeated claims that it did not engage in price fixing, the Department of Justice brought suit in 1931, arguing that many of the institute's practices violated the Sherman Antitrust Act. The trial was one of the longest and most expensive on record—the government had two lawyers and a 520-page brief, while the defense had fifty lawyers and a one-thousand-page brief—and by some accounts, it was the most important antitrust case since the Standard Oil case of 1911. In the end, a federal judge found the Sugar Institute guilty of restraint of trade. He did not order the dissolution of the institute but did impose about forty-five restrictions on its activities. After a unanimous Supreme Court upheld this decision, the Sugar Institute disbanded. Based on an analysis of the private notes kept by a major sugar refiner, economic historians have argued that the Sugar Institute did not fix prices directly. Rather, it enforced a set of rules that increased the probability of secret price cuts and thus facilitated collusion within the industry.[63]

In 1943, even as sugar rationing was in place (Figure 6), the producers and processors of cane and beet sugar formed the Sugar Research Foundation, a nonprofit organization designed to help stabilize demand in the postwar period. Under the leadership of Joseph F. Abbott, head of the ASRC, the Sugar Research Foundation gave $125,000 to the Massachusetts Institute of Technology for research in such areas as the role of sugar in human metabolism and new uses for sugar and related carbohydrates. The funds could also be used to train young scientists in the field of carbohydrate chemistry and to prepare them for service in the industry. The foundation also gave money to the National Academy of Sciences, and its

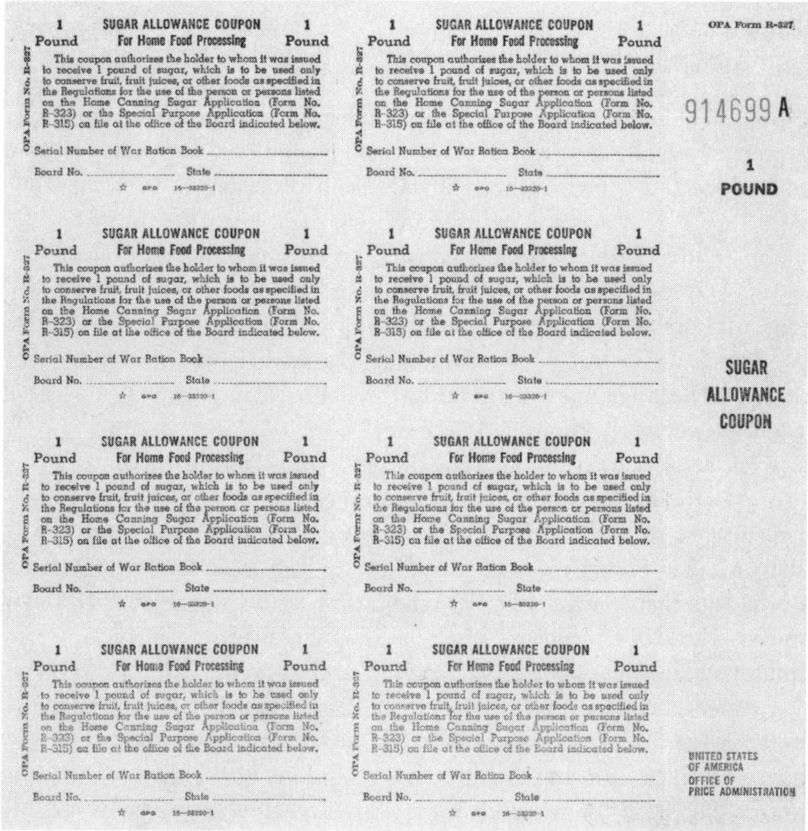

Figure 6: Recognizing that wartime activities would lessen the availability of sugar, the federal Office of Price Administration instituted rationing in May 1942 and distributed ration-allowance coupons to Americans across the country. Most families were allowed half a pound of sugar per person per week, or about half of what they were using before the war, but those who processed fruits and other perishable foods at home could have more. Courtesy Political Sciences Collection, National Museum of American History, Smithsonian Institution.

National Science Fund awarded substantial prizes to stimulate research into the use of sugar. The foundation became the Sugar Association in 1947. One arm, Sugar Information Inc., aimed to convince American consumers that sugar provided ready energy and contained only eighteen calories per teaspoonful. The other, the Sugar Research Foundation, sponsored scientists looking into the possibly harmful effects of saccharin, cyclamate, and other nonnutritive sweeteners.[64]

Cuba

American sugar traditionally came from Barbados, Jamaica, and other European colonies in the West Indies. In the mid-nineteenth century, after a political revolution destroyed the sugar plantations and mills on Saint Domingue and after British sugar interests had decamped to India, Cuba became the primary source of American sugar, and the United States became Cuba's primary trading partner. Americans purchased sugar plantations and mills on the island, worked in Cuba's sugar and ancillary industries, and provided many of the tools and much of the machinery used to produce Cuban sugar. Most of the machetes and knives wielded by cane and sugar beet workers were made by the Collins Company in Connecticut and were popularly known as "Collinses."

The reciprocity treaty signed with Hawaii in 1876 gave the United States access to the harbor at the mouth of the Pearl River and allowed Hawaiian sugar to enter the United States duty free, enabling Claus Spreckels, a refiner in San Francisco, to control the sugar market in the West. Refiners in the East then sought a similar relationship with Cuba. Henry O. Havemeyer, who owned land in Cuba as well as shares in several Cuban sugar mills, was quoted as saying, "Cuba's the island we want."[65]

American intervention in the Cuban Revolution of 1898, which formed part of what would be known as the Spanish-American War, paved the way for a special relationship that, in theory, would enhance the economies of both nations. Cuban president Tomás Estrada Palma favored a reciprocity treaty of some sort. So, too, did Theodore Roosevelt, whose Rough Riders had stormed San Juan Hill. American cane and sugar beet growers demurred, arguing that they would suffer grievously if Cuban sugar were to come into the country duty free. The ASRC tried to conceal its hand, but word leaked out that Havemeyer had paid a lobbyist $2,500 to promote duty-free raw sugar from Cuba. In 1903, after lengthy and bitter debates, after the United States had gained access to a naval and coaling station at Guantánamo Bay, and after the ASRC forced the beet-sugar interests to withdraw their opposition, Congress agreed to lower the tariff by 20 percent on sugar and tobacco coming from Cuba into the United States in exchange for Cuba's agreeing to lower its tariffs on a host of American-made goods.

The Cuban revolution that brought Fidel Castro to power in early 1959 did not immediately affect the sugar business, but by summer the ASRC was warning stockholders that the island's new agrarian reform laws might have "far-reaching effects on the future of our Cuban investment." A year later, after President Dwight D. Eisenhower banned the importation of Cuban sugar into the United States, the Cuban government appropriated thirty-six American-owned sugar mills and two million acres of American-owned sugar-cane land. The ASRC, which had spent $10 million to buy a 110,000-acre sugar plantation with a modern sugar mill in the Camaguey province in 1919, as well as an additional $10 million to build a large mill on its Canagua estate the following year, estimated its loss at $14.9 million and immediately increased its purchase of sugar from Mexico and the Dominican Republic.[66]

The End of an Era

Facing a plethora of problems, the leaders of the sugar-refining industry did what other business leaders were doing. That is, they turned to reorganization and diversification. The first step came in 1963, when the American Sugar Refining Company merged with the Spreckels Sugar Company to become the American Sugar Company. Press releases issued by the new firm mentioned "a new and cumbersome United States sugar law which effectively tied the United States price for sugar to the more volatile world market," as well as the "generally higher and erratic prices for refined sugar" that tended to "encourage the use of substitute and artificial chemical sweeteners." In 1970, after acquiring Duff-Norton (a firm that made equipment for light and heavy industry) and Hoffmeister (a firm that made disposable paper products) and after establishing a Food Service Division (which sold individual services of salt, pepper, mustard, ketchup, jams and jellies, relishes, and salad dressings), the firm changed its name to Amstar, because American Sugar "just doesn't describe us anymore." Expanding still further in 1973, Amstar purchased a high-fructose corn syrup factory in Dimmitt, Texas.[67]

In a 1984 management buyout, Amstar was sold for $428 million to Amstar-Holdings, an organization formed by the investment-banking firm Kohlberg, Kravis, Roberts & Company. Amstar-Holdings then took the company private and liquidated several of its divisions. The Dimmitt corn division was sold to American Fructose Corporation for $50 million in cash and assumption of liabilities. The Spreckels Sugar Company and its industrial products group fetched $170 million. The Amstar Sugar Corporation brought $305 million. Thus, in 1989, then simply a power tools and electronics company with piles of cash, Amstar bought $175 million worth of junk bonds from Essex Industries, "a heavily indebted maker of nonresidential door locks and exit devices."[68]

Amstar Sugar was bought by Tate & Lyle, the great British sugar concern, and in 1991 it was renamed Domino Sugar Corporation to capitalize on the firm's well-known brand. Ten years later, Tate & Lyle sold Domino to a conglomerate of Florida sugar-cane growers and refiners led by Alfonso and J. Pepe Fanjul, Cuban immigrants with major sugar holdings in Florida and the Dominican Republic.[69]

The Fruits of Their Labors

Like others of ample means, the sugar barons shared some of their wealth with the wider community. The Stuart brothers gave each employee a copy of the scriptures and presents during the holiday season. They purchased and retrofitted a magnificent house for James McCosh, president of Princeton College and the leading Presbyterian scholar in the country. They also provided substantial funding for Stuart Hall at the Princeton Theological Seminary as well as for two Presbyterian churches on Fifth Avenue in Manhattan. Robert Stuart intended to give his American paintings to the Metropolitan Museum of Art in New York, but he died before making this donation. His widow, upon learning that the Metropolitan would remain open on Sundays, bequeathed the collection to the Lenox Library (now the New York Public Library), to be exhibited in its own room free of charge.[70]

Henry O. Havemeyer was already a collector when he married Louisine Elder (a daughter of George W. Elder, an original partner in the firm), and together they became important patrons of the arts. While their tastes were eclectic, they are particularly remembered for having brought impressionism to America. Most of the Havemeyer collection is now in the Metropolitan Museum of Art and the National Gallery of Art in Washington, DC. The American folk art collected by their daughter, Electra Havemeyer Webb, now resides in the Shelburne Museum in Vermont.[71]

chapter 2

MOLASSES

Molasses is what remains behind when crystallized sugar is made from cane juice. It contains some sugar as well as vitamins, minerals, and interesting flavors. Rum is molasses fermented and distilled. The relationship between these products and the English colonists in North America began with three key dates: 1627, when a few English men and women settled on Barbados, a small island on the eastern fringe of the Caribbean; 1637, when sugar was first planted on Barbados; and 1647, when the governor of the Massachusetts Bay Colony, John Winthrop, learned from his friend Richard Vines that settlers in Barbados were so intent on sugar that they would rather "buy foode at very deare rates than produce it by labour, soe infinite is the profitt of sugar workers." As governor, Winthrop clearly understood the significance of this news for his local economy. The early New England settlers had sent beaver pelts to England in exchange for the English goods they needed to maintain their way of life. As the beavers became scarce, however, the colonists had nothing to offer the mother country. Provisioning the settlers on the sugar islands would offer another line of credit. So, Winthrop exclaimed in his diary, "it pleased the Lord to open us a trade with Barbados and other islands in the West Indies, which . . . proved gainful," noting that the resultant bills of exchange were a good help in discharging "our engagements in England."[1]

Trade between New England and the sugar islands was soon a booming business. Ships bound for the Caribbean carried fish and cattle for food, horses for powering the sugar mills and for land transportation, and timber for barrels and buildings; they returned home with sugar, molasses, rum, and bills of exchange. Colonists farther south grew tobacco, cotton, and other crops that found a ready market in England, but even they traded with the West Indies. George Washington sent fish, flour, and a wayward slave from Mount Vernon to Barbados, receiving in return rum, molasses, and fruits. New Yorkers produced pig iron, fire bricks, and sugar boilers suitable for use in the West Indies.[2]

Molasses was used in such dishes as brown bread, gingerbread, Indian pudding, shoofly pie, and apple pan dowdy. John Adams favored boiled puddings made of corn and molasses and claimed that George Washington "always asserted and proved that Virginians loved molasses as well as New Englanders did." Baked beans flavored with molasses and animal fat, put into the oven on Saturday and ready for eating on the Sabbath, may have been a staple of the New England diet, and the colonists possibly learned of this famous dish from Native Americans. There is, however, reason to question the standard story. One careful historian of material culture found but few mentions of beans in colonial probate inventories and no mention of a bean pot until 1811. One account of sweetened beans dating from 1825 states that this "favorite yankee dish may be wonderfully improved by add-ing a table spoonful of molasses to each quart of beans at the time of putting them into the oven." The earliest public account of "Boston baked beans" appeared in 1847.[3]

While middle- and upper-class Americans ate molasses on occasion, lower-class Americans ate it daily. This correlation of consumption with class was already evident in 1728 when Boston newspapers published three budgets for families of a "Middling Figure," and each included a line item for molasses or small beer, a common molasses-based drink of low alcohol content. In the first debates over the tariff that would provide funds for the new federal government, Fisher Ames and Benjamin Goodhue described molasses as one of the "necessaries of life," especially for "the poorer class of the people" who use it "instead of sugar." Elbridge Gerry, also of Massa-chusetts, went on to say that many fishermen and their families consumed some thirty gallons of molasses a year to sweeten their Indian meal and produce their beer.[4]

Poor people "feed their children on bread and molasses," exclaimed one do-gooder in 1872, adding that "the little ones smear molasses all over the house, and it is usually swarming with flies." A *New York Herald* reporter, wondering how a shirtmaker could live on her meager earnings, found that most days she subsisted on bread and molasses and tea. A *New York Times* reporter quoted a bootblack who refused to live in the Newsboys' Home because "they don't give them nothing but bread and molasses, and that ain't enough for nobody to live on. Why, there was one feller went to a doctor, he was so sick, and the doctor said he had consumption, all from eating bread and molasses."[5] This practice of serving molasses to inmates of almshouses and prisons was probably under way in America long before the expatriate scientist Benjamin Thompson (Count Rumford) introduced it to Europe in the 1790s, noting that a ten-pound gallon of molasses sold in the United States for a mere twelve to fourteen pence.[6]

Most enslaved Americans—including the young men who worked in Thomas Jefferson's nail factory at Monticello and the many men and women who labored on plantations in the Deep South—lived on molasses, cornmeal, and fatty meat, and they would continue to do so long after emancipation. A dietary study undertaken under the aegis of the U.S. Department of Agriculture (USDA) in the 1890s found that African American families in Alabama were substantially malnourished: not only did they not get enough protein, but they also consumed "an undue proportion, but not amount," of starch and sugar. Dr. Joseph Goldberger argued in the 1920s that the deficiencies of this diet caused pellagra, a disease devastating many Americans, blacks especially but many whites as well.[7]

Rum

The early history of rum is somewhat hazy, but we do know that the stuff was produced in the West Indies by the early 1640s. For Giles Silvester, an Englishman living in Barbados, rum was "a hott hellish and terrible liquor." But when planters learned to produce a smoother drink, consumption increased. In his monumental study of rum in colonial America, John McCusker estimates that nearly 7.52 million gallons of rum were consumed in 1770. If each American slave received 1 gallon of rum a year, the average white inhabitant must have consumed a phenomenal 4.2 gallons.[8]

Relatively inexpensive compared with brandy and wine, rum was, according to a 1740 history of Jamaica, drunk by "Servants and the Inferior Kind of People."[9] Increase Mather, a Puritan cleric who would enforce morality in Massachusetts, was particularly concerned with drunkenness, especially among the lower orders and especially as caused by rum. As he sermonized in 1686, "It is an unhappy thing that in later years, a kind of strong drink hath been common amongst us, which the poorer sort of people, both in town & country, can make themselves drunk with, at cheap & easy rates." Cotton Mather followed his father in thundering against the abuse of rum, fearing that it would prove "more pernicious to us than any French invasion, that we have yet any apprehension of." Mather recognized that much of the rum in New England came by "Fleets of our Beloved Friends," and for the prosperity of those voyages, he prayed most heartily.

That prosperity was apparently very great, as the ships brought fifteen hundred hogsheads of rum to New England each year—where a hogshead barrel held over fifty gallons. If this should not "satiate the inordinate craving of too many of our poor people after it," Mather said, "an unknown quantity is prepared here at home, to supply the unhappy market." James Oglethorpe, the British social reformer who established the colony of Georgia, found that the men hired to build a lighthouse worked only one day in seven because a day's wages would buy a week of inebriation.[10]

The harm possibly done to the colonists, of course, paled in comparison to that done to Native Americans. The Englishman Ned Ward, in a book now known to be fictitious, claimed that the land for Boston had been bought for a bushel of wampum and a bottle of rum. More to the point, John Joselyn faulted the French and English for having introduced Indians to that cursed liquor, thereby inducing inebriation and preventing the spread of Christianity. Arriving in America in 1683, William Penn remarked that the Dutch, Swedes, and English had by brandy and rum "almost debauched the Indians all." Increase Mather noted with alarm that some wicked persons had given or sold strong liquor to the Indians and made them drunk. Eleazer Wheelock, the Congregationalist minister who established Dartmouth College in the late eighteenth century, is remembered for having carried five hundred gallons of New England rum with him when he went to convert Indians to Christianity.[11]

Rum also formed part of the triangle trade that brought so many enslaved Africans to the West Indies and the colonies on the American continent. This trade began when Europeans brought guns, gold, and brandy to West Africa, but by 1725, British traders at Sierra Leone were reporting that no slave trade could be made without rum. Rhode Islanders would later boast that their merchants had introduced rum to Africa in 1723 and that the large consumption of rum in Africa had deprived France of the sale of an equal quantity of brandy. Bostonians likewise boasted that a considerable branch of their trade was with Africa and that many of the slaves purchased there were sold in the West Indies. Thus, although John McCusker has shown that Americans' role in the triangle trade was relatively small, it was not insignificant.[12]

Most of the rum transported from America to Africa was made in New England. Emmanuel Downing, John Winthrop's brother-in-law, erected the first still in Massachusetts in 1647. He intended to produce whiskey but, like many others who followed in his wake, found molasses to be more plentiful than grain. In 1676, the governor of New York prohibited the distilling of any grain unless it was unfit to be milled into flour. This action, it was said, had the effect of lowering the price of bread, promoting trade with Barbados, enriching the Duke of York through import duties, and stimulating the production of rum.[13]

In 1661, the General Court of Massachusetts declared the overproduction of rum "a menace to society." This menace would expand as the colonists realized that, with abundant forests for fuel and mechanics to build stills and keep them in repair, they could produce rum more efficiently and economically than could their counterparts in the West Indies. By the mid-eighteenth century, rum had become the leading manufactured product in the northern colonies. It was not, however, very good. In the words of a contemporary observer, Bostonians were "more famous for the quantity and cheapness than for the excellency of their rum."[14]

Looking past our Victorian heritage, historians have shown that candidates for office in the eighteenth century relied on the "persuasive powers of food and drink dispensed to the voters with open-handed liberality." In 1758, when George Washington was running for a seat in the Virginia House of Burgesses, his agents plied voters and their families with twenty-eight gallons of rum, fifty gallons of rum punch, and substantial amounts of wine, beer, and hard cider. Historians have also shown that rum, or more probably rum punch, was the favored drink of colonists who gathered in taverns to debate topics of local and national interest. A wonderful relic of these times is the silver punch bowl that Paul Revere made for the Sons of Liberty in 1768, now in the Museum of Fine Arts in Boston. James Madison would later argue that the corrupting influence of liquors was inconsistent with the purity of moral and republican principles, but Washington, who had his own still at Mount Vernon, suggested that public stills be erected in the several states.[15]

Rum and Revolution

The thirteen continental colonies differed from one another in many ways and even contained internal differences; so, historians have been able to identify a large number of causes of the American Revolution. But John Adams, who came from Massachusetts, had no doubt that molasses was "an essential ingredient in American independence," for "many great events have proceeded from much smaller causes."[16]

The connection between molasses and independence may be said to have begun in 1689 when England banned the importation of French brandy. When France retaliated by banning the importation of rum, the French colonists in the West Indies found themselves awash with rum and molasses that they were willing to sell at cut-rate prices. The British sugar interests responded by asking Parliament for protection, arguing time and again that Barbados and Jamaica contributed much more to the British economy than did the continental colonies. The "least Sugar Island we have," said one, "is of ten Times more consequence to Great-Britain than all Rhode Island and New England put together." Another suggested that if the northern colonies were prohibited from trading with the French, the French colonies would simply throw their molasses away. Yet another suggested that the British sugar colonies should be appeased because they would never seek independence, while the New England colonies, with their prosperous farms, fisheries, and industries, would surely seek independence some day soon.[17]

Parliament responded in 1733 with the Act for the Securing and Encouraging the Trade of His Majesty's Sugar Colonies in America. The Molasses Act, as it was generally known, imposed a stiff tariff on foreign rum, molasses, and sugar brought into Britain or any of her colonies. As it was costly to staff customs houses and police a three-thousand-mile coastline riddled with secluded coves and harbors, however, Britain turned a blind eye to smuggling, bribery, and other forms of evasion.[18]

In the midst of the French and Indian Wars, Prime Minister William Pitt enjoined colonial governors to enforce the prohibition against commerce with the enemy. That is, they should stop American merchants from trading with the French West Indies. To facilitate enforcement, King George II authorized the issuance of writs of assistance (i.e., search warrants). James Otis, the Boston lawyer hired by the merchants of Massachusetts to plead their cause, lost in court, but his arguments against the writs would form the basis of the Fourth Amendment to the U.S. Constitution.[19]

In 1763, as Britain came to recognize the magnitude of its national debt, Parliament considered renewing the Molasses Act. Alarmed by this news, Massachusetts merchants formed the Society for Promoting Trade and Commerce within the Province and set forth several reasons why any action that operated to the "disadvantage, discouragement, and consequent weakening" of the northern colonies must also "strike at the well-being of the sugar colonies" and thus tend to enfeeble Britain itself. If New England merchants could not trade freely with foreign islands, they would cease their trade with the West Indies. This, in turn, would destroy a great part of the northern colonies' strength and riches, meaning that the colonists would no longer be able to afford the prodigious quantity of "luxurious British imports" that they had come to enjoy. It would also destroy the livelihoods of those who fished for cod and whales, those who built ships, and those who provided lumber and provisions for the West Indies. Finally, by raising the price of New England rum, the molasses tax would interfere with the steady supply of slaves so needed to produce sugar in the islands.[20]

The merchants of Rhode Island repeated these arguments in their 1764 Remonstrance against the Sugar Act, adding that molasses was the engine that enabled them to pay for British manufactured goods. In the past year alone, Rhode Island ships had brought fourteen thousand hogsheads of molasses from the West Indies, only twenty-five hundred hogsheads of which had come from the English islands. Much of this molasses had been turned into rum in "upwards of thirty Distill houses," the principal materials of which had come from Britain. Much of this rum had been exported to the coast of Africa to purchase the slaves needed in the English colonies in Carolina, Virginia, and the West Indies.[21]

Despite these arguments, Parliament passed a Sugar Act that reduced the duty on foreign sugar, molasses, and rum, presumably in order to diminish the temptation to smuggle, but increased the penalties for avoidance.[22] This, in turn, sparked another spate of protests. Stephen Hopkins, the governor of Rhode Island, argued that even a tax of three pence per gallon was much higher than a gallon of rum could bear and so would end the colonies' export trade to the French and Dutch sugar colonies, the vibrant cod fisheries, and the many costly distilleries that made possible the rum trade to the coast of Africa, throwing these business opportunities into the hands of the French. It would also substantially lessen colonial consumption of British manufactures.[23]

The Continental Congress that convened in Philadelphia in September 1774 voted to ban the importation of British goods as well as the exportation of all merchandise and commodities to Great Britain, Ireland, and the West Indies until the several American grievances were redressed. Britain refused to budge, but British planters in the West Indies realized that it was to their advantage to make common cause with the continental colonists.[24]

Following the Declaration of Independence, the Continental Congress faced the problem of provisioning American troops on land and sea. To this end it established quotas for the several states. Those for 1780 called for 195,628 gallons of rum from Massachusetts, 120,000 gallons from South Carolina, 100,000 gallons from Virginia, and lesser amounts from Connecticut, New Hampshire, Pennsylvania, and Rhode Island. To satisfy the central demand, several state governments banned the export of rum, molasses, cattle, and other articles needed to supply their own armies and inhabitants. As commander in chief, Washington worried whether his men had their daily rations of rum and extra rations on holidays; he had done the same while serving as an officer in the British army in the 1750s.[25]

Molasses and the Tariff

When Congress convened in 1789, its first major problem was finding funds to settle the war debt and run the government. There was general agreement that there should be a tariff on imported goods but not about the details. The debate over molasses was particularly lengthy and acrimonious. The merchants and traders of Portland, a town then in Massachusetts, argued that the proposed duty of six cents per gallon of molasses would "operate injuriously" on the New England states and have "pernicious consequences" for their manufactures, and they prayed that molasses might remain "free from all imposts and duties whatever."[26] Benjamin Goodhue, a Harvard graduate and representative from Massachusetts, emphasized the connection between the molasses trade and fisheries, noting that a large proportion of the fish caught by New Englanders were of an inferior quality that could not be sold anywhere but the West Indies. Any interruption of these New England fisheries would turn men away from the sea, and thus the United States would have no seamen for its navy, should it wish to have one.

In the end, the tariff signed into law on July 4, 1789, imposed a modest two and a half cents on each gallon of molasses coming into the country and between one and three cents on each pound of sugar, depending on degree of refinement. It also included a substantial tax of ten cents per gallon on "distilled spirits of Jamaica proof." This protection, although welcomed by the American distillers, did not enable rum to maintain its favored position in the national economy. The rum business was also beset by the increasing production and consumption of beverages made from American grains, a growing distaste for products grown by slave labor, and the rise of other and more lucrative American industries. There were also matters of quality and cost. In his *Report on Manufactures* presented to Congress in 1791, Alexander Hamilton noted that an increased demand for molasses in the United States, together with the recent revolution in Haiti, had led to a substantial increase in the price of molasses; thus, American distillers found it difficult to compete with the much better rum brought from the West

Indies. Nonetheless, several American rum makers remained in business well into the twentieth century.[27] Ironically, as per-capita consumption of rum decreased, "rum" became the generic term for intoxicating liquors, demon rum became the poster child for the American temperance movement, and rumrunners brought alcohol to dry locales. During Prohibition (1920–1933), oceangoing ships would sail to the rum line—that is, the edge of U.S. territorial waters, three miles from the coast—and wait there to transfer their foreign contraband to smaller, faster vessels. In 1924, having found that perhaps 95 percent of all contraband was coming ashore and that many contraband ships were registered abroad, the United States and Great Britain signed a treaty moving the rum line to an hour's steaming distance from the coast.[28]

Transporting and Marketing Molasses

Molasses was traditionally packed in barrels from which grocers would scoop out the amount each customer desired. Attempting to improve this process, inventors introduced a host of mechanical devices for drawing viscous liquids from casks or barrels. The first patent was issued in 1866 to Daniel Drawbaugh, a Pennsylvania inventor who would be known as the Thomas Edison of Cumberland County. By the 1880s, the Enterprise Manufacturing Company of Philadelphia was selling a Champion Molasses Gate that would not leak when in use or break when removed from the barrel. This firm also offered a self-priming and -measuring pump with a total-registering device, promising grocers a saving in time, labor, and molasses, as well as an absence of flies and dirt.[29]

Pumps were also important features of the containerized transportation that reduced the cost of cooperage and handling as well as loss through leakage. The *Novelty*, an iron brig launched at Boston in 1869, held ninety thousand gallons of molasses, and her steam pump could discharge her cargo in six hours. By the 1890s, most of the molasses that the United States imported from Cuba was carried in bulk in steamships that carried petroleum as return cargo. The *Amolco*, a steel ship launched with great fanfare in 1913, could hold eight hundred thousand gallons of molasses and boasted a pumping system designed for viscous liquids.[30] Molasses tanker cars for trains were used in Louisiana by the end of the nineteenth century. In yet another experiment, a Philadelphia man used a three-inch wrought iron pipe and a steam pump to move molasses from a tank alongside the Delaware River to the Grocers' Sugar House more than a mile away.[31]

In 1908, as molasses was becoming a branded product available in family-sized containers, several Louisiana canners were said to be spending "more money each year *in advertising alone* than was spent formerly in the general operations of a large molasses business." Many of these advertisements aimed to attract middle-class customers. While writer Harriet Beecher Stowe had described slave boys "with hands and faces well plastered with molasses" playing under the table in her 1852 novel *Uncle Tom's Cabin*, Penick & Ford boasted "clean sanitary cans" that were airtight and germproof and "guaranteed to keep in any climate for any length of time."[32]

Penick & Ford also boasted that its open-kettle plantation molasses was "made in the old time primitive way"—not in modern mills that removed most of the sugar from the cane juice, leaving behind a molasses that was not very sweet. The company chemist, however, let it be known that that was not the case with Penick & Ford's product.[33]

Penick & Ford introduced two new brands in 1907: Aunt Dinah and Brer Rabbit. Although portrayed as a black mammy, Aunt Dinah seems somewhat less demeaning than Trixy, the wild-haired poster child for the "Louisiana old style cooking molasses" sold by D. B. Scully of Chicago. Brer Rabbit was the character from the *Tales of Uncle Remus* who, since the first edition of the book in 1881, had been shown sitting on rock, a jug of molasses at his feet, talking to five small bunnies.[34] The elderly white woman shown on the labels of Grandma's Old Fashioned Molasses probably appealed to customers with different sensibilities.

Penick & Ford distributed such booklets as *Something Every Mother Should Know* (1928), *Old Fashioned Molasses Recipes* (1934), and *Brer Rabbit's Modern Recipes for Modern Living* (several editions since the 1930s). Similarly, the American Molasses Company distributed *Forty Ways to Use Molasses* (1923), *One Hundred and Fifty Practical and Tasty Recipes Using Grandma's Old Fashioned Molasses* (1928), and *Grandma's Recipes for Mother and Daughter* (1950 and 1955), as well as a film, *Six Ways to Use Unsulphured Molasses*.

Molasses and the Federal Government

Packaged foods, developed originally for soldiers and wealthy explorers, reached middle-class American homes in the latter half of the nineteenth century, bringing fruits and vegetables out of season, perishable fish, and condiments such as catsup. Vinegar was the first important packaged food, other than rum, made from molasses. Postum, a drink promoted as a healthy alternative to coffee, was made from molasses and wheat berries.[35] After an 1877 patent described a method of canning beans and pork, Boston baked beans became widely available around the country and the world. Novelist William Dean Howells described grocers' clerks in Italy who were ambitious to read the labels on cans of Boston baked beans, and the Massachusetts delegation to Congress raised a ruckus when the Senate's restaurant served Boston Canned Beans rather than what they considered to be the real thing.[36]

As packaged foods and pharmaceuticals became ever more prevalent and commercial producers became ever more distant from their consumers, Americans became concerned about the quality of what they put into their mouths. As early as 1899, Penick & Ford boasted that its molasses was "guaranteed pure" and "the kind you used before the days of adulteration." In 1907 the firm announced that its trademark and government serial number 3799 guaranteed that its molasses was "pure within the meaning of the Pure Food and Drugs Act,"[37] a reference to the landmark law enacted by Congress on June 30, 1906, that laid the groundwork for the eventual Food and Drug Administration. The original law gave the secretary of agriculture responsibility for developing standards of purity and definitions of adulteration, for ensuring that labels on packaged foods and drugs were not fraudulent or misleading, and for policing food and drugs intended for interstate or foreign commerce or manufactured or sold in the District of Columbia or any American territory.[38]

Secretary of Agriculture James Wilson delegated much of this responsibility to Harvey Wiley, a chemist of enormous ego and energy long active in the pure-food movement. In April 1907, however, after Wiley's uncompromising stances, abrasive personality, and sometimes irrational decisions had provoked conflicts with powerful business interests, Wilson established a three-man Board of Food and Drug Inspection that could hold hearings and make recommendations. Wiley chaired the board, but with clearly diminished authority. The actual decisions were signed by the secretaries of agriculture, the Treasury, and commerce and labor.[39]

Long skeptical of food additives, Wiley was able in 1902, with congressional funds in hand, to establish what he called a "hygienic table" and what the *Washington Post* termed a "poison squad." For this test of how harmful the most common additives actually were, Wiley recruited a dozen healthy, young white men who agreed to eat and drink only what they were given and to submit to regular medical examinations. One year the poison of choice was sodium benzoate, which proved less toxic than Wiley had anticipated. Another year it was sulfurous acid (H_2SO_3), and when massive doses produced some unpleasant side effects, Wiley felt free to argue that foods with even trace amounts of sulfur be vigorously discouraged, if not entirely forbidden.[40]

As it happened, most makers of wine and dried fruits saw sulfur as a residue of their basic production processes. So did those who clarified their sugar with lime (calcium carbonate) and sulfurous acid. In the fall of 1907, some of these sugar men invited Wiley and his assistants to participate in a trial at a Louisiana mill: one day they would use no sulfur, the next they would use the normal amount, and on the third day they would use half the normal amount. Wiley came away convinced that the unsulfured molasses was just fine, but the brokers and dealers at the New Orleans Sugar Exchange saw it as a hard sell—it was, they said, "almost unmerchantable."[41]

The Louisiana sugar men also countered Wiley's hygienic table experiment with a clinical test of their own. Scientists affiliated with the Louisiana Sugar Planters Association and the Louisiana Board of Health selected a dozen healthy, young black men serving short sentences in the parish prison and transferred them to the New Orleans House of Detention. While Wiley's men knew they were testing substances that might be poisonous, the Louisiana men were told simply that they were testing various syrups and molasses. While Wiley's men ate sulfur straight, in capsule form or in water solutions, the Louisiana men ate it in molasses. Further, while Wiley's men received set amounts of sulfur, the Louisiana men could eat as much molasses as they desired, which was apparently quite a lot. At the end of five weeks of nourishing food and regular exercise, there was no change in the prisoners' vital signs. In a similar experiment, a doctor in California found no ill effects in healthy young men who ate fruits processed with sulfites.[42]

Faced with conflicting evidence, Secretary Wilson crafted a compromise. Food Inspection Decision 76 stated that there would be no prosecutions with regard to sulfur "as usually employed in the manufacture of food products" if there were no more than 350 milligrams of sulfur per kilogram or liter of product and the product carried a label stating that the food was preserved with sulfur dioxide. An accompanying memorandum, which Wiley did not sign, explained that there was insufficient evidence to condemn sulfur dioxide as presently used and that the standard was based on a large number of analyses of typical food samples.[43]

Food Inspection Decision 89, which modified the earlier ruling, said that minute amounts of sulfur dioxide would be allowed pending determinations of the Referee Board of Consulting Scientific Experts, which President Theodore Roosevelt had authorized in January 1908. Under the leadership of Ira Remsen, a chemist then serving as president of The Johns Hopkins University, the referee board would conduct yet another set of experiments pertaining to food additives. With regard to sulfur, the board decided to focus on dried fruits, reasoning that most Americans were likely to encounter sulfur in this form and that the physiological effects of the sulfur in fruit would probably resemble those in other products. They then conducted three lengthy and independent human trials, one at the medical school at Northwestern University, one at the medical department of the University of California, Berkeley, and one at the Sheffield Scientific School at Yale. They also made elaborate chemical analyses of foods and field trips to food processors. In the end they found that small doses of sulfur dioxide (up to 0.3 grams per day) combined with the sugars of foods had not shown "deleterious or poisonous effects and had not been found injurious to health." The same generally held true for larger doses (up to one gram per day), and in the few instances of ill health, the symptoms disappeared as soon as the intake of sulfur was discontinued. The board submitted its report in February 1913.[44] By then Wiley had resigned from the USDA (and switched his political allegiance from Republican to Democrat), which caused the question of sulfur and sulfites to lose its urgency. The unsulfured molasses that was introduced in 1923 and remains available today came from experiments to clarify cane juice using mechanical rather than chemical means.[45]

Another problem with molasses pertained to adulteration. In 1908, federal agents in Tennessee seized twenty-six barrels of Penick & Ford molasses from Louisiana, charging that it was actually a mixture of molasses and glucose (or corn syrup) and that the label should be amended accordingly.[46] A third problem pertained to the appearance of the name "New Orleans" on many packages of molasses, regardless of their relation to the Crescent City. After inviting molasses men to Washington for a hearing on the subject and polling grocers as to what they understood the name's inclusion to mean (the answers were far from unanimous), Wiley found how difficult it could be to associate names and foods. With no clear answer in sight, another compromise was in order. Acknowledging that the public understood "New Orleans molasses" to refer to any molasses made in Louisiana, Food Inspection Decision 134 accepted this definition with the proviso that labels might bear further information about quality.[47]

Molasses and Big Business

Like their counterparts in other industries, molasses men recognized the costs of competition and the benefits of association. While an 1899 attempt to form a $10 million molasses trust came to naught, a New York dealer named Noah Taussig succeeded in bringing several firms together under the auspices of the American Molasses Company. Understanding the advantages of working with Washington, DC, Taussig joined a committee of the U.S. Food Administration during World War I and fostered a campaign urging Americans to save sugar by eating more cane syrup and molasses (Figure 7). His son, Charles William Taussig, became an early member of Franklin Roosevelt's brain trust and served on the President's Advisory Council for the Government of the Virgin Islands. He also convinced two other brain trusters, Rexford Tugwell and Adolph A. Berle Jr., to leave the government and join American Molasses. By the late 1930s, the American Molasses Company had factories in New Orleans, Montreal, Boston, and Wilmington (North Carolina), as well as a new sugar refinery in Brooklyn. Its SuCrest division supplied molasses and syrups to industrial and consumer markets throughout the United States, and its advertisements in trade literature during the 1970s explained that molasses added color, moisture, and flavor to all sorts of processed foods and increased their shelf life.[48]

In 1940, with sugar shortages again on the horizon, Penick & Ford provided funds so that scientists at the Massachusetts Institute of Technology and the Harvard Medical School could conduct research to prove that New Orleans molasses contained more iron than any food except beef extract. When sugar rationing was imposed, Penick & Ford promoted molasses as a healthy supplement for children and a solution to the "periodic problem" that "all women share."[49]

Blackstrap Molasses

As sugar mills became increasingly efficient, they extracted more of the sucrose from the cane juice, leaving behind large amounts of low-grade molasses for which there was little demand. According to one report, Louisiana sugar men dumped this molasses into bayous and rivers and willingly gave it to anyone who would furnish the barrels and pay the freight to take it away.[50] In time, however, as livestock appeared to thrive on molasses mixed with grain, demand grew for blackstrap molasses in animal feed.[51]

Figure 7: World War I Poster: "Save Sugar by Using Best Louisiana Molasses and Sugar Cane Syrup." Princeton University Poster Collection. Courtesy Archives Center, National Museum of American History, Smithsonian Institution.

The glut of blackstrap molasses in the 1890s also coincided with a growing demand for industrial alcohol. The two forces came together when the Old Colony Distillery in New Haven, Connecticut, announced plans to distill alcohol from raw or blackstrap molasses.[52] The U.S. Industrial Alcohol Company (USIAC) was formed in 1907 after Congress agreed that (poisonous) denatured alcohol could be manufactured and sold in the United States. The firm soon had factories in New York, New Orleans, Buffalo, Baltimore, and Boston and was making alcohol for smokeless powder, fuel and antifreeze for motor vehicles, and numerous other purposes.[53] The USIAC gained notoriety on January 15, 1919, when a storage tank on the Boston waterfront came apart, spewing 2.3 million gallons of blackstrap molasses. This accident killed 21 people, injured 150, and covered several blocks of downtown Boston to a depth of several feet.[54]

Gayelord Hauser's best-selling advice book *Look Younger, Live Longer* (1950) touted the advantages of blackstrap molasses, as did his endorsement of Plantation blackstrap molasses. Regular consumption, he said, would provide enough vitamins and minerals to alleviate or prevent a host of ailments and might even extend life for several years. Blackstrap molasses today is regularly found on the shelves of health food stores.[55]

chapter 3

CANE SUGAR IN LOUISIANA

The sugar enterprise in Louisiana began around 1700 and gathered steam when French Jesuits brought cane from the island of Saint Domingue to the mouth of the Mississippi River. The next step came in 1796 when a Louisiana planter named Étienne Boré, with the help of a sugar maker from the West Indies, managed to get his cane juice to granulate. That is, by making sugar and not just syrup and then clearing $12,000, Boré showed that Louisiana sugar could be a profitable business. The Louisiana Purchase of 1803 was followed by Louisiana statehood in 1812, an influx of northern money and entrepreneurs, and the introduction of ribbon cane, a variety that matured rapidly and was relatively resistant to frost. By the 1830s, there were over six hundred sugar plantations and mills in the southern part of the state, and sugar had become the dominant crop in the region.

In 1830, Congress asked the secretary of the Treasury to commission "a well digested manual, containing the best practical information on the cultivation of sugar cane, and the fabrication and refinement of sugar, including the most modern improvements." This task eventually fell to Benjamin Silliman who, as professor of chemistry and natural philosophy at Yale College and founding editor of *The American Journal of Science and the Arts*, was one of the best-informed men of science in the country. Silliman was interested in practical as well as abstract matters, believed that science could serve practical purposes, and welcomed what were then considered "professional" opportunities to supplement his academic salary.[1]

As Silliman reported in his *Manual on the Cultivation of the Sugar Cane and the Fabrication and Refinement of Sugar* (1833), the techniques used in Louisiana were similar to those used around the world for hundreds of years. The harvested cane was brought to a nearby mill to be crushed by heavy cylinders moved by animal power. Bagasse, the spent cane from which the juice had been expelled, was put aside, perhaps to be burned for fuel when it had become sufficiently dry. After passing through a filter that removed the larger bits of remaining cane, the juice flowed into a large kettle known as the *grande* where, with the addition of lime and heat, further extraneous matter floated to the top and was skimmed off. The juice was then ladled sequentially into a series of other kettles—the *propre*, the *flambeau*, the *syrup*, and the *battery*—where, with further heating and skimming, it became increasingly transparent. When the time was right, the clarified syrup was ladled into flat pans, where it would crystallize. The resultant sugar was then placed in barrels with holes in the bottom so that the molasses could drain away. The sugar could be consumed as it was or sent to the refinery, perhaps in New Orleans, to be refined.

Silliman explained further that Louisiana cane had to be cut before the first frost, whereas cane in the West Indies could grow until the sugar had reached its peak. He intimated that sugar produced from cane grown in one locale, especially when refined, was indistinguishable from that produced from cane grown elsewhere—or, as we might say, cane has no *terroir*. He also hinted that sugar beets grown in northern temperate climes might provide serious competition to tropical cane.

Silliman was well aware that the industrial revolution that had begun in Europe in the late eighteenth century was impacting the production of sugar in various ways and that a few Louisiana sugar men were taking advantage of these new techniques. Some used steam engines to power their mills and heat their cane juice; some used vacuum pans to boil the juice at a lower temperature than was possible with the open kettles. Silliman was particularly impressed with this technology, noting that it represented "the most perfect method ever yet devised for completing the concentration of the syrup."[2]

Harper's New Monthly Magazine reported in 1853 that sugar demanded substantial sums of money. The cheapest sugarhouse in Louisiana cost $12,000, while a well-equipped one might cost twice as much. A planter who wanted to optimize this investment needed several hundred acres of cane, as well as land on which to grow food for his family and slaves, fodder for his animals, and wood to fuel the sugarhouse and build the barrels that would hold the sugar. Despite these initial and ongoing costs, however, sugar production could be extremely lucrative.[3]

Leon Godchaux

The Louisiana sugar industry was worth some $200 million when Abraham Lincoln was elected president in 1860; through the emancipation of slaves and destruction of property, however, it lost about 90 percent of its value during the Civil War. This situation, so disastrous for antebellum planters, presented a golden opportunity for Leon Godchaux, a New Orleans merchant with cash on hand. Godchaux bought his first plantation in 1862. At his death in 1899, he was the wealthiest man in Louisiana and the largest sugar producer in the country. Personal qualities such as energy, ambition, intelligence, and a head for business contributed to Godchaux's success. So, too, did money and timing. As the sugar industry recovered and then modernized, he was able to provide his facilities with the best of everything. In the words of the *Louisiana Planter*, Godchaux "introduced the latest and most scientific principles of sugar making in his factories and so tended to advance the art in our state."[4]

Godchaux hailed from Herbeville, a hamlet in the Lorraine province of France. His granddaughter Alma remembered him as spunky and independent and, although illiterate, able to learn what he needed to know. Thirteen years old and penniless when he landed in New Orleans in 1837, he quickly found a merchant willing to advance him notions and other goods desired by plantation families who could not easily shop in the Crescent City. Godchaux carried his pack on his back until he could afford a horse and wagon. He opened a small shop in the town of Convent in 1840, then a larger one in New Orleans in 1844, and was soon able to send for his mother and siblings.

Family played another key part in the story. Godchaux's wife, Justine Lamm, the well-educated daughter of a French schoolteacher, eventually bore ten children. As each son came of age, he was brought into the business. Paul Leon (who studied in Strasbourg) worked in the store, as did Albert (who studied at Yale). Walter, Edward (who studied at Tulane), and Jules (who studied mechanical engineering at the Massachusetts Institute of Technology) supervised the three sugar factories. Charles was the overall manager of the family's sugar interests. Emile (a graduate of the Yale Law School) was the family's lawyer.[5]

The *Times Picayune* described Leon Godchaux as a "progressive citizen, who believed in New Orleans, and invested largely in the improvement of the metropolis," while the *Times Democrat* noted that he was a member of Temple Sinai and always a "liberal giver" to such charities as the Touro Infirmary (also known as the Hebrew Hospital of New Orleans), the Jewish Orphans' Home, and the Young Men's Hebrew Association.[6]

Godchaux bought Reserve, his first plantation, from the widow of French printer Antoine Boudousquie. This land was located in St. John the

Baptist Parish, on the east side of the Mississippi River forty miles north of New Orleans. Godchaux first visited Reserve in 1837 and, when he had fallen ill, was given a bed until he regained health. Perhaps remembering this kindness, he contributed generously to the building fund of the new opera house in the French Quarter of New Orleans, a project spearheaded by Charles Boudousquie, Antoine's son, in the late 1850s. After acquiring the plantation, he invited the elderly Madame Boudousquie to remain in the French Creole manor house, rent free.

Godchaux's first tasks at Reserve involved extensive repairs to the fields and factory. We do not know who did the actual work, but with the local sugar enterprise in tatters, many talented mechanics and sugar makers must have been looking for a man with deep pockets who intended to make a go of the business. Once Reserve was up and running, Godchaux bought several other plantations. One was Souvenir in nearby Ascension Parish—Godchaux had never forgotten that the antebellum owners had chased peddlers from their lands. The most important purchases, however, were located along Bayou Lafourche on the west side of the Mississippi River: Elm Hall in Assumption Parish and Raceland in Lafourche Parish.

Louis Bouchereau, a statistician who initiated a series of annual reports on the sugar crop of Louisiana, gave voice to the issues facing the resurgent sugar enterprise. So did J. Y. Gilmore, editor and proprietor of the *Louisiana Sugar Bowl*, a weekly newspaper with articles in English and French. These issues included the intricacies of scientific farming and modern mechanics. Finance was also problematic, for costs were high and credit was tight. Not, however, for Godchaux, who enjoyed the profits of his store and real estate holdings. Indeed, as early as the 1850s, he was lending money to extravagant but cash-strapped planters who put up their land as collateral. In 1877, he announced his intention to conduct the sugar factorage and commission business on an extensive scale and to offer planters liberal advances on their crops. Gilmore described Godchaux's commission house as "one of the staunchest" in the state, adding that "as it does business on better terms than usual, it is securing a large and safe business."[7]

The Louisiana Sugar Planters' Association (LSPA) was formed in late 1877, its stated aim being to "develop the culture of the sugar cane, the manufacture of sugar therefrom in all its branches, to furnish such statistics and facts as will justify favorable legislation on the part of the United States Congress in behalf of this great industry, and to harmonize and concentrate, for the above purpose, the efforts of all those engaged in the cultivation, manufacture, and handling of the sugar products of this State, and of those who are engaged in the manufacture of machinery therefore." Godchaux may not have attended any meetings of the LSPA, but he clearly supported the organization's aims, understanding that the information it generated and disseminated would benefit the engineers, chemists, sugar makers, and other scientific and technical men in his employ.[8]

The LSPA pressured Congress to maintain the sugar tariff, and it convinced the U.S. Department of Agriculture (USDA) to provide the sort of aid to Louisiana cane that it was already providing to sorghum in the Midwest. It also created a sugar experiment station in Kenner, the *Louisiana Planter* (Godchaux was one of the thirty-eight original stockholders of this weekly newspaper), the Audubon Sugar School (Godchaux's $500 check for the building fund was the largest contribution for that purpose), and the Louisiana Sugar Exchange to facilitate transactions between sugar producers and purchasers (in 1888, Godchaux sent a keg filled with the best Louisiana open-kettle molasses to a Mr. J. Görz of Berlin "in appreciation of his prompt reply to an inquiry cabled by the Louisiana Sugar Exchange").[9]

In its first volume, that of 1888, the *Louisiana Planter* noted that there had always been much confusion in Louisiana with regard to the economics of the business. A few weeks later, when Godchaux presented a detailed account of the various costs in his sugarhouses, the paper announced that he had done great service to the industry and suggested that planters who compared these figures with their own work might attain the "effective and economic conditions" reached by Godchaux. Another year, when Raceland produced a mixture of sugar, syrup, and molasses, observers noted that this businesslike procedure aimed "to get the maximum dollars and cents out of a ton of cane, rather than the greatest number of pounds of sugar."[10]

Many less tangible factors also caused comment. One visitor to Raceland (perhaps inspired by Jane Austen) remarked that everything about this plantation "indicates taste, good judgment, ample means and thrift." A visitor to Elm Hall found that "as usual everything was in apple-pie order and not a speck of dirt to be seen." Yet another commented on the extreme cleanliness that prevailed overall, noting that the greatest precautions were taken against fire. But even with precautions, sugarhouses were risky businesses, and so the Godchauxs paid into the Sugar Planters' Mutual Insurance Company and were covered when Souvenir burned to the ground and when Elm Hall was hit by a hurricane and later destroyed by fire.[11]

The McKinley Tariff of 1890 removed the tax on foreign sugar and instituted a bounty of two cents per pound on sugar produced in the United States. As the bounty applied, however, to sugar that polarized at 90 percent or above—that is, sugar of quite high purity—it only benefited those who had fairly modern facilities. Godchaux and his sons received a bounty of $248,800 one year and $468,900 another. Planters with less modern equipment got nothing from the deal, and many were forced out of business all together.[12] This windfall enabled the Godchauxs to purchase a number of other plantations. Madewood had sold for $300,000 in 1853, but the Godchauxs got it for a mere $30,000. They were not particularly interested in the manor house, built in the 1840s for the wealthy and prominent planter Col. Thomas Pugh and now a National Historic Landmark described as "one of the finest and purest examples of Greek Revival style architecture in a plantation home." They were, however, interested in the fields and set about building ramps and a floating bridge to bring Madewood cane across the bayou to Elm Hall.[13]

As the efficiency of their sugarhouses increased, the Godchauxs bought cane grown by other planters. In the late 1880s, the *Louisiana Planter* urged the Godchauxs to establish a tramway that would run through the farms surrounding Raceland and encourage their neighbors to plant cane instead of the rice that had exhausted their fields. If the Godchauxs offered a reasonable price, this arrangement would undoubtedly "prove mutually profitable to the grower of the cane and to the manufacturer." Some years later, as the boll weevil was devastating local cotton, the Godchauxs provided seed cane to "former but now weevil-disgusted cotton planters along the Atchafalaya River."[14]

Since roads in southern Louisiana tended to be poor, if not impassable, the Godchauxs relied on barges that floated along the bayous, as well as narrow-gauge railroads pulled by mules and later by steam locomotives.[15] They also invested in mechanical carriers that lifted cane from the fields into the barges and train cars and then from these conveyances into the mill, finding that these devices were efficient and reduced the demand for labor. In 1912, the *Modern Sugar Planter* observed that the Godchaux organization had been "developed to the most perfect degree attainable, both for the delivery of cane from the many hundreds of tributary fields and in the daily handling of the huge tonnage at the mills."[16]

Water overflowing the rivers and bayous had always presented a challenge to Louisiana planters and became increasingly problematic as swamps were reclaimed for agricultural purposes. Like other planters, the Godchauxs paid serious attention to this matter. In 1893 they spent over $100,000 to close a break in the levee to keep the Mississippi from flooding the Reserve plantation and other properties in the area. Water was also a problem at Raceland. After purchasing this forested wetland for a song, the Godchauxs hired a plantation engineer and spent around $27 an acre to prepare the land for cane. They felled trees, many of them laden with Spanish moss, dredged great drainage canals, and built levees on all sides to keep back the water. They also installed an immense drainage station, allegedly the largest in the state, and equipped it with two Connersville pumps, each with a guaranteed capacity of one million gallons per hour. When the levee along Bayou Lafourche gave way in 1899, flooding many prosperous plantations, the Godchaux lands remained dry.[17]

Chemistry and Quality Control

Because the sugar tariff represented such an important part of the federal budget, the government was understandably concerned about the many attempts to disguise the quality of the sugar and molasses coming into the country. Benjamin Silliman was aware of this problem but could offer no solution. Indeed, the best this Yale chemistry professor could say was that the saccharometer, or Baumé hydrometer, was often used to learn the "saccharine richness of cane liquor" but did not afford "a sure criterion" for customs purposes.[18]

In the 1840s, as the problem continued to fester, another secretary of the Treasury turned to Alexander Dallas Bache, superintendent of the U.S. Coast Survey and thus de facto head of the federal Office of Weights and Measures, asking him to undertake "scientific investigations in relation to sugar and sirups, and the saccharine matter contained in them." Richard McCulloh, the young Philadelphia chemist to whom Bache delegated the task, reported that hydrometers were widely used to determine the proof of alcohol but not for sugar analysis. More to the point, McCulloh called attention to the saccharimeter, an optical instrument recently developed by scientists and artisans in Paris as an aid to the beet-sugar industry. Judah P. Benjamin, a Yale-educated lawyer with a sugar plantation in Louisiana and a wife in Paris, provided McCulloh with important information about his French saccharimeter and wrote an effusive account for *DeBow's Review*, a New Orleans journal with wide circulation that drew attention to the agricultural, commercial, and industrial progress in the South.[19] While it is unlikely that many Louisiana planters followed Benjamin's lead in the antebellum period, sugar was increasingly bought and sold on the basis of polariscopic analysis.

As the South began to recover from the Civil War, progressive planters came to see the value of keeping close tabs on the physical and chemical properties of the cane as it was turned into sugar. Opticians in New Orleans advertised sugar thermometers, vacuum pan thermometers, and hydrometers along with spectacles and opera glasses. Henry Studniczka, a sugar chemist from Vienna who settled in New Orleans and styled himself a sugar planters' "advisor and supplier," sold French and German polariscopes and gave lessons in their use.[20] The Louisiana Sugar Chemists' Association, formed in 1889, published its *Report on Methods of Sugar Analysis*.[21]

While the names of most sugarhouse chemists are absent from the historical record, we do know that various Louisiana plantations employed Studniczka throughout the grinding season of 1877. Moses Trubek, a Jewish chemical engineer from Latvia, worked for the Godchauxs during several sugar campaigns in the 1890s; he would later join the American Chemical Society and establish his own chemical firm in New Jersey. J. M. E. Stow was at Elm Hall from 1906 to 1909. G. L. Klein, a man "of large experience in sugar house laboratory work," was at Raceland in 1909.[22]

Labor

Silliman said nothing about the labor involved in cultivating, cutting, and milling the cane and moving the syrup from one kettle to another, although he surely knew that slaves of African descent performed these tasks (Figure 8). Indeed, historian John Rodrigue has argued that sugar "transformed southern Louisiana from a society with slaves into a slave society"—that is, a society whose economy depended on slaves. Recognizing that slaves in southern Louisiana and the West Indies had a much lower life expectancy than those working in other areas of the Americas, Michael Tadman has posited a demographic cost to cane culture and identified several factors that contributed to this appalling situation. The work was hard and hazardous and had to be done in a timely manner. Since the birth rate of Louisiana slaves was low and the demand for them rose precipitously after sugar became established, many slaves were brought from afar—from Africa and the West Indies and, after the 1808 prohibition against importing enslaved Africans, from plantations in the Upper South. Many died before they could adjust to their new environment. However expensive new slaves might have been, planters focused on recouping the substantial capital investments they had made in their mills.[23]

Following the partial abolition of slavery in Louisiana in 1861, the Emancipation Proclamation of 1863, and the adoption of the Thirteenth Amendment to the Constitution in 1865, many black men and women

Figure 8: African Americans cutting cane and loading it into a cart. This image appears on a $1 banknote issued by the Central Bank of Tennessee in Nashville, ca. 1855. Courtesy Numismatics Collection, National Museum of American History, Smithsonian Institution.

celebrated their release from the cane fields. In time, however, recognizing the limited scope of their opportunities, a large number returned to the plantations. Louis Bouchereau, a spokesman for planters, wrote about "the objectionable characteristics of the present system of labor, which gives no effectual means for coercing the performance of the contracts entered into by the freedmen. Notwithstanding the good treatment, the comfortable quarters, the high wages punctually paid them, and the regular rations of food with which they are supplied, the negroes are so indifferent about hiring, that it may be safely stated that not more than two out of twenty sugar planters have their full complement of laborers." Bouchereau also noted with annoyance that "women are nowhere to be found now in the fields"—but illustrations tell a different story (Figure 9).[24]

John Rodrigue has written that most of the sugar lands in Louisiana were worked by "gangs" of men and women who worked from daylight until dark and slept in cabins on the plantation, some of them "lying on the floor and getting their rest as best they [could]." There is scant indication of peonage in the Louisiana cane fields but ample evidence of workers in debt to planters and of planters agreeing, among themselves, to hold the line on wages. The situation came to a tragic head in the Thibodaux Massacre of November 1887, when planters, or goons in their employ, shot workers who, with the support of the Knights of Labor, refused to cut the cane. The official death toll stood at thirty. The Godchauxs must have been aware of this situation as the town of Thibodaux was but a few miles from Raceland.[25]

Figure 9: "Cutting Sugar Cane in Louisiana." Photograph by William Henry Jackson, late nineteenth century. Detroit Publishing Company Collection. Courtesy Prints & Photographs Division, Library of Congress.

During the economic depression of the early 1890s, the wages of field hands were scaled back to between forty and sixty cents per day. As the economy recovered, the Godchauxs advanced these wages by 16 percent, understanding that other planters would follow suit. The skilled laborers employed in the sugarhouses were paid according to the market price of sugar and, accordingly, saw an increase in wages of between 20 and 30 percent over the previous year.[26]

Eager to break their dependence on African Americans, some Louisiana planters hired local Creoles as well as Chinese, German, and Italian immigrants. The latter were reported to work well but tended to send money to the old country rather than spend it in the state. Elm Hall had fifty Italians "who gave satisfaction as hoe hands" in 1891 and one hundred Germans "cutting cane in addition to the usual force" in 1896. The Godchauxs also advertised in Midwestern papers for white farmers willing to become tenants at Elm Hall, leasing their lands and receiving a fixed price for the cane they delivered to the central. Other planters, hoping to attract a "new and better class of Western agriculturist," sold land to their tenants.[27]

In 1920, the *Louisiana Planter* called attention to "the hegira of colored farm labor and even white labor from the sugar district, brought on by the demand for it during and ever since the war." Recognizing that tractors offered a new way to circumvent this problem, one Godchaux brother was quoted as saying that "for breaking lands, one tractor and two men could do as much work in a day as six men and fourteen mules." Tractors, moreover, attracted "a better class of labor" than did mules. Because the exodus of farm labor was even greater during World War II, Louisiana planters made use of German prisoners of war housed in the state, and as late at June 1946, the American Sugar Cane League was asking the War Department to allow these men to remain in Louisiana a little while longer.[28]

Sugar Centrals

Industry leaders were well aware that Louisiana sugar men must upgrade their operations and take advantage of economies of scale. Five-roller mills extracted more juice from the cane than did three-roller mills, and six-rollers were more effective still. Vacuum pans and multiple-effect evaporators reduced the cost of fuel, as did centrifugal machines that spun the water out of the syrup. But since this modern machinery was expensive, specialization was imperative. Planters with modest resources should stick to cultivation and sell their cane to those better equipped to manufacture sugar. The smaller mills should be shut down, and the larger ones should be turned into central factories (or centrals) that would process cane grown over as wide a region as possible.[29]

Appreciating the logic of this argument, the Godchauxs worked steadily to expand and improve their facilities. They spent $75,000 on Elm Hall in 1896, raising its daily grinding capacity from eight to twelve hundred tons of cane. To increase the capacity still further, they awarded a prize to the central that ground the largest amount of cane each year. Elm Hall won it in 1910, with its mills "running as they never ran before" and processing 136,000 tons of cane, an amount that had been exceeded only twice in Louisiana, both times by Reserve. Sterling won the prize in 1912 by grinding almost 150,000 tons of cane, while Reserve ground 145,000 tons and Elm Hall slightly less.[30] Raceland today, now operated by M. A. Patout & Sons Ltd., has an annual grinding capacity of 1.25 million tons.

Other techniques for processing the cane juice received equally serious attention. When Godchaux ordered thirteen large presses made by the German firm of Webelin & Hübner, the *Louisiana Planter* noted that as he "generally knows what improvements are most likely to add to the profits of sugar making, his selection of this filter press is a high recommendation."

John Brewer, manager of the Raceland central, would later develop, patent, and install a mechanical filter that he hoped would revolutionize the filtering of cane juices throughout the world.[31]

Equally important was the Rillieux multiple-effect evaporation system, so called because it was devised by Norbert Rillieux, a quadroon from New Orleans, and because it used the energy from heating the cane juice over and over again. After studying in France and teaching applied mechanics at the École Centrale in Paris, Rillieux returned to Louisiana in the early 1830s, patented his invention, and convinced a few progressive planters to invest in his ideas. In the late 1850s, as the freedoms enjoyed by free persons of color became ever more restricted, Rillieux returned to France. By the 1880s, sugar makers around the world and engineers involved with other processing industries were adopting his ideas. The Godchauxs knew about Rillieux because Elm Hall had an original Rillieux apparatus from the late 1840s. The double-effect evaporator that the Godchauxs installed at this plantation in 1891 was said to be the largest in the state, if not the largest ever made; with pans measuring ten feet across, it could handle 150,000 gallons of syrup in twenty-four hours.[32]

With an assortment of new technologies, sugar centrals could produce a very high grade of sugar, thereby blurring the once clear distinction between raw and refined. Yet making the best sugar possible was not always the best business decision. A chemist who worked at Reserve in the 1890s recalled that Charles Godchaux's "genius" was his ability to evaluate a constantly fluctuating market and decide each day the optimum output of sugar destined for a refinery and that destined for direct consumption.[33]

The Leon Godchaux Company Ltd.

The *Louisiana Planter* detailed the advantages of corporations for the ownership of sugar plantations and central factories in 1894, noting that this form of organization was common in the North but relatively unknown in Louisiana. The Leon Godchaux Company Ltd. was formed two years later. It aimed to secure the best men and the best machinery, and with $2.5 million in capitalization, it managed to do just that. Each Godchaux central was soon equipped with large six-roller mills, massive vacuum pans, and equally impressive double-effect evaporators.[34] The expanded facilities attracted a host of visitors. One such was President William Howard Taft, who, while en route from St. Louis to New Orleans to promote traffic along the Mississippi River, stopped for a photo opportunity at Reserve, bringing an entourage that included 25 governors and 177 congressmen.[35]

Reserve refined ten million pounds of sugar in 1911, the bulk of it coming from Cuba. This was, said the *Modern Sugar Planter*, the first time that a Louisiana sugar factory had "been employed in the off season to refine a large quantity of outside raw supplies." The LSPA trekked to Reserve in 1911, for its first meeting outside of New Orleans, to see and learn about the new machinery and techniques.[36]

Tariffs and Trusts

In 1884, when Congress was considering abolishing the sugar tariff, Leon Godchaux and other leading Louisiana planters decided that their voices must be heard. To this end, they organized a Convention of the Representatives of the Louisiana Protected Industries and resolved to demand that cane sugar and rice continue to receive tariff protection. A decade later, when Democrats abolished the sugar bounty without reinstating the tariff, Louisiana sugar men discussed the possibility of finding a Democrat who would vote with Republicans on tariff matters or, if that were not possible, quitting the Democratic Party all together.[37]

Obviously pleased with the Dingley Tariff of 1897 reinstating the tax on imported sugar, Louisianans were dismayed by the Underwood Tariff, enacted in 1913 after Democrats had again regained national power, which again eliminated the tariff on sugar. Indeed, while the Underwood bill was under consideration, Jules Godchaux and other members of American Cane Growers' Society went to Washington to urge lawmakers to recognize that the "prosperity of Louisiana was at stake." When the beet-sugar men joined, this organization became the American Sugar Growers' Society.[38]

In their struggle to maintain the tariff, the Louisiana sugar men came into direct conflict with the Sugar Trust, as the American Sugar Refining Company (ASRC) was generally known. Since the ASRC aimed to control the entire sugar business in the United States, it promoted a low tariff on raw sugar and a reciprocity treaty that would admit Cuban sugar into the country duty free. It also obtained rebates from railroads and offered unsustainably low prices for Louisiana sugar.

The ASRC took over the two largest refineries in New Orleans in 1887 and consolidated their operations into one. Dismayed by this move, the *Louisiana Planter* urged the Louisiana centrals to upgrade their facilities. Such a move, it said, would bring higher prices and avoid the necessity of dealing with the trust. A few years later, the Louisiana legislature raised the license tax of the ASRC refinery from $2,500 to $37,500 a year. In retaliation, the ASRC closed its Louisiana refinery, throwing some six hundred men out of work; the refinery reopened when the courts overturned the tax.[39]

The Louisiana sugar men also made several efforts to establish their own refinery in New Orleans, but each came to naught.[40] A retired planter would later testify that "abject fright" of the trust was the real cause of the failure of the planters' refinery. A leader of the trust would suggest that the planters' effort should be seen as a "combination in restraint of trade."[41] In 1913, the Godchaux brothers joined with fourteen other Louisiana sugar men who hoped to form a $60 million holding company that would introduce economies of scale and modern business practices, as well as offer serious competition to the trust. When that scheme failed, the Louisiana legislature declared sugar refineries to be public utilities and thus subject to state regulation. This bill, enacted at a special session in early 1915, was quickly overturned by the district court in Atlanta and then the Supreme Court in Washington, DC.[42]

Godchaux Sugars Inc.

Godchaux Sugars Inc. was chartered in New York in 1919 and, with a capitalization of $7 million, further expanded the operation. At Reserve, a new furnace produced the boneblack used to decolorize and remove impurities from sugar. At Elm Hall, a new retort produced Norit, a vegetable-carbon mixture that served the same purpose. The Godchauxs also began using electricity, where it was possible and economical, for power and oil, which was cheap in Louisiana, for fuel.[43]

By 1929, with its machines running round the clock, Reserve was refining two million pounds of sugar each day. To attract domestic customers, the firm published such pamphlets as *Recipes for Using Godchaux's Cooking Sugars* (1920), *Famous Recipes from Old New Orleans* (from the 1930s), and *The Story of Godchaux's Pure Cane Sugar* (1935), which offered tested recipes using granulated, brown, and confectioners' sugar, as well as household hints that would do Martha Stewart proud. Most of the sugar, however, probably went for industrial purposes. To that end, for instance, Godchaux salesmen targeted the many firms that manufactured carbonated beverages.[44]

Godchaux Sugars erected model towns for the employees at Reserve and Raceland, Elm Hall having been destroyed by fire in 1922. Clearly impressed, the editors of the *Louisiana Planter* noted that the Godchauxs were "known for the thoroughness of all their various projects and the sugar planters of the State will be interested watchers of this latest progressive move on their part." These new towns, although racially segregated, boasted houses that the workers could afford and community centers with a swimming pool, tennis courts, and movies. There were hot meals in the factory, health and life insurance policies, a new school, and presents at Christmas. Three steam-powered generators produced the electricity and purified the water used in the refineries, as well as free electricity and free potable water for the residents. A monthly paper was known as *The Blue Band*, the title coming from the design of packages of Godchaux sugars (Figure 10). This civic involvement was nothing new. Indeed, as early as 1887, Leon Godchaux had paid for the stained glass rose window and much of the interior paneling for the new St. Peter Catholic Church in Reserve.[45]

Figure 10: The Reserve factory town. Photograph by Frances Benjamin Johnston for the Carnegie Survey of the Architecture of the South, 1938. Courtesy Prints & Photographs Division, Library of Congress.

While many men and women spent their whole working lives in the Godchaux fields and factories, philanthropy and paternalism could not compel docility. One employee, Edward Hall, helped organize a union in the Reserve refinery and served on the negotiating committee until his retirement in 1970. Hall also filed a federal lawsuit challenging the discriminatory procedures of the registrar of voters in St. John Parish and helped establish the St. John branch of the National Association for the Advancement of Colored People.[46]

The National Labor Relations Board heard at least two cases involving Godchaux Sugars in the early 1940s. In one, a local union affiliated with the American Federation of Labor (AFL) argued unsuccessfully that management had exerted undue influence over the Godchaux Sugars Employees Labor Council at Reserve. In the other, a local union affiliated with the Congress of Industrial Organizations (CIO) argued successfully that a unit composed entirely of supervisory employees could engage in collective bargaining, even if that unit was affiliated with the production workers' union.[47]

The federal minimum-wage law enacted in the 1930s applied to workers in the sugar mills. Field hands, however, were covered by the 1937 Sugar Act, which stated simply that, in exchange for price supports, planters were expected to be good employers. That is, they would not hire children and would provide, free of charge, a habitable house, medical attention, and similar traditional perquisites. Wages were expected to be fair and reasonable and would be determined at annual meetings of local planters and representatives of the USDA. Following the 1951 meeting, many Louisiana field hands earned $700 a year, the minimum that planters could pay if they were to receive the one-cent-per-pound federal subsidy for domestic sugar. A small number of skilled and semiskilled workers in their prime might earn as much as $1,800. No rules governed the length of the working day.

In 1952, as many Americans were seeking their fair share of the American dream after so many years of economic depression and war, the National Agricultural Workers Union, aiming to represent cane field workers, male and female, black and white, established Local 317 with headquarters in Reserve. After the annual meeting, the union prepared a joint wage claim involving several hundred of its members, demanding more than $32,000 in back wages. The following year, with no resolution in sight, about three thousand cane field workers went on strike against the principal plantations in nine Louisiana parishes, demanding permission to organize under the aegis of the union. The strike leader argued that the issue was not wages but rather "the right, guaranteed under our Constitution, of agricultural workers to have an organization that will represent their interests." Leon Godchaux II responded by saying, "If this country is to survive, the unscrupulous methods and irresponsibility of unionized field hands will not be tolerated."[48]

While asking President Dwight D. Eisenhower and the Department of Labor to intervene in this job action, the Negro Labor Committee noted that the plantation owners threatened to evict poorly paid workers from company-owned homes and intimated that they would import "hundreds of Jamaican or illegal Mexican immigrants to break the strike." *Time* also took account of this "Cane Mutiny," reporting that "most workers live rent-free in company-owned houses, some of them hovels, some adequate. Most workers trade in company stores and are completely dependent on the plantation owners. About two-thirds are illiterate; some know no English, speak only a Cajun patois. Four-fifths are Negroes." The union's cause was severely damaged when a Louisiana court issued an injunction requested by the American Sugar Cane League that prevented farm workers from picketing factories. The union appealed to the Supreme Court but was out of funds and could not continue the strike. The Supreme Court finally set aside the injunction in December 1955, but by then it was too late. After studying this matter in detail, historian Thomas Becnel concluded that the strike was effective only insofar as it exposed the dreadful conditions in the Louisiana cane fields.[49]

The American Sugar Cane League also argued successfully that the strike was illegal as the federal Wagner Act and state mediation law expressly excluded agricultural workers. Indeed, as explained by Socialist leader Norman Thomas, the AFL-CIO had mounted a strong campaign to repeal the right-to-work laws for industrial workers in Louisiana, and to get the large sugar, rice, and cotton makers to support this cause, it had thrown the agricultural workers "to the wolves." That is, the union omitted agricultural workers from their purview and stretched the definition of "agricultural" to include labor "engaged in the earlier stages of processing" these crops.[50]

In 1955, about fifteen hundred factory workers, under the umbrella of the United Packinghouse Workers Union, went on strike against Godchaux Sugars Inc. and the Colonial Sugar Company, also of Louisiana, and organized a "don't buy" campaign (Figure 11). The workers sought a raise of ten cents in hourly wages plus four cents in hourly fringe benefits. Management counteroffered with a five-cent raise across the board, bringing wages into the range of $1.39 to $2 per hour. After publishing a few comments

Figure 11: Poster promoting the strike of 1955. The text on the back reads, "Your support can help win—a decent standard of living for southern workers and bring prosperity to the south." Collection of the author.

from management, the mainstream media reported that the details of the final settlement had not been disclosed. The *Pittsburgh Courier*, however, described the Godchauxs as a "feudal barony," and the *Chicago Defender* described Dixie as a "bitter foe of labor." The Negro press also reported that while management tried to keep things moving, a Godchaux locomotive hit a truck hauling pilings and killed two people: it was the only recorded accident on the line.[51]

At the time of the strike, there were some 650 workers in the Reserve factory: 25 black men who did the heavy lifting, 580 white men (including hourly workers and such salaried workers as supervisors, scientists, and foremen), and 45 white women who worked in the Small Packaging Department putting granulated sugar into two-, five-, and ten-pound sacks and brown and confectioners sugar into one-pound boxes. Men tended the machines that filled the twenty-five- and one-hundred-pound bags. During the grinding season, from October to December, these numbers rose to 750 whites and 100 blacks.[52]

In January 1956, just weeks after the strike came to an end, William Zeckendorf, the New York real estate tycoon who served as president of Webb & Knapp, announced his intention to purchase a controlling interest in Godchaux Sugars Inc. for $15.5 million. His stated purpose was to develop some of the land for commercial and industrial purposes and to exploit its mineral resources. The Godchauxs were well aware of these resources. Annual reports had mentioned oil developments since 1938, and in recent years the firm had leased sixteen thousand of their thirty-three thousand acres to oil companies that had capped three gas-distillate wells on the property.[53]

Uninterested in sugar, Zeckendorf sold the Reserve refinery and the Raceland central to the National Sugar Refinery of New York, the nation's second-largest sugar refiner. After the Federal Trade Commission saw this deal as a violation of the antimerger law, the Godchaux sugar properties were sold to Julio Lobo, a sugar baron from Cuba. When Lobo went bankrupt a few years later, the properties were sold to Southern Industries in Mobile, Alabama. That conglomerate, in turn, sold the properties to the Hunt Brothers, who also owned the Great Western Sugar Company of Denver, Colorado. When the Hunts declared bankruptcy in 1975, the Raceland central was purchased by Patout, a firm that has been producing sugar in Louisiana since 1825, and the Reserve property defaulted to the Port of South Louisiana. The plantation house at Reserve is now on the National Register of Historic Places.[54]

chapter 4

CANE SUGAR IN FLORIDA

Spanish adventurers planted the first cane grown within the bounds of what would become the continental United States at St. Augustine in 1572. Native Americans would later grow cane in the area, as would the farmers from Georgia and Alabama who moved south after the United States acquired Florida in 1818. The first sugar mill in Florida was erected at New Smyrna in 1830. The success of the industry, however, would await the draining of the Everglades, the infusion of northern capital, and the development of disease-resistant cultivars.[1]

The great swamps that covered the southern tip of the Florida peninsula became known as the Everglades in the early 1820s. The U.S. Senate discussed the Everglades in 1848, three years after Florida became a state, noting that this land was "now nearly or quite useless to the United States; and will so remain until reclaimed by draining by means of canals." The cost of this work, it was believed, would probably not exceed $500,000.[2] Although no action was taken, the Swamp Lands Act of 1850 deeded such lands to the several states with the proviso that the funds from the sale of these lands be used for their reclamation. Florida then organized an Internal Improvement Fund to manage its twenty million acres of public lands, most of them in the Everglades.

In 1881, the governor of Florida approached Hamilton Disston, a wealthy northerner who enjoyed fishing in southern waters, about filling the depleted coffers of the Internal Improvement Fund. After agreeing to purchase four million acres of the Everglades for $1 million, Disston formed the Atlantic and Gulf Coast Canal and Lake Okeechobee Land Company and began dredging a canal between the great lake and the Gulf of Mexico. The company also established an experimental farm and showed potential customers which crops could be grown on reclaimed land. After samples of Florida cane were shown at the Cotton Centennial Exposition in New

Orleans in 1884 and 1885 and judged superior to cane grown in Louisiana, Mexico, and Cuba, Disston erected a mill and planted four hundred acres of the crop.[3]

Prospects for the new venture were soon looking rosy. The *Chicago Tribune* reported that Florida might be expected "to produce 60 per cent of the whole amount of sugar we consume." The *Atlanta Constitution* reported that, thanks to Disston's drainage operations, "thousands of acres of once submerged lands are now bearing upon their fertile surface immense fields of sugar cane and groves of tropical fruits." The *New York Times* reported that the St. Cloud Sugar Plantation, now owned by Disston, was expected to make three million pounds of sugar a year and obtain a federal bounty worth $60,000. The article went on to say that the plantation was equipped with the newest and most expensive machinery as the margin in sugar production was "too small to allow of any old-fashioned methods."[4]

Harvey Wiley, the chief chemist of the U.S. Department of Agriculture (USDA), was equally enthusiastic, noting that the prospect of success in the production of sugar in Florida was so great that the land should be "recovered from overflow and developed." Moreover, because of climatic factors, the manufacture of sugar in the area around Lake Okeechobee would eventually surpass that in Louisiana. In 1891, after Congress appropriated $50,000 for sugar experiments, Wiley established an agricultural station on the Runnymede Plantation, planted some eighty varieties of cane, and built a model mill. Much to his dismay, however, Grover Cleveland's secretary of agriculture pulled the plug on the project.[5] This decision, along with the economic depression that descended on the country in 1893, followed by Disston's death in 1896, spelled the end of the first phase of large-scale sugar production in Florida. But small-scale production continued apace, with numerous farmers producing syrup for domestic use and local distribution.[6]

Local politicians, however, never lost sight of the bigger picture and, seeing themselves as progressives, provided the state funds that, they argued, would make worthless marshland useful for growing cane, citrus, and various other fruits and vegetables. William Jennings, who became governor of Florida in 1900, proposed that reclamation of the Everglades was both feasible and practicable and would create "the most valuable agricultural land in the Southern States." Under his watch, Florida received title to the Everglades from the federal government. Two firms were then given concessions to drain the area and convert it into "the greatest sugar producing territory in the world."[7] Gov. Napoleon Bonaparte Broward promised to turn the "fabulous muck" of the Everglades into a productive resource, and as drainage canals proliferated, promoters announced that these lands might produce as much sugar as was then imported into the United States.[8] Acknowledging that Florida "used to be considered a malarial swamp," Gov. Albert Gilchrist boasted that dredges had turned Florida into one of the healthiest states in the Union. In 1912 he opened the Gulf-to-Atlantic Canal, the first of five great canals that he hoped to see cross the Everglades.[9] At his inauguration in 1917, Gov. Sidney Catts announced that the "most important thing facing Florida today is the drainage of the Everglades" and that bonds would be floated to finance the project. Begun in 1917, the Tamiami Trail opened automobile traffic from Tampa to Miami. By 1922, over $8 million had been spent to drain one million acres of swamp, and more was yet to come.[10] The USDA established a research station at Canal Point, on the south shore of Lake Okeechobee, and the Florida legislature authorized an Everglades agricultural experiment station nearby at Belle Glade.

With this infrastructure in place, Florida attracted a new wave of entrepreneurs. The Florida Sugar and Food Products Company transported a secondhand mill by barge to Canal Point, and the Pennsylvania Sugar Company built a $1.5 million mill close to Miami. While this latter firm failed, its mill was taken over by the Southern Sugar Company, and that in turn was taken over by Bror E. Dahlberg, a handsome, ambitious, and well-capitalized immigrant from Sweden.[11]

Dahlberg was not a sugar man per se but a businessman who banked on the mutually profitable relationship between sugar and Celotex, an insulating building material made from the leftover fiber of the cane. Dahlberg had incorporated Celotex in Delaware in 1920, established headquarters in Chicago, and built a factory in Marrero, Louisiana, across the Mississippi from New Orleans. As the demand for Celotex exceeded the local supply of bagasse, Dahlberg contemplated a factory in Florida; before that factory could be built, however, the cane would have to be grown. So, Dahlberg hired agricultural experts and drainage engineers to evaluate the land around the southern edge of Lake Okeechobee and was pleased to learn that the "rich muck lands of the Everglades awaited only the touch of modern science to bring them into unique fertility."[12]

Dahlberg was also familiar with the new cane cultivars developed at the Proofstation Oost Java, the Dutch agricultural research station in Indonesia, which E. W. Brandes, the senior plant pathologist in the sugar division of the USDA, had brought to the United States. Some of these cultivars crossed domesticated Javanese canes with semiwild ones from the slopes of the Himalayas in India. More to the point, some proved resistant to the sugar-cane mosaic virus that had devastated much of the cane crop in the United States and around the Caribbean. When some of the new cultivars produced a phenomenal sixty to eighty tons of cane per acre in southern Florida, compared to an average of twenty-one tons in Cuba and thirty tons in Louisiana, Dahlberg announced that the "Alchemy of Muck and Sunshine" provided ideal conditions for his new enterprise.

Aiming to create a sugar empire in Florida, Southern Sugar soon owned extensive acreage purchased primarily from small investors who had been caught up in recent buy-land-and-get-rich-quick schemes. It was also involved in what historians have described as one of the basic conflicts in American agriculture—that is, "the clash between the goal for family-sized farms and the economies and efficiencies of mechanized, large-scale agricultural production."[13]

In an effort to control the water levels in the lake and on the land, Southern Sugar began dredging canals and installing pumps, eventually boasting about the millions of dollars that had been spent, the millions of cubic yards of material that had been moved, and the vast amount of saw grass land that had yielded to the plow. It called attention to the number of acres planted with cane, applauded the modern highways and railways that were helping transform the region into a section of "prosperous activity," and located its headquarters in Clewiston, a town designed by John Nolen, a Harvard-educated urban planner.[14]

Southern Sugar broke ground for the Clewiston Sugar House in 1927, letting it be known that this $3.5 million facility would be the largest, most advanced, and most efficient mill in the continental United States. With Gov. John Martin officiating at the groundbreaking ceremony, Dahlberg predicted that the sugarhouse would produce fifteen hundred tons of raw sugar daily from cane grown in drained land in the "formerly useless swamps of the Everglades." Then, with new governor Doyle Carlton starting the machinery in January 1929, Dahlberg predicted that Florida would soon constitute the "sugar bowl of the nation." Just months later, Dahlberg announced a $6 million expansion to increase acreage, enlarge the drainage system, and augment mill capacity.[15]

Determined to keep abreast of scientific progress in cane production and sugar manufacture, Dahlberg hired Benjamin Bourne, a plant pathologist from Barbados who had studied in the United States and conducted cane-breeding experiments at the USDA station at Canal Point. Bourne's new job was to build and equip facilities for breeding and testing varieties of cane that would flourish under local conditions and to experiment with fertilizers and other methods of soil amendment.[16]

Southern Sugar valued technological solutions to agricultural problems and aimed to minimize the use of human and animal labor in its fields. Accordingly, it announced that it had a "completely motorized plantation—perhaps the first in existence to depend wholly upon power machinery." There were seventy-five tractors "plying away unceasingly day and night breaking up the drained swamps of Florida," as well as machines to carry seed cane to the fields and plant it in the ground and mechanized wagons to carry the mature cane from field to factory. Recognizing that most cane was still cut by hand, Southern Sugar was particularly proud of its harvester, describing it as "the outstanding mechanical invention for agricultural purposes since that of the McCormick reaper." This harvester, which did the work of 150 skilled cane cutters working a ten-hour day, had been designed by Carl Muench, the inventor who had developed Celotex.[17]

When the Muench harvester failed to perform as promised, Dahlberg asked Allis-Chalmers, the Milwaukee firm that made its tractors, to build fourteen harvesters according to the design of Ralph S. Falkiner. Each machine would cost between $15,000 and $20,000 but would harvest some thirty to forty tons of cane an hour, the rate depending on the acreage yield, thereby supplanting two hundred laborers. Moreover, each machine would cut the cane, separate the debris from the stalks, and discharge the cane in mill lengths into carts. Evidence suggests, however, that these new harvesters fared no better than those designed by Muench.[18] Falkiner, incidentally, hailed from Australia, a country where expensive white labor cultivated and cut the cane, and where management demanded high levels of efficiency. Australia was also the home of the several harvesters that would prove most effective in Florida some fifty years later (Figure 12).[19]

In 1929, Dahlberg exuberantly announced the formation of Dahlberg Sugar Cane Industries, an umbrella organization that would ensure safe and efficient management of his several enterprises and continue their tradition of "exhaustive investigation, scientific research, and far-sighted planning."[20] But his timing was bad, with the stock market about to crash and a depression about to descend. A glut of sugar on the world market and a consequent slump in prices further increased the riskiness of Dahlberg's situation. This problem, which had begun during World War I when the decline of European beet-sugar production had led growers in Java and Cuba to increase their acreage, was enhanced by the resumption of European production at the end of the war. In an effort to control the damage, Cuba

Figure 12: "Harvesting Sugarcane, United States Sugar Corporation, Clewiston, Florida." Farm Security Administration photograph, 1939. Courtesy Prints & Photographs Division, Library of Congress.

limited its production of sugar in 1926, hoping that other sugar-producing nations would follow suit—but to no avail.

Dahlberg believed that a higher tariff would enable the American industry to compete with the lower-cost production in Cuba and Java. The American Sugar Cane League agreed, and like lobbyists for other industries, it covered itself with the flag. Pointing to the thousands of family farmers who grew small amounts of cane, the nine corporations that dominated the industry argued that "farm relief ha[d] become the most insistent issue of the day, made so by the American conscience and the American sense of justice." Americans should "discard every consideration except the patriotic necessity of making our country self-sustaining in the matter of sugar," and national pride "will impel us to lift our sugar industry to the level of our other industries."[21] The Smoot-Hawley Tariff of 1930 provided some protection to the industry but not enough to keep the Dahlberg enterprise from falling into receivership.

Led by Charles Stewart Mott, a General Motors executive who held a large block of Southern Sugar stock, the firm was reorganized as the U.S. Sugar Corporation. Managed by Clarence Bitting, a specialist in financing and managing industrial properties, the firm would become the largest sugar producer in the world, and Florida would surpass Louisiana as the most productive cane-sugar state in the nation.[22]

U.S. Sugar took advantage of several New Deal programs. The Reconstruction Finance Corporation lent the funds that enabled the firm to erect a two-million-gallon molasses tank and make the Clewiston Sugar House into the largest and most efficient mill in the world. In 1936, U.S. Sugar boasted that its sugarhouse employed more than three thousand people and used about forty-five hundred tons of cane each day. The Agricultural Adjustment Administration gave U.S. Sugar more than $1 million for not planting cane on some of its land and another $80,822 for diverting land from soil-depleting to soil-conserving crops and soil-building practices. This latter sum was larger than that received by any other farm participating in the program and led Bitting to praise the secretary of agriculture for his "dramatization of soil conservation." With cash in hand, U.S. Sugar built one hundred cottages, twelve homes, and a termite-, wind-, and fireproof hotel of eighty rooms, describing this expansion as a jobs-creation program.[23]

Maintaining a vigorous commitment to research, U.S. Sugar rehired Benjamin Bourne who, after the failure of Southern Sugar, had earned a PhD from Cornell University. In 1942, having substantially increased crop yields and introduced canes not affected by freezing weather, Bourne was promoted to vice president in charge of research. U.S. Sugar also employed hydraulic engineers, soil chemists, plant pathologists, geneticists, agronomists, nutritionists, entomologists, livestock experts, and scientists who "check[ed] the top two feet of soil on each crop unit, and prescribe[d] what is needed to restore the plant food balance of the nitrogen-rich, mineral-poor Everglades soil." The scientific team also developed several new crops, including new varieties of sweet potatoes (grown largely for cattle feed), ramie, pasture grasses, lemon grass, and other vegetables.[24]

The amazing success of U.S. Sugar further exacerbated the problem of overproduction caused by similar productivity gains made by other firms elsewhere in the world. The first serious international effort to address this problem was the Chadbourne Plan of 1931, according to which Cuba agreed to limit its production and Java agreed to limit its exports, but these quotas were too high to accomplish much in the face of declining world demand.[25]

American efforts to balance production and consumption began with the 1934 Jones-Costigan amendment to the Agricultural Adjustment Act, which gave the secretary of agriculture authority to establish quotas for how much sugar could be produced in the United States and how much could be imported from specific foreign countries. The initial domestic quotas, based largely on previous production, specified 1.5 million tons of beet sugar and 260,000 tons of cane sugar. As would be the case with every subsequent piece of sugar legislation, the Jones-Costigan bill juggled several objectives. One was protecting the domestic sugar industry, which included that in Hawaii, the Philippines, Puerto Rico, and the Virgin Islands. Another was assuring adequate supplies of sugar for American consumers, industrial as well as domestic. A third was supporting sugar production abroad so that Americans could have access to this sugar when needed and foreign producers would have the funds for purchasing other American goods.[26] The Supreme Court struck down the Agricultural Adjustment Act in 1936, but Congress quickly passed a Sugar Act that extended the quotas for another three years.

With its expansion well under way, U.S. Sugar began chafing under restrictions that limited Florida production to 6 percent of the domestic market. Clarence Bitting argued that Florida was "capable of filling the Nation's sugar bowl at lower cost than it has ever been filled, and, at the same time, pay[ing] higher real wages than have ever been paid in any other sugar-producing area of the world." And if the federal government would lift the quotas, U.S. Sugar could hire more people and thus ease the problems of unemployment. Claude Pepper, the junior senator from Florida, took a similar tack, asking if his state was "forever to be put in a straitjacket and to have no hope of getting out of the class of those discriminated against; or are we to be able to hope that as soon as possible these ironclad restrictions will be removed and we shall have the right to produce sugar according to our natural right?" Following suit, the Florida Chamber of Commerce sponsored what the *Washington Post* described as one of the "most glittering junkets . . . in the long American tradition of free travel." A luxury train carried a host of senators, representatives, government officials, and their wives to Miami for a day of play, then to Clewiston and Canal Point.[27] Despite this effort, the domestic quotas were reinstated in 1940, lifted during the war, and remained in effect until 1974.

The Federal Writers' Project described U.S. Sugar as "a romance of large-scale ingenuity and effort," noting that the firm had ten plantations stretching along fifty-two miles of the shore of Lake Okeechobee. Each plantation had about four hundred "Negro laborers" and a number of white overseers, assistants, timekeepers, and storekeepers. Each sported houses, dormitories, mess halls, bathhouses, churches, schools, first aid stations, recreation facilities, and company stores that sold merchandise at cost.

From a similarly positive perspective, the *Wall Street Journal* reported that the wages paid in the U.S. Sugar mills were higher than those paid by sugar facilities in other countries. With high-quality housing provided rent free, free medical care, fuel, ground for vegetable gardens, and educational facilities, cleanliness and contentment prevailed in its plantation villages. Also, no one who shopped at the company stores was allowed to fall into debt.[28] Critics, however, would point out that U.S. Sugar, like other major corporations, provided these amenities in order to keep its employees on the job throughout the year and from one year to the next, as well as to control their behavior off the job as well as on (Figure 13).

U.S. Sugar was less caring of the several hundred temporary workers who cut its cane. Evidence gathered by the Federal Bureau of Investigation indicated that the firm worked the men long hours, paid them poorly, provided minimal medical assistance when they were cut by machetes or by the cane itself, supervised them closely, dismissed them when they could not meet their quota of up to ten tons of cane per day, provided dismal

Figure 13: "USSC [U.S. Sugar Corporation] Village for Negro Workers in Cane Fields, Clewiston, Florida." Photograph by Marion Post for the Farm Security Administration, 1939. Courtesy Prints & Photographs Division, Library of Congress.

barracks and disagreeable food, and prevented them from leaving before the harvest was done. When these allegations came to light, U.S. Sugar was indicted for peonage. The firm got the indictment quashed on a technicality but did not dispute the findings.[29]

In 1943, as large numbers of Americans joined the military services and war-related industries, Washington began to import large numbers of people from other countries to work in American agriculture. Sugar was not the only beneficiary of this emergency measure, but it was one of the largest. By 1943, Mexicans were harvesting American sugar beets, and West Indians, most of them from Jamaica, were harvesting American cane. Although these immigrants were eager to escape the dire economic conditions at home, they were appalled at the situations they encountered in the United States. For the Jamaicans working for U.S. Sugar, that meant terrible food and living conditions, lower wages than expected, and Jim Crow laws and customs.[30]

One might have expected these temporary-worker programs to expire at the end of the war, but that was not to be. Indeed, temporary agricultural workers would receive H-2 visas as defined under the Immigration and Naturalization Act of 1952. Although the Mexican Bracero Program ended in 1964, the West Indian program remained in place until 1992.

The Cuban Revolution of 1959 and subsequent nationalization of the island's sugar industry led Congress to ban the importation of Cuban sugar and raise the acreage allotments on American cane and beet fields. Within a decade, sugar had become the most important crop in Florida, second only to citrus. U.S. Sugar took the lead, erecting a new sugar mill and, in conjunction with the Savannah Sugar Refining Company, building a refinery nearby. About fifty-one independent farmers organized the Sugar Cane Growers Cooperative of Florida, planted cane on twenty-two thousand acres of land on the southeast edge of Lake Okeechobee, and built a modern mill just outside the small town of Belle Glade.[31] The Florida Sugar Cane League, an association of growers and processors, was formed in 1964. Cuban immigrants jumped at the chance to start again in Florida.

Fernando de la Riva, once the second-largest sugar producer in Cuba, arrived in Florida "with little money but excellent credit" and, according to his Yale-educated son, easily raised the $10 million needed to start the Talisman Sugar Corporation. As most of the farmland around Lake Okeechobee was already under cultivation, Talisman bought eighteen thousand acres of land further south in the Everglades and built a mill at Belle Glade. He hired Cuban immigrants to clear, dredge, and plant the land.[32]

Alfonso Fanjul, a wealthy Spaniard whose family owned a large sugar-trading company, had married Lillian Gomez-Mena, the only child of an equally wealthy family that had been growing cane in Cuba since 1850. With a substantial amount of American real estate in his portfolio, Fanjul was financially well-off when he fled to southern Florida and brought three secondhand sugar mills from Louisiana to Lake Okeechobee. Fanjul's sons— Alfonso Jr. (Alfie), Jose (Pepe), Andres, and Alexander—would continue the businesses, operating under an umbrella organization known as Flo-Sun. Purchase of Gulf & Western's sugar enterprise in 1984 made the Fanjuls the largest sugar producers in Florida and the Dominican Republic, and the Dominican Republic was the foreign country with the largest quota for selling sugar to Americans.[33]

Even as the sugar business flourished in Florida, critics charged it with unfair labor policies, undue political influence, and blatant disregard of the environment. Labor problems came to light in 1972 when 250 Cuban American truck drivers, cane haulers, and handymen employed by Talisman objected that their conditions—twelve-hour days, seven days a week, for $2 an hour with no overtime or fringe benefits—were un-American. The media took notice when a college girl distributing union literature was killed by a company truck operated by a Jamaican cane cutter who had been brought in from the fields to substitute for the strikers.[34]

Representatives of the United Farm Workers were then in Florida trying to unionize workers in the citrus orchards. When they learned that there were ninety-five hundred West Indian cane cutters in the state, they asked the district court in Miami to set aside the federal certification that allowed these temporary workers into the country, arguing that these men were alien, poor, and afraid and that the contracting companies had not made any reasonable efforts to find Americans to handle this admittedly difficult, dangerous, and seasonal job. The judge denied the plea but did note some irregularities in the Department of Labor's worker-recruiting procedures.[35] Yet what Cesar Chavez called the "inhuman treatment" of guest workers continued. A Gulf & Western vehicle overturned in a ditch in 1974, killing one man and injuring eighty-six. The company had a policy of only 80 men per vehicle, but this aluminum van carried 130 men and had no windows, no seats, and no inside lighting. Florida Rural Legal Services Inc. brought another suit in 1975, noting that the West Indians averaged $3.03 an hour for doing what a Labor Department official called "the most arduous and unpleasant task in agriculture." Americans, it was said, would do the work if wages were raised.[36]

In 1983, the House Subcommittee on Labor Standards conducted an investigation of cane cutters in Florida in order to determine why so many guest workers were brought in when there were so many unemployed Haitians living in the United States. They found that the nine thousand Jamaicans in Florida were remarkably adept at this "hard, dirty, and dangerous" work and that management had a decided preference for this "elite corps" of experienced men who could not organize, strike, or effectively protest. The Jamaicans earned at least $4.73 an hour, which was substantially higher than the federal minimum wage. The subcommittee then pointed to piece-rate quotas, industry blacklists, and other barriers that kept the Haitian immigrants and other Americans from availing themselves of their legal right to jobs as sugar-cane cutters. The issue came again to the fore in 1986 when 353 foreign workers were sent packing after they protested about the wages paid by one of the Fanjul operations in southern Florida.[37]

In his powerful exposé, *Big Sugar: Seasons in the Cane Fields of Florida*, Alec Wilkinson argues that by providing an "endless supply of West Indian cutters," the H-2 program protected the sugar-cane industry from competition that might cause it to raise wages, improve working conditions, adopt attainable standards of production, and concern itself with safety and that this amounted to a "hidden subsidy of considerable magnitude." Colman McCarthy, a liberal columnist writing for the *Washington Post*, observed that cutting cane surpassed coal mining and meat packing "as the nation's most dangerous and grueling occupation" and that the sugar companies wanted "docile workers grateful to be earning $6 an hour when relatives back home [were] earning $3 a day, if that." The Department of Labor, he said, needed twelve years of litigation to get toilets into the fields worked by Americans and had even less concern for these non-Americans. West Indian governments, however, appreciated the money earned in the American fields.[38] Writing from a management perspective, the *Wall Street Journal* described "muscular men" from the West Indies doing a job "so grueling" that even Haitian refugees and illegal immigrants would have nothing to do with it. The *Journal* went on to say that, although the cutters were often injured on the job, they could earn about $4,500 to $6,500 a season, which was high by island standards.

The longevity of the temporary-worker program in the Florida cane fields can be explained in part by the strength and skill of the West Indian cutters and in part by the difficulty of developing a machine that could do the work. Because cane is sweetest at the bottom, it must be cut close to the ground, but machines that cut low tended to hit rocks and other debris.

Because cane has little sucrose at the top, this material is best left as waste in the field, but machines did not have good discrimination in this regard. Because cane is very heavy, it is tempting to cut the stalk in several sections, but each cut increases deterioration and thus the speed with which the cane must be transported to the mill. Finally, the boggy nature of the Florida muck prevented the adoption of the mechanical harvesters used in Louisiana, Texas, and Hawaii.[39]

In the end, social factors forced Florida growers to replace their human cane cutters with machines. In 1991, following yet another congressional report on immigrant cane cutters, growers complained that labor litigation had become so costly that they were being "herded toward the use of expensive harvesting machinery that sometimes cuts cane inefficiently and damages the fragile soil." By then, U.S. Sugar was harvesting more than half its crop by machine. Joe Klock, general counsel for Flo-Sun, observed, "Machines can't discriminate between good and bad cane. . . . But the fact is that machines don't sue."[40]

The second serious challenge to Big Sugar pertained to the complicated federal programs needed, the sugar companies believed, to maintain an American industry in the face of unfair international competition, as well as a balance within the United States between cane and beet. Opponents of these programs charged that American consumers paid too much for sugar—often twice as much as consumers in other countries—and that American taxpayers subsidized wealthy agribusiness rather than small family farmers. The problem began with the above-mentioned Jones-Costigan quota plan of 1934. Although the program died in 1974, the tariff was still in effect, and the secretary of agriculture still had authority to maintain price supports. In 1977, U.S. Sugar received $11 million, owing to the federal plan that called for a maximum two-cents-per-pound payment when sugar prices fell below 13.5 cents per pound. In 1978, Charles Stewart Mott, generally described as a philanthropist rather than a sugar magnate, urged Congress to raise price supports to seventeen cents per pound, an action that the administration estimated would cost consumers $1.2 billion.

The sugar program was resurrected in 1981 as part of a complex negotiation in which "boll weevil" Democrats agreed to support Ronald Reagan's budget and tax measures, and Reagan agreed to include sugar in the omnibus farm bill. Rep. John Breaux of Louisiana spoke for many when he said, in this regard, that his vote could not be bought, but it could be rented.[41] The new bill gave sugar producers all the benefits as before and introduced commodity loans so that farmers could grow the crops in question. Since the loans were actually made to the producers who passed the money on to the farmers, the price supports gave the landowners an incentive to repay these loans.

The issue came up again in 1996 when Congress swept aside a number of farm programs in effect since the New Deal. It did not, however, cut price supports for sugar (or peanuts). Congressman Mark Foley, a Florida Republican whose district encompassed the sugar mills around Lake Okeechobee, said at the time, "While free markets sound good on the surface, we're quickly finding that the free market doesn't lead to the best end result." The Fanjuls, who received some $65 million a year from price supports, also played a role in this decision. Alfie Fanjul was a major contributor to the Republican Party and especially generous to Bob Dole, a senator who voted consistently to protect the sugar lobby. Recognizing that business trumped politics, however, Fanjul also gave large sums to the Democratic Party and candidates. The following year, facing another effort to kill the price supports, Alfie and Pepe Fanjul explained to readers of the *New York Times* that the United States' sugar policy "prevents the dumping of foreign sugar, just as its trade policy prevents the dumping of foreign-subsidized cars" and that they were proud "to employ more than 3,000 Floridians making an average of $31,000 a year."[42]

Conservation versus Reclamation

Herbert Hoover visited Florida in February 1929, a few months after the great hurricane that, with a death toll of over two thousand, is remembered as one of the worst natural disasters to hit the country. Appalled by the devastation, he declared the reclamation of the Everglades to be one of the great problems facing American engineers. Accordingly, the massive $145 million Rivers and Harbors Act that he signed in 1930 included $9 million for the Army Corps of Engineers to build a navigable waterway across Florida and an eighty-five-mile dike along the southern and eastern shores of Lake Okeechobee to contain its hurricane-whipped waters. Hoover returned to Clewiston in 1961, when the latter project was renamed the Hoover Dike.[43]

While this reclamation work was under way, Hoover's secretary of the interior was supporting plans for a national park that would save at least some of the Everglades from development. Because one of his concerns was maintaining a flow of fresh water into the rivers at the southern tip of the region to avoid the encroachment of salt water, he argued that no cross canals should interfere with the general drainage of water from Lake Okeechobee.[44] Established in 1947, the Everglades National Park, besides generating education, recreation, and inspiration for visitors, has provided a laboratory for scientists monitoring the health of the ecosystem. National park status, however, neither mitigated the damage that artificial drainage had done to the ecosystem nor prevented further modifications.

Marjorie Stoneman Douglas brought public attention to the ecological problems of the region in her best-selling book *The Everglades: River of Grass* (1947), which opened with the words, "There are no other Everglades in the world," and went on to explain the damage caused by private, state, and federal reclamation projects. The muck was drying and shrinking, and its nutrients were being destroyed. Because of subsidence, much of the land no longer drained toward the sea, and so artificial irrigation was needed in many areas. As the rocks at the edge of the Everglades were dynamited so that canals could pass through, salt water was able to invade the land, seeping into wells, fields, and orchards. Many native plants and animals struggled to survive. For all her energy and eloquence, however, Douglas could not halt the momentum of those who believed that what they called the "Empire of the Sun" must be made safe for agriculture, industry, and human habitation.[45]

In 1948, after severe hurricanes flooded the rapidly urbanizing areas of southern Florida, Congress authorized $200 million so that the Army Corps of Engineers could embark on an ambitious water-management project that would, in time, be seen as one of the most colossal environmental mistakes in American history. On the south side of Lake Okeechobee, the corps would create a web of levees, canals, pumping stations, and reservoirs. On the north side of the lake, it would reroute the meandering one-hundred-mile-long Kissimmee River, forcing it into a straight fifty-two-mile-long ditch.[46]

Conflict between conservationists and reclamationists would flare up, time and again, over the course of the next half century. In 1963, scientists at the Everglades National Park reported a severe man-made drought caused by the Army Corps of Engineers. In 1982, following a "mercy killing" of some 750 deer stranded on barren islands, the ninety-two-year-old Marjorie Douglas was quoted as saying, "I have an increasingly profound conviction that man makes great mistakes" by tampering with the Everglades. Business interests argued that the problem of high water levels had been caused by unseasonably heavy rains, while environmentalists alleged that the water levels were kept artificially high for the benefit of those who grew cane and other agricultural products. Years later, citing the need to restore habitats for panthers and other endangered species, Florida began restoring the natural flow of water by putting the kinks back into the Kissimmee River.[47]

Water quantity was not the only problem in the area. In 1993, scientists working at the Loxahatchee Wildlife Refuge reported that cattails and other nonnative plants were choking out the native plants needed by fish and other animals, and they attributed the rapid growth of these invasive plants to phosphorus from the fertilizers used for cane and other crops. Denying responsibility for the damage, the sugar producers filed dozens of lawsuits to block plans to reduce chemical runoff. Yet another environmental challenge came from the *Bufo marinus*, a South American toad let loose in Florida in 1945 in an effort to control sugar-cane pests. By the 1990s, these large and voracious animals were said to be killing dogs and other pets that touched their poisonous skin.[48]

Secretary of the Interior Bruce Babbitt stepped into the fray in 1994, announcing the largest and most expensive restoration project in the history of the national park system. It would cost $465 million, $322 million of which would come from growers. It would extend over twenty years and put right the quantity and quality of water flowing into the Everglades National Park. After lengthy negotiations, Flo-Sun agreed to pay $100 million to help meet water-quality standards and to withdraw its support for lawsuits that would block federal and state plans for the recovery of the Everglades. U.S. Sugar, however, resisted Babbitt's proposal, even though it had earlier pleaded guilty to eight counts of violating federal hazardous-waste-disposal laws at one of its mills and agreed to pay a $3.75 million fine, the largest hazardous-waste fine in the nation until then.[49]

On President's Day 1996, Vice President Al Gore announced a complex plan to protect the Everglades by taking some farmland out of production and restoring some of the natural flow of fresh water. Of the $1.5 billion price tag for this project, $900 million would come from a twenty-five-year tax of one cent per pound of sugar produced in Florida. Later that day, while President Bill Clinton was in the Oval Office telling White House intern Monica Lewinsky that he no longer felt right about their intimate relationship, he took a call from Alfie Fanjul, presumably about this matter. The tax was presented as a Florida ballot initiative in November, and after millions of dollars were spent on each side of the issue, it went down to defeat.[50]

As it became clear that the 1996 plan would not solve the problem, the Army Corps of Engineers devised a scheme to capture fresh water and sequester it in massive lagoons and underground aquifers. Environmentalists were in favor of taking some two hundred thousand acres of land out of production. The sugar industry, however, recognizing the inevitability of the situation, convinced the corps to reduce the amount of affected land to sixty thousand acres.[51] With an eye toward the 2000 presidential election, Congress included this scheme in an omnibus and bipartisan bill calling for the "restoration, preservation and protection of the South Florida ecosystem." The plan would cost a whopping $7.8 billion, extend over the course of thirty-six years, and be hailed as the largest restoration project in the nation's history. Clinton signed the bill into law on December 11, 2000, an hour or so after the Supreme Court heard oral arguments in *Bush v. Gore*.

chapter 5

BEET SUGAR:
PROFITABLE AND PATRIOTIC

The sweetness of root vegetables had long been known in the kitchen, but it became a scientific fact when the French agronomist Olivier de Serres reported in *Le théâtre d'agriculture et mesnages des champs* (1600) that boiled beet juice was similar to sugar syrup. The next step came in 1747 when a German chemist named Andreas Margraff told the Berlin Academy of Sciences that he had extracted sugar from beets and thus people in temperate lands could break their dependence on tropical cane. Margraff's student, Karl Franz Achard, displayed beet sugar to men of science and heads of state in 1799 and convinced Frederick William III of Prussia to establish a sugar beet farm and beet-sugar factory in Silesia. The British blockade of Continental Europe, together with Napoleon's ban on commerce with the enemy, prevented Europeans from getting sugar from cane grown on the British island of Barbados and stimulated further experimentation with beets. In 1811, after receiving two loaves of domestic sugar, Napoleon decreed that thirty-two thousand hectares of France should be planted with beets and that specialized schools should give instruction "conformably to the processes of chemists."[1] By midcentury, sugar beets were grown on a large scale in most European countries, and beet sugar was beginning to offer serious competition to cane sugar on the world market.

In 1811, while serving as the American minister to France, Joel Barlow sent Dolley Madison "the oddest present that you will receive from France if not of the least value." It was, he said, "a beetroot, of that sort that they make so much noise about as cultivated for sugar," and the First Lady should put it in her garden for seed, "not that I think it worth our while to make sugar of them, but to eat & feed our sheep & cattle."[2] We do not know what became of this project, but we do know that word of sugar beets and beet sugar was soon widespread in the United States. In 1824, in his great oration protesting a protective tariff, Daniel Webster mentioned the attempts "to make sugar from common culinary vegetables, attempts which serve to fill the print shops of Europe, and to show us how easy is the transition from what some think sublime to that which all admit to be ridiculous."[3] Despite fulminations of this sort, Americans were becoming interested in the cultivation of beets and the production of beet sugar, both for individual profit and for national sugar independence.

Joseph M. White, a delegate from the Florida Territory who hoped to see an expanded sugar industry, presented a resolution to Congress in 1830. This would have the president cause to be procured "such varieties of the sugar cane, and other cultivated vegetables, grains, seeds and shrubs, as may be best adapted to the soil and climate of the United States."[4] White's resolution passed with but minor objections, and in 1836, the task was given to the newly appointed commissioner of patents, who, in addition to his other duties, was responsible for America's federal agriculture activities. Of the seventy thousand packages of seeds that the Patent Office distributed in 1848, some were for sugar beets.

James Smithson was an English man of science who died in 1829 and left his estate to the American people to found in Washington, DC, an institution for the "increase and diffusion of knowledge." Since Smithson's language was so vague, suggestions for what to do with his bequest were soon forthcoming. Charles Lewis Fleischmann, a graduate of the Royal Agricultural Institute of Bavaria who worked as a draftsman in the Patent Office, thought it should be used to establish an agricultural school in Washington, replete with fields, workshops, and a beet-sugar manufactory. In a subsequent petition to Congress, Fleischmann traced the history and progress of beet sugar in Europe, suggested that this industry should be seen as of the "highest importance to any and every government charged with the duty of promoting the great interests of a nation," and recommended that an agent be sent to Europe to examine the recent discoveries and improvements in this field. Congress printed five thousand copies of Fleischmann's report but would not send him abroad as "it has been hitherto the policy of the National Government to leave pursuits of this character to the sagacity and ingenuity of individual enterprise."[5]

In a similar vein, John Fitz Randolph, a Whig representative from New Jersey, presented a petition from citizens asking Congress to "foster and aid" the culture of mulberry trees and sugar beets. With encouragement from the Committee on Agriculture, Randolph then canvassed farmers and enthusiasts around the country. Discussing this matter in 1838, Congress recognized that Americans were paying immense sums of money to foreign countries for silk and sugar, that it would be important to encourage the culture of these products in the United States, and that five thousand copies of Randolph's report should be published as a means of correcting the deficiency of "practical intelligence" on the subject. Congress also considered authorizing the president to lease some public land for the cultivation of these crops, but concern with constitutional prohibitions against direct support for agriculture and industry carried the day.[6]

Randolph's report mentioned the Beet Sugar Society of Philadelphia, an organization that had sent the Anglo-American agriculturist James Pedder to Europe, published his *Report . . . on the Culture, in France, of the Beet Root* (1836), and distributed some five hundred pounds of French seeds to farmers around the United States. This society was led by three progressive entrepreneurs: John Vaughan was an English wine merchant who had come to America with letters of introduction from Benjamin Franklin and served as librarian and treasurer of the American Philosophical Society; James Ronaldson was a type founder who served as the founding president of the Franklin Institute and later wrote *Observations on the Sugar Beet and Its Cultivation* (1840); Jacob Snider Jr. was a merchant.[7]

Henry Clay, a Whig politician who favored federal protection of industry and agriculture, had some of this French seed planted at Ashland, his home near Lexington, Kentucky. When it proved more productive than any he had tried before, Clay became convinced that the American climate and soils were better adapted to the growth of beets than those of France. Since the establishment of beet-sugar manufacturing in the United States would redound to the common benefit, information on this matter clearly deserved the "liberal patronage of Government." Aware that Congress would not approve such a bill, however, Clay urged the Philadelphians to turn to state governments. As it happened, Snider had already asked the Pennsylvania legislature for $3,000 to introduce and disseminate information pertaining to the manufacture of beet sugar. This measure passed in the Senate but failed in the House.[8]

Organizations similar to that in Philadelphia were formed in Baltimore, Maryland, and in White Pigeon, Michigan.[9] Jesse Buel, editor of *The Cultivator* and a proponent of scientific agriculture, published a letter from James Le Ray de Chaumont, a French aristocrat whose father had supported the American Revolution, who grew sugar beets on his estate in France and whose sugar had received a gold medal from the French government. Le

Ray noted that labor was more expensive in the United States than it was in France, but land and fuel were less dear. Moreover, recent discoveries and improvements should make the fabrication of the sugar easier and more profitable.[10]

Beet-sugar culture in Massachusetts began with John Prince, a Roxbury farmer who imported seeds from Paris and distributed them through the office of the *New England Farmer*.[11] In 1831, after the Massachusetts Society for Promoting Agriculture offered $20 for the greatest quantity of sugar beets grown on an acre of land, a Charlestown farmer raised almost enough beets on one acre to produce six pounds of sugar.[12]

Maximin Isnard, the French vice consul in Boston, informed the *Daily Advertiser* and the Massachusetts Society for Promoting Agriculture that the manufacture of beet sugar had ceased to be an object of ridicule and that France drew palpable and great advantages from it. Because of beet sugar, he said, France was no longer paying $10 million a year to foreign nations and was spending this amount on her farmers, mechanics, and laboring classes. Isnard had served as superintendent of the beet-sugar school at Strasbourg until allied forces had destroyed this facility in 1816, and he was confident that he could plan and supervise a beet-sugar factory in the United States. Before doing so, however, he would need to obtain French seed and examine the latest machinery and processes.[13]

Isnard hit pay dirt in November 1836 when he addressed a large meeting in Northampton, a Massachusetts town with a large and active liberal community. The Northampton Beet Sugar Company was then formed and raised enough money to send Isnard to France and to import high-quality seed for sale at cost to farmers of the Connecticut River Valley.[14] Edward Church, the spokesman for the company, had encountered beet sugar while serving as the American consul in France and planned to grow beets on his property near Paris. Following the political upheavals of 1830, Church returned to the United States fully persuaded that beet sugar was an "eminently important discovery" that must eventually "prove of inestimable value to our country." To this end he prepared *Notice on the Beet Sugar* (1837), a compilation of English translations of several French publications.[15] He was probably also familiar with the English edition of L.-J. Blanchette, *Refining Sugar from Beets* (1836). Both Pedder and Church reported that the mature beets were grated and the resultant mass squeezed with a hydraulic press. Then, after adding lime (calcium carbonate) to remove many of the impurities—a process also used with cane—the beet sugar would crystallize and be ready for refining.

In March 1837, Church let it be known that, with capital of $200,000, the Northampton Sugar Beet Company was sure to lead all American establishments in the business. In May, the company offered Sicilian seed for forty cents per pound and promised to purchase beets for $5 a ton. One local commentator argued that, unless "trammeled by the prejudice of making innovations in our ordinary crops," American sugar beets would be as successful as those grown in France, and even if the sugar did not live up to expectations, the beets would make splendid fodder for cattle, sheep, and horses. Another warned, however, that "prudence, caution and skill" were as needed as were enterprise and industry. The *Maine Farmer* expressed confidence that the agricultural and mechanical skill of New England was equal to that of France, noting that if beet sugar could be manufactured with profit in Massachusetts, it could be done also in Maine.[16]

After the state legislature offered a bounty of $30 for every hundred pounds of "merchantable sugar" made in Massachusetts from beets raised within the commonwealth, C. A. H. of Northampton reported that he had made some experiments with beet sugar and, despite various difficulties, obtained satisfactory results. "The chemical part of the process, which is thought to be the most difficult, is in reality very simple," he said, "and if the Peasantry all over France have skill enough to go through the required manipulations, I am very sure that the American farmer, if he be not more than half a 'yankee,' can readily acquire the necessary information and can successfully compete with them in the manufacture of this article." When pressed by the editor of the *Northampton Courier*, C. A. H. refused to provide details of his methods. He did say, however, that he had found the business to be not remarkably profitable, and as a private individual, he had done his share "pro bono publico."[17]

Into this situation stepped David Lee Child, an ardent abolitionist who looked forward to the day when Americans would enjoy sugar made from beets grown and processed by free labor rather than from cane grown by slaves. Child had visited beet farms and sugar factories in France, Belgium, and England on behalf of entrepreneurs in Alton, Illinois, who envisioned a beet-sugar business along the Mississippi River. Visiting Northampton in March 1838, Child let it be known that the production of beet sugar abroad was probably more complicated than necessary and that "the industry of a Frenchman is quite a different thing from that of a Yankee."[18]

As a proprietor of the Northampton Beet Sugar Company, Child showed beet sugar at the Massachusetts Anti-Slavery Fair in 1839. He won a silver medal from the Massachusetts Charitable Mechanics Association and $100 from the Massachusetts Society for Promoting Agriculture; he also wrote *On the Culture of the Sugar Beet* (1840). He then turned to other matters, presumably finding that, during a time of economic depression, beet sugar could not pay the bills.[19]

Several members of the American scientific and technical community showed interest in the subject in the 1840s. The Franklin Institute published English translations of some French accounts of beet-sugar science and technology, one of which said that, with 292 factories in operation, the French beet-sugar enterprise could sustain competition from colonies in the tropics. Richard McCulloh, a chemist studying sugar analysis for the federal government, reported that scientific inquiry had so improved the cultivation of sugar beets and the manufacture of beet sugar that cane sugar would soon be facing serious competition.[20]

The final antebellum effort to make beet sugar came from Mormons who had been chased out of the Midwest and were eager to become economically independent of the rest of the nation. Shortly after arriving in Utah in 1847, Brigham Young asked John Taylor, an Apostle then on a mission abroad, to investigate industries that would serve the Saints. After visiting beet farms and sugar factories in France, Taylor raised funds from new converts and organized the Deseret Manufacturing Company. He also purchased sugar-making machinery and shipped it across the Atlantic to New Orleans; it then traveled on boats up the Mississippi to St. Louis and down the Missouri to Ft. Leavenworth, then on wagons across land to Salt Lake City. By the time the Deseret Manufacturing Company had produced some molasses using beets grown from seeds distributed by the Patent Office, it was out of funds. Although the Mormon Church assumed the assets and liabilities and built a sugarhouse, it could not keep the venture alive. Fred Taylor, historian of the later, successful Utah-Idaho Sugar Company, believed that his predecessors lacked the necessary "knowledge of the many delicate chemical processes involved in extracting the juice from sugar beets and converting its liquid sugar to solid crystals."[21]

In the early 1850s, now serving as U.S. consul in Stuttgart and encouraging Germans to immigrate to America, Charles Lewis Fleischmann argued that no crop could be cultivated "more easily and at less cost" than the beet and that "in time, the West [would] produce more sugar than the entire United States require[d], and that without slave labor, but solely with the labor of whites."[22] Indeed, many of the German farmers who came to the United States brought their knowledge of sugar beet cultivation.

Beet Sugar in the Late Nineteenth Century

Established in 1862 when the ascendancy of the Whig Party led to greater acceptance of federal support for internal improvements, the U.S. Department of Agriculture (USDA) was soon planting various varieties of beets and analyzing their sugar content.[23] The commissioner of agriculture reported on the history and current production of beet sugar in 1868, calling attention to the government's encouragement without which this sugar "might not now be numbered among the industries which bless the world," and suggesting that Congress encourage the industry so far as "official means and opportunities may permit." He included an account of beet-sugar manufacture in Europe in his annual report for 1869 and another of the American industry—there were then small factories in several states— in 1870. Unlike such sweet stuff as maple, honey, sorghum, and cane syrup, beet sugar was never a cottage industry.[24]

The Morrill Land Grant Act of 1862 (providing funds for state colleges that emphasized such practical subjects as agriculture and engineering) and the Hatch Act of 1887 (authorizing the establishment of agricultural experiment stations in the several states) provided employment for a cadre of agricultural chemists, many of whom worked to improve the culture of sugar beets in their several locales. One such was Charles A. Goessmann, a German immigrant with a PhD from the University of Göttingen who specialized in sugar chemistry and, as a professor of agricultural chemistry at the Massachusetts Agricultural College (now the University of Massachusetts, Amherst), conducted an extensive course of experiments on sugar beet culture and sound agricultural principles. William Smith Clark, the president of the Massachusetts Agricultural College, supported this work, telling audiences that Massachusetts spent $5 million for foreign sugar each year and asking if it would not be better to "retain our gold at home" and spend it raising sugar beets and building factories.[25]

The Universal Exposition held in Paris in 1867 included an extensive display explaining the complexity and cost of the industry: farmers must coordinate with factories, and factories demanded enormous capital for their erection as well as for equipment and maintenance. For Americans who could not travel abroad, the U.S. commissioners produced an informative and well-illustrated report.[26] Because the science and technology were changing so rapidly, the USDA sent its chief chemist to Europe in 1878, and Congress paid for the publication of his report: 15,200 copies for the House of Representatives, 3,800 for the Senate, and 1,000 for the USDA.[27]

From funds of $10,000 for experiments in the manufacture of sugar from various plants, the USDA promised $1,200 awards for the two best reports concerning beet cultivation.[28]

When he became the chief chemist of the USDA in 1883, Harvey Wiley was tasked with developing a domestic sugar industry. He spent most of his time and attention on sorghum and cane but always understood that beets were the better bet. To this end, he distributed seeds, quizzed growers about their methods of cultivation, analyzed sugar content, and determined which seeds produced beets of optimum sugar content and purity and under which soil and climatic conditions they grew most readily. E. H. Dyer, head of the Standard Sugar Manufacturing Company in Alvarado, California, informed Wiley that beets raised in California yielded "as many tons per acre and [were] as rich in saccharine matter as any produced in Europe." Dyer also argued that the success of this industry depended greatly, if not wholly, on encouragement from the federal government.[29]

Wiley's student, Guilford Spencer, was in Europe in 1884, learning about the very latest beet-sugar techniques. The USDA then bought some new machinery, and Wiley and Spencer tested it on sorghum and cane. They must have been pleased when Congress decided that foreign machinery for beet-sugar production could come into the country duty free.[30] The most important machinery related to diffusion, a process that had originated in France and been perfected in Austria. Here, the beets were cut into thin slices, and these *cossettes* were put into a series of warm-water tubs known as a diffusion battery.

Wiley's 1890 report on *The Sugar Beet Industry: Culture of the Sugar Beet and Manufacture of Beet Sugar* was available at nominal cost, went through several editions, and would be credited with having "formed the basis of most of the work undertaken by private enterprise in this important industry." Under the leadership of James Wilson, who served as secretary of agriculture during the administrations of William McKinley, Theodore Roosevelt, and William Howard Taft, the USDA sent field agents to inspect sugar beet farms and beet-sugar factories across the country and to interview growers and manufacturers.[31] Lewis Sharp Ware, a wealthy Philadelphian who had studied chemistry in Paris, was dismayed that Americans paid an annual tribute of $80 million to foreign countries and foreign labor for sugar and that sugar was "the most costly, and the most bulky single article which we import from abroad." To rectify this situation, Ware wrote *The Sugar Beet* (1880) and established and edited a bimonthly journal of the same name. He also disseminated information about beet culture and sugar manufacture from around the world, distributed tons of beet seeds to farmers around the United States, planted hundreds of acres of sugar beets in New Jersey, New York, and Pennsylvania, and conducted thousands of tests in his private laboratory.[32]

Beet Sugar Becomes Big Business

Nebraska was among the first states with a thriving beet-sugar industry, and numerous people—including scientists, boosters, investors, farmers, farm laborers, and politicians eager to garner public funds for local enterprises— contributed to its success.

Hudson Nicholson was teaching physical sciences at the Nebraska State Normal School when, in 1882, he was tapped to become the first professor of chemistry at the University of Nebraska. Rachel Lloyd, the first American woman to earn a PhD in chemistry—she had gone to the University of Zurich as a widow in her mid-forties—was named associate professor of analytical chemistry at the University of Nebraska in 1887, becoming the first woman anywhere in the world to become a professor of chemistry other than at a women's college. Working together, and using the resources of the local agricultural experiment station, Nicholson and Lloyd analyzed soils, measured the saccharin content of beets, and issued a series of reports on *Experiments in the Culture of the Sugar Beet in Nebraska*. Nicholson and Lloyd also gave lectures on sugar chemistry, the technology of beet-sugar manufacture, the culture of the sugar beet, and the theory of light and the use of the saccharimeter. The course opened with twenty-five students in January 1892. The following month, the Beet Sugar Convention in Nebraska asked Congress for $50,000 to support this program in the hope that it would develop into a full sugar school to meet the demand from within the state and around the country. The federal appropriation did not come through, but the course continued.[33]

Harry Koenig was president of the Citizens National Bank in Grand Island, a Nebraska railroad town along the Platte River, and an enthusiastic booster of the local economy. Envisioning an American beet-sugar enterprise like those he had known in his native Prussia, Koenig had the local soil tested and distributed seeds to farmers, many of whom were German immigrants already familiar with the crop. After sending some local beets to the University of Nebraska and to the USDA in Washington, DC, he was pleased to learn that their sugar content was high. Koenig and his neighbors then raised $100,000 and signed a contract with Henry T. Oxnard, an American of French descent looking for a place to build a beet-sugar factory. Further enticements included a large land grant, liberal tax concessions, a state bounty, and Oxnard's Harvard roommate, who lived in the area.[34]

The Oxnard family had recently sold its Fulton Sugar Refining Company for $750,000 worth of certificates in the Sugar Trust, yielding a profit of over $500,000. Thinking it would be "profitable and patriotic" to introduce beet-sugar production to the United States, Henry T. Oxnard had gone to Europe to learn about the business.[35] By early 1890, he was boasting that capital and experience would enable him to succeed where most other Americans had failed. To ensure efficient production, he imported machinery from France and hired French engineers to set it up. To ensure a steady stream of 350 tons of beets every twenty-four hours, he signed contracts with local farmers who controlled two thousand acres and would cultivate beets using seeds imported and methods approved by the firm. To keep his energy costs in line, he contracted for coal "at a fixed rate for ten years ahead."[36] Oxnard was soon the biggest employer in the area, hiring men to build and work in the factory and paying beet farmers some $40 to $60 an acre, about three times as much as they could get from corn.[37]

A Sugar Palace erected in Grand Island in 1890 proclaimed, "The sugar-beet is king and Nebraska his kingdom" (Figure 14). Local papers described the sugar factory and industry as godsends to Nebraska and testimony to what could be produced in the Great American Desert. The Nebraska building at the 1893 Columbian Exposition held in Chicago featured an exhibit on the production of beet sugar.[38]

The Nebraska delegation to Congress proved understandably supportive of the new industry. Under the leadership of Algernon Paddock, the Senate Committee on Agriculture and Forestry produced a report on the production of beet sugar to accompany a bill authorizing funds for the encouragement of sugar beets and beet sugar. Rep. George Washington Emery Dorsey requested $100,000 for an agricultural experiment station to be located near Grand Island to test every phase of the industry from seed to sugar manufacture. The delegation also supported the printing of 21,500 copies of a report on the Bohemian sugar beet industry.[39]

Oxnard and his brothers built two more beet-sugar factories in 1891, one in Norfolk, Nebraska, and the other in Chino, California, east of Los Angeles. The *Los Angeles Times* described the Chino operation as the biggest ever started in the state and the largest such factory anywhere outside of Germany. It also predicted that sugar beets would prove more profitable than any crop yet grown in Southern California, that Southern California

Figure 14: Grand Island Sugar Palace, late nineteenth century. Courtesy Stuhr Museum of the Prairie Pioneer, Grand Island, Nebraska.

would become the greatest manufacturing country in the West, and that sugar factories would "give us our start." The *Los Angeles Times* elsewhere described the building of the Chino factory as the most important event in the productive history of the region "since the first carload of oranges was shipped east." In 1898, when the Oxnards built an even larger factory in Ventura County, in a location soon to become the town of Oxnard, the *Los Angeles Times* described it as the "greatest enterprise upon the Pacific Coast."[40] While these predictions may have been a bit over the top, beet sugar did become an important factor in the California economy and would remain so until the early twenty-first century, when it faced drought, disease, urbanization, competition from other crops, and a plummeting per-capita consumption of sugar.

After the Oxnard brothers merged their four factories into the American Beet Sugar Company, Henry T. Oxnard informed Congress that New York bankers had insisted on putting some $15 million worth of "water" into that $20 million corporation and that he and his brothers had made about $1 million by selling their shares of this "watered stock." When asked about the morality of this action, Oxnard replied, "You know Wall Street's conscience."[41]

Claus Spreckels, a German immigrant who had gained control of cane-sugar production in Hawaii and sugar refining in California, turned his attention to beets in the late 1880s. After a trip abroad to learn about the industry, he imported German seed and growers, as well as German builders, machinery, and machine operatives, and established the Western Beet Sugar Company at Watsonville, south of San Francisco. His second, larger factory was built nearby in Salinas, an area that would become known as Spreckels.[42]

By 1896, seven considerable beet-sugar factories operated in the United States: two in Nebraska, three in California, and one each in New Mexico and Utah. In addition, there was a small factory in Virginia and another being built in Wisconsin. Just six years later, the country had forty-one factories, with several more under construction. Colorado had the most, but California ranked first in output.[43] By the late 1920s, sugar beet production had become one of Michigan's stable industries, providing a cash crop for farmers and alleviating the economic hardships that followed the rapacious exploitation of the area's once vast timber resources. Sugar beets were among the principal money crops in several counties in Nebraska. By 1961, they were grown on twenty-four hundred farms on over one million acres in 269 counties.[44]

Tariff Protection and Other Government Controls

Henry T. Oxnard boasted in 1913 that he had spent an average of between $10,000 and $20,000 a year for the past twenty-three years to develop a "friendly attitude on the part of the public and public men toward the sugar industry." Moreover, the American Sugar Beet Association, of which he was a founding member and the first president, had spent an additional $5 million for this purpose over the same period, and "none of this money had been spent improperly."[45]

Oxnard and others in the business were especially concerned with the protective tariff, believing, together with Harvey Wiley, that the industry must be fostered until its growth was assured. The Nebraska governor pointed out that his was an agricultural state and its people, who were now establishing beet-sugar industries, "want[ed] every encouragement the Government [could] give them." He went on to say that the sugar industries of Germany, France, and Austria had been developed under protective tariffs, "and it seems impossible for us to hope to obtain the same results in any other way."[46] Congress finally decided that raw sugar could come into the country duty free, but American producers would receive two cents for every pound of sugar made from domestic sources. The bounty was rescinded in 1894, but the Dingley Tariff of 1897 restored sugar duties to their previous high levels. Economists would long debate the merits of the matter, but beet-sugar men would long insist that the tariff had stimulated the growth of the industry.

As beet sugar prospered, the Sugar Trust slashed the price of refined cane sugar in the Missouri River territory to make it competitive with beet sugar.[47] It also promoted a reciprocity treaty that would allow raw sugar from Cuba into the country duty free. Oxnard, in turn, argued that Cuba did not exclude the Chinese, saying, "Our people cannot compete with such cheap labor." He had taken a similar tack against the annexation of Hawaii in the 1890s, arguing that Hawaii employed "coolies and the cheapest kind of Chinese labor" and sent its sugar to the Pacific coast "to compete against the beet sugar produced at American labor wages."[48]

The election of 1912 gave Democrats control of the White House and both houses of Congress, thus the ability to enact a tariff that reduced sugar duties by 25 percent and mandated their extinction within a few years. Then, because of the "numerous, industrious and insidious lobby" that aimed to influence the legislation—the words are those of President Woodrow Wilson—the Senate formed an investigative committee and found that the beet-sugar lobby had coddled lawmakers on both sides of the aisle, bought publicity, and engaged Harvey Wiley to deliver fifty lectures promoting beet sugar and the politics that made it possible.[49] The committee was especially interested in *Sugar at a Glance* (1912), a booklet published at congressional expense on the assumption that it consisted primarily of statistical charts that Henry Cabot Lodge of Massachusetts had displayed on the Senate floor. Before the charts went to press, however, Truman Palmer, the Washington representative of the U.S. beet-sugar industry, had added a lengthy account of "The Influence of Sugar-Beet Culture on Agriculture and Its Importance in Relation to National Economics." The Government Printing Office eventually printed some 1.5 million copies of this protariff publication, and Palmer convinced Lodge to send out 320,000 copies under his official frank. When called on the carpet, Palmer observed that Congress had also authorized the publication of the antitariff booklet *Sugar at a Second Glance* (1913).[50]

The tariff pendulum swung back the other way in the 1920s, aided in large part by the U.S. Beet Sugar Association, which had spent $500,000 since the Republicans had been returned to power.[51] Reed Smoot of Utah, the first

Mormon in the Senate, was a great advocate of protective tariffs in general and of sugar tariffs in particular. Because beet sugar was the most important agricultural industry in the intermountain West in general, particularly in Utah, it was said that "Senator Smoot venerates unspeakably the sugar beet." Following the enactment of the Smoot-Hawley Tariff of 1930, which contained the highest duties ever imposed by the United States, Smoot was quoted as saying, "Without sugar beets . . . the sites of hundreds of thousands of happy contented homes would see little but the prowling coyote and the skulking timber wolf stalking wild deer and smaller game. No one need feel concerned with respect to the increased duties on sugar. . . . They are a national blessing."[52]

In 1934, with sugar production greatly exceeding consumption, Congress limited the amounts of cane and beet sugar that could be produced in the United States or imported from other countries. Since this basic system remains in place today, the beet-sugar lobby now works with the cane-sugar lobby to maintain commodity price supports and limit the amount of foreign sugar in the country.

Labor in the Beet Fields

In 1933, *Fortune* described beet sugar as an American industry with un-American working conditions, noting that the cultivation of sugar beets required vastly more labor than any other common crop in God's green United States. *Sugar* agreed that the industry had been "shackled to 'stoop' labor, hired in battalions to work on hands and knees or stooped doubled all day long at thinning, weeding, topping and loading sugar beets" and acknowledged that it "was almost the last American form of agriculture to leave the age-old peasant level of farming."[53]

The basic tools in American beet fields were no different from those used elsewhere in the world. Weeding was done with hoes, either long handled or short, and the beets were dug and topped with cane knives— essentially machetes with short blades that were squared off at the end and provided with a hook. Bare hands pulled weeds and threw beets onto wagons. On his trip to France in the 1830s, David Lee Child had noticed that weeding was done mostly by women and children. Harvey Wiley noted that the thinning and bunching of the plants was a "laborious" activity that teenage girls and boys could perform.[54]

Beet cultivation on small farms might be done by family and friends, but as fields expanded in response to factory demand, growers turned to others. Chinese workers predominated in California until the Exclusion Act of 1882 halted immigration from that country. Japanese workers then took to the fields in large numbers, and although the Gentlemen's Agreement of 1907 limited immigration, workers of Japanese descent would long remain a vital part of beet cultivation on the West Coast, as would East Indians and Filipinos.[55]

Russian immigrants of German descent—that is, ethnic Germans who had spent some time in Russia before coming to the United States—tended to work in family units as they had in the old country, cultivating a certain number of acres for a fixed amount of money.[56] This situation eventually caught the attention of reformers who believed that children should be in school and that their hours and conditions of work should be regulated. Under the aegis of the National Child Labor Committee, social worker Edward Clopper and photographer Lewis Hine visited the beet fields of Colorado in 1915. There they found some five thousand children aged seven to thirteen years, many of them handling twelve to fifteen tons of beets daily. Few attended school, but all contributed to the family welfare. Subsequent studies, some of them conducted for the U.S. Children's Bureau, found similar situations in Nebraska and Michigan. Newspapers described these children's conditions as "even worse than in the textile mills and in the tenement workshops of the most backward states," noting that sugar beet cultivation provided the most egregious example of the exploitation of children for agricultural labor.[57] The Jones-Costigan Act of 1934, known primarily for initiating crop quotas, included a provision giving the government the power to limit or regulate child labor in the fields. Although the act constituted the first federal intrusion into agricultural labor, it was much weaker than the House version decreeing that crop restrictions on sugar beets "may contain provisions which will eliminate child labor and fix minimum wages for workers" (Figure 15).[58]

Betabeleros were sugar beet workers from Mexico or of Mexican descent. Their numbers, small at the turn of the century, increased dramatically when World War I disrupted immigration from Europe and Asia and when military service and war-related industries attracted many young Americans. Although Congress had enacted a strict immigration law, the secretary of labor relented somewhat: Mexicans could come into the country without passing a literacy test or paying the $8 head tax—if they stayed for no longer than six months and worked only in agricultural fields.[59] The program was terminated in December 1918 but reinstated soon thereafter solely for employment in sugar beet production.[60] The depression of 1921,

Figure 15: Children in the sugar beet fields, Hall County, Nebraska. Photograph by L. C. Harmon in Records of the Office of the Secretary of Agriculture, 1940. Courtesy National Archives and Records Administration.

which threw many Americans out of work, put another end to the program but not to the use of Mexican labor in the industry. Indeed, one scholar has shown that by the late 1920s, "between 75 and 90 percent of beet workers in the Midwest were Mexicans, a yearly total of 15,000 to 20,000 workers" (Figure 16).[61]

Growers who relied on *betabeleros* eventually recognized that stable families ensured a stable workforce that would be available when needed. Accordingly, the Oxnards urged California growers to furnish each Mexican worker with living quarters for himself and his family, which he could call his home and where he could live the year round. The Oxnards went on to recommend adobe, noting that it was inexpensive and the Mexican could build the house himself when not working in the fields.[62]

As in other industries, ethnically diverse workforces could lead to tense situations. Laboring men in Nebraska referred to boys from the Genoa Industrial Training School for Indians as alien labor. "Native" Californian

Figure 16: "Field Worker with Knife Used in Topping Sugar Beets, Adams County, Colorado." Photograph by Arthur Roth-stein for the U.S. Farm Security Administration, 1939. Courtesy Prints & Photographs Division, Library of Congress.

settlers worried that Russian immigrants would lower wages in their area, and white Americans clashed with Mexicans in Michigan.[63] Sometimes, however, workers managed to overlook their differences. Thus, in 1903, the Japanese Mexican Labor Association was able to maintain wage rates in the beet fields around Oxnard, California.[64] World War II caused a substantial increase in demand for sugar and a substantial decrease in labor available to work in the fields. Even as plans for the Japanese internment program were being finalized, farmers, businessmen, and civic leaders appreciated the contributions that Japanese people displaced from their homes could make to their agricultural and public works projects. In the end, thousands of Japanese immigrants and Japanese Americans worked on beet farms in Oregon, Utah, Idaho, and Nebraska.[65] After the expulsion of Japanese workers from California, the governor asked the federal government to take measures to supply Mexican workers in order to avert a situation potentially disastrous to the entire victory program. Following the enactment of the Mexican Farm Labor (aka Bracero) Program in August 1942, some fifteen hundred Mexican men went to California for the beet harvest. The Farm Security Administration provided for their transportation, subsistence, and medical care. Ads for California sugar were soon announcing that "smiling Mexicans—'Good Neighbors' from south of the border—came by the hundreds to work in our sugar beet fields." By 1945 the quota for Mexican agricultural workers was fifty thousand.[66]

A German sugar expert who visited Nebraska in the early 1890s noted that American wages were high compared to those in Germany, but so was American productivity. He also noted that the American farmer was "unacquainted with institutions that oblige[d] him to pay dues on insurance against sickness and accidents, pensions for old persons and invalids and obligations toward laborers disabled in his employ." In the 1920s, leaders of organized labor argued repeatedly that there would be no shortage of labor in the beet industry if growers would pay living wages to American workers. Management responded that with a commodity like sugar, wages reflected the international market.[67]

Machines would eventually displace most of the people who labored in the beet fields. The USDA had begun promoting mechanized sugar beet cultivation in the 1890s, but the first robust machines were developed in the 1930s under the aegis of California's agricultural experiment station and the U.S. Beet Sugar Association. The war-related labor crunch and rising demand for sugar forced the issue, but while various designs were available for production, the needed raw materials were not.[68] As the war drew to an end, the industry formed a development foundation with the express aim of improving mechanization and making stoop labor a thing of the past. With the return of peace and prosperity, beet farmers purchased tractors and various machines that sheared multigerm seed balls into smaller units, planted the seeds, and blocked, harvested, and topped the plants.[69] Furthermore, the introduction of pelleted, and later monogerm, seeds substantially reduced the need for hoeing and the annual scramble for itinerant labor.[70] Thus, when the Bracero Program ended in 1964, these workers were no longer needed in such large numbers.

Irrigation

Since it takes eight hundred pounds of water to produce one pound of beet sugar, sugar beets were known as water hogs, and farmers who grew this crop in arid regions readily resorted to irrigation.[71] A historian of the beet-sugar business in Colorado has explained that the Great American Desert's irrigation potential could not be fully realized until Colorado abrogated the riparian rights common in the East, replacing them with the so-called Colorado Doctrine according to which the first settlers in an area could divert waters for their own uses, be they mining or farming, even though this would diminish the flow downstream. Put in place in 1876 and copied in other states, this doctrine revolutionized methods of water control throughout the West. Congress weighed in on the issue in 1894 by passing the Carey Act, which stated that proceeds from western land sales would be given to arid regions, states, and territories for public irrigation projects. The Newlands Act of 1902 went a step further, providing federal funds for reclamation purposes. By 1930, the United States had invested $100 million in the construction of reservoirs, canals, and water-distribution systems.[72]

Sugar Beet Seeds

Because American farmers relied on European sugar beet seeds and European farmers tended to get the choicest ones, many Americans pushed for a domestic source of seed. Harvey Wiley noted in 1898 that growing seeds was "one of the most intricate features of the whole enterprise, requiring a large investment of capital and the application of considerable scientific knowledge"; thus, "it will be some years before the United States will have fully established a safe and reputable sugar-beet seed production."[73]

The world war sparked by the assassination of Archduke Franz Ferdinand in Sarajevo in June 1914 caused the price of beet seed to skyrocket and the supply of seed in warehouses along the eastern seaboard to dwindle. Growers in Utah, having learned of the impending trouble from Mormon missionaries in Europe, had stocked up on seeds. Others, however, had to scramble. The *Los Angeles Times* reported that, despite curtailed cable and telegraphic facilities and paying $40 a bag (about five times the prewar cost), the Anaheim Sugar Company managed to secure four hundred of the nine hundred bags of sugar beet seed known to be "high and dry above the great tidal wave of war now sweeping Europe."[74]

As the crisis escalated, the federal government decided that international law did not ban the importation of sugar beet seeds, dyestuffs, and chemicals from Germany, especially if these goods were shipped from a neutral country. In January 1915, the government facilitated the purchase, from factories in Rotterdam, of 115,000 bags of German seed worth $635,000 in gold and convinced Britain to refrain from attacking ships carrying seed. When the United States entered the war, the German government cancelled all sales, forcing Americans to look to Russia for seed. The armistice of November 1918 made German seed again available to American growers.[75]

The modest program to develop American sugar beet seeds, begun in 1902 at the USDA in Washington, DC, and at several agricultural experiment stations in the West, expanded dramatically in 1916 when Congress appropriated $10,000 to aid "in the development and improvement of American strains of sugar beet seed and especially for the establishment of a permanent sugar beet industry in the United States."[76] One result of this program was a seed that could be grown in one year instead of two. Another was a monogerm seed that produced one beetroot rather than several, thereby eliminating the need to thin plants in the field. Yet a third was a seed that would produce beets resistant to curly top, a disease that curled the leaves, withered the root, and threatened the survival of the industry.[77]

A plant pathologist with the USDA examined twenty-three theories about the cause of curly top and was satisfied with none. The director of the agricultural experiment station in Utah found a relationship between the disease and the tiny insect known as the leafhopper. A scientist at the Bureau of Plant Industry offered substantial proof that a virus transmitted by the leafhopper from one plant to another caused curly top.[78] In 1929, after enormous losses to farmers and factories, the USDA received a whopping $300,000 for work related to curly top and leafhoppers. An entomologist working for the Spreckels Sugar Company followed the insects to their winter lairs and sprayed them with poison. The better solution, however, would come from beets resistant to the disease altogether. The first of these, known as U.S. No. 1, became commercially available in 1933.[79]

Selling Beet Sugar

A marketing study commissioned by the U.S. Beet Sugar Association in the mid-1940s found that while industrial users understood that sucrose was sucrose, whatever the source, many housewives thought beet sugar had an unpleasant odor. The study went on to explain that cane-sugar refineries operated continuously and consistently, but beet-sugar operations were seasonal, and so the product was less consistent. A massive campaign was soon under way to convince consumers that beet sugar was now as good as cane sugar (Figure 17).[80]

HERE'S A VEGETABLE
That Writes Insurance

Buy sugar and you buy *sucrose*, compounded by Nature in numberless plants and trees. Buy beet sugar and you buy *sucrose* plus an "insurance policy" that guarantees the United States an internal supply of an essential food, no matter what may happen to supplies from overseas areas. Buying domestic beet sugar is good sense, it's good business. And, besides, there's no better sugar!

More than assuring a domestic sugar supply to thirty million Americans, the sugar beet itself is a form of crop insurance to farmers who grow it in a third of our states from Michigan to California. Deep-rooted beets suffer less from hail and pests than many other crops. Beets are a dependable cash crop, not competing with other major crops. Beets demand diversification—not all the farmer's eggs in one basket. Other crops show higher yields when properly grown in rotation with beets. Beet by-products, fed to livestock, improve soil fertility. Beet farms are better farms.

One hundred thousand efficient American farmers grow sugar beets on nearly a million fertile acres. In field and factory the industry gives employment to tens of thousands of workers, and its income energizes the economic life of a hundred communities.

The engaging story of the benefits which the sugar beet confers on business and agriculture is told in a booklet, 'The Silver Wedge,' sent on request.

An industry engaged in developing American natural resources, improving American agriculture, and supplying American markets with an all-American food product

UNITED STATES BEET SUGAR ASSOCIATION

344 GOLDEN CYCLE BLDG.   COLORADO SPRINGS, COLO.

Figure 17: Advertisement touting the importance of sugar beets for the national economy. From *Time*, July 12, 1937. Collection of the author.

Figure 18: The American Crystal Sugar Company in Rocky Ford, Colorado, neglected to mention that the "fine granulated" sugar in this cotton bag came from beets. Collection of the author.

Leading the charge was Irene McCarthy, a home economist who had compiled menus and commissary lists for the Japanese internment camps. Under the name "Nancy Haven," McCarthy ran a test kitchen for Western Beet Sugar Producers Inc., answered letters from the public, spoke to women's organizations, and wrote pamphlets on the use of beet sugar. Reinhold Starke, pastry chef at the famed Mark Hopkins Hotel in San Francisco, was quoted as saying that there was no longer any difference between beet and cane sugar. Other chefs, however, disagree with this assessment, and while observant consumers can opt for "sugar" (presumably from beets) (Figure 18) or "pure cane sugar," the production and consumption of beet sugar remains high in the United States and throughout the world.[81]

chapter 6

CORN, CHEMISTRY,
AND CAPITALISM

Sucrose is a complex sugar, or disaccharide. Each molecule contains twelve atoms of carbon, twenty-two of hydrogen, and eleven of oxygen; thus, scientists express its formula as $C_{12}H_{22}O_{11}$. With the addition of a molecule of water, sucrose can be split into two simple sugars, glucose and fructose. Even though these monosaccharides have the same formula, $C_6H_{12}O_6$, because their atoms are arranged somewhat differently, they have somewhat different properties. A key difference pertains to taste: fructose is somewhat sweeter than sucrose, and glucose is somewhat less sweet. While both can be found in nature, glucose can be produced by treating a starch with acid, and fructose can be produced by treating glucose with an enzyme. Glucose was popularly known as starch sugar or grape sugar, and that made from corn was known as corn sugar or corn syrup.

The glucose story began in 1811 when Gottlieb Kirchhoff, a German chemist working in St. Petersburg, added sulfuric acid to starch and obtained a sweet substance—not sucrose, but a boon nonetheless for those whose access to cane sugar was limited by the continental blockade of goods from Britain and her colonies in the West Indies. In 1838, after other chemists found that Kirchhoff's starch sugar seemed identical to the sugar in grapes and in the urine of diabetics, J. B. Dumas coined the term "glucose." William Allen Miller introduced the term "sucrose" in 1857 to refer to the sugar derived from cane and beets; Francophones preferred the term "saccharose," introduced by Marcellin Berthelot in 1860. By midcentury, the production of glucose from potatoes was widespread, especially in France and Germany. According to one report, more than one hundred thousand pounds of European glucose entered the port of New York each week in 1874.[1]

American involvement with glucose began in 1831 when Samuel Guthrie, a chemist, physician, and inventor living in Sackets Harbor, a small New York town on the shores of Lake Ontario, informed Benjamin Silliman, professor of chemistry at Yale College and editor of the *American Journal of Science*, that he had made sugar from potato starch. With its "delicious sweet" taste, this stuff could be used on the table in lieu of honey, and as it fermented readily, it could be made into beer. He had not, however, been able to get it to crystallize. Silliman tasted a sample of Guthrie's molasses, deemed it "nearly as rich as that from the sugar maple," and was sure that crystallized sugar from potatoes would soon be forthcoming.[2]

With corn so plentiful and sugar so dear, Americans were easily attracted by the possibilities of producing glucose from corn. This business began in the 1860s. By 1875, the profits connected with the business were enormous, and investors could expect an annual return of at least 50 percent. By 1880, there was some $30 million in the enterprise, with ten factories in operation, employing around twenty-one hundred men, and nine others under construction. In 1881, eleven million bushels of corn would be processed, and there was every reason to expect this amount to double in 1882. Glucose cost about one cent per pound to manufacture, and it sold for three to four cents per pound.[3] Much of this business was shrouded in secrecy, but its rapid success was clearly enabled by the chemists who improved the production techniques and helped convince the public that the product was as healthy as other sweeteners. The capitalists who raised funds and greased the skids for factory production also played an important role, as did the farmers who grew the corn, the men and women who staffed the factories, and a culture that favored industrial advancement over worker safety and environmental preservation.[4]

Patents and Privacy

Frederick W. Goessling, a Buffalo man known variously as a professor or "German chemist," obtained three patents in 1864. The first described a method of preparing syrups from Indian corn and from beets, then mixing them together to form a sugar "superior to the best cane-sugar." The second was much the same but claimed that either the corn or beet syrup would make "a good quality of sugar if either [was] taken and treated separately." The third described an "improved compound sugar" made of corn and cane syrups.[5] Goessling's patents were so loosely written that one might imagine he had produced sucrose from corn, but scientists knew better.

Orange Judd, editor and proprietor of *American Agriculturist*, noted with some derision that starch sugar was an old story, but making cane sugar from starch was another matter altogether. Charles A. Joy, the scientific expert at *Scientific American*, agreed. If Goessling "has discovered a cheap process of making cane sugar from corn," he said, "he has made one of the greatest chemical discoveries of the age, but if he is merely changing starch into grape sugar he is accomplishing nothing more than has been done ever since the origin of the art of making fermented liquors from grain."[6]

Others, however, were more credulous, especially when Goessling intimated that he could obtain as much molasses from a bushel of corn as a still could make whisky, and headlines proclaimed, "A new era in the manufacture of sugar. A promised revolution in commerce. A golden road to wealth." Thus, a group of New Yorkers, led by an eminent sugar refiner named Adolphus Ockershausen, organized the Union Steam Sugar Refinery, hired Goessling, and bought his patents for the astonishing sum of $600,000. Goessling received three more patents and saw his corn syrup win a silver medal at the American Institute Fair. Then he died, and when others could not make glucose—Goessling's patents clearly did not explain the key elements of the process—the investors threw in the towel. The glucose on hand, which had turned into a solid, waxy mass, was sold to the *New York Tribune* and allegedly used to make ink rollers.[7]

Goessling's story would be repeated time and again. Patents were issued, but few provided enough detail for others skilled in the art to produce glucose at all, let alone efficiently or cost-effectively. The same could be said of 1881's *Practical Treatise on the Manufacture of Starch, Glucose, Starch Sugar and Dextrine*, an apparently informative book—its fifty-eight engravings covered almost every branch of the subject, "including examples of the most recent and best American machinery"—edited by the proprietor of the Philadelphia Starch Sugar Works and based on the German text written by a professor at the Royal Technical High School in Budapest. The press reported repeatedly that glucose production was surrounded by mystery that the manufacturers "sought in every way to maintain and magnify."[8]

Glucose Production in Buffalo

Arthur W. Fox and Horace Williams were brewers and vinegar makers in Buffalo who, having obtained rights to some of Goessling's early patents, announced in 1870 that they were manufacturing grape sugar for the use of winemakers, confectioners, and brewers.[9] In 1874, with help from a venture capitalist named Cicero Hamlin, Fox and Williams formed the Buffalo Grape Sugar Company. The firm was immediately prosperous, paying dividends of $38,000 after just six months. By 1880, Fox was dead, Williams had gone onto other things, and Hamlin's $20,000 investment was yielding weekly profits of $30,000 to $40,000.[10] By 1884, Buffalo had several glucose factories, glucose production was among the most important interests in the city, and some one thousand men were employed in the business.[11]

Hamlin shunned everyone who might discover and reveal the details of his machinery, methods, or business practices. At most an enterprising reporter could learn that the main building of the Buffalo Grape Sugar Company stood seven stories tall, covered an area of eighty by one hundred feet, and was "fitted up in the most elaborate manner to accommodate the enterprise which it occupie[d]." The corn was taken from the railroad cars to the top floor, where it was cleaned, crushed, and, through chemistry, converted into "a gummy semi-liquid." Further processes and filtration changed this stuff into a "pure, limpid thick syrup," then into a product that looked "not unlike refined mutton tallow, or paraffine in cakes" and, while not sweet enough to take the place of cane sugar for family use, was "of the greatest utility for many manufacturing purposes." The factory produced three hundred barrels of syrup daily, but its facilities were "taxed to their utmost capacity; so much so, indeed, that its work is uninterrupted, day and night, from one month's end to another."[12]

Hamlin and his sons owned the rights to several patents, but it is not clear which, if any, were important to their production techniques. It is clear, however, that there was much patent litigation. One fairly egregious case involved William Jebb and his son Thomas, who together would own some thirty patents on subjects as diverse as churns, railroad tickets, and methods and apparatus for recovering starch from grain. The Jebbs also had a large stake in the American Grape Sugar Company, which they sold to the Hamlins for $80,000. The Jebbs then bought several patents pertaining to processes and machinery for manufacturing starch, assigned them to yet another firm (the New York Grape Sugar Company), and brought a $5 million patent-infringement suit against the Hamlins. An arbitration panel decided that the Jebbs were due a royalty of twenty-five cents on every bushel of corn used in this way, but a U.S. circuit court judge disagreed, noting that several glucose firms—including those run by the Hamlins and by the Jebbs themselves—had used the techniques in question for years. That is, the Jebbs sought to make the Hamlins "pay the profits which were created by their own acts of infringement."[13]

Glucose production required lots of water; so, when facing a hike in the Buffalo water rates that would cost them an additional $42,000 a year, the Hamlins approached communities offering concessions to attract business. In Des Moines, Iowa, for instance, the Board of Trade proposed exempting Hamlin's firm from all taxes for ten years.[14] In the end, however, the Hamlins bought a glucose company in Leavenworth, Kansas.[15]

Prussian-born physician and inventor Joseph Firmenich's Sugar Refining Company in Buffalo provided employment for 350 men. In time, however, the Firmenichs moved to Marshalltown, Iowa, "where water [could] be obtained for the mere pumping."[16] In due course they were sued for being a noxious neighbor. One case alleged that the factory and cattle sheds released large amounts of "acids, poisons, manure and other filth" into a creek, making the water unfit for stock or domestic purposes. Another case sought to recover damages for injuries to health and depreciations in value in various properties caused by "foul and offensive odors emanating from the factory." The jury in the latter case found for the Firmenichs, leading the industry to anticipate that the verdict would "shut off much pending and prospective litigation." In another case, however, the judge found that "the discharge from the sewer was as 'rank' as Richard's 'offense' is abundantly shown." When the Firmenichs objected that they were in a different county from the plaintiffs, the judge found "sufficient warrant for the assumption that the ordinary law of gravitation is in full force in both counties, and that water at the mouth of the sewer, whether foul or pure, finds its way down stream without regard to county lines." When Marshalltown had no adequate means for the "proper or scientific disposal" of its sewerage, however, the Firmenichs lent the city $25,000 to erect a suitable plant for this purpose.[17]

Glucose work was probably safer than work in heavy industry, but there were risks nonetheless. One case involved a man who worked and died in a room containing eighteen large shaker boxes as well as "belts, pulleys, a revolving shaft, tanks, pillars and other things occupying much space." When the family sued the Firmenichs, the court decided that the man's death resulted from his own negligence. Since he had worked in the factory for three months, it could be assumed that he "did not exercise the care which the instincts of nature demanded for his safety."[18] The biggest danger, however, came from the starch dust that tended to explode when heated.[19] An 1894 fire in Buffalo killed twelve men, most of Polish descent, and hastened the Hamlins' exit from the business.[20]

Glucose Production in Chicago

Franz O. Matthiessen would eventually obtain more than thirty patents pertaining to the process and machinery of sugar and glucose manufacture. He had apprenticed in Hamburg before immigrating to the United States, joining with William Wiechers, and establishing a sugar refinery in Jersey City, New Jersey. Family connections were also important: his uncle, Henry O. Havemeyer, was the wealthiest and most powerful sugar refiner in the country.[21]

Becoming interested in glucose, Matthiessen and Wiechers hired Arno Behr, a young German chemist with a PhD from the University of Heidelberg, and set him to work on corn. As Behr's patents began piling up, Matthiessen and Wiechers organized the Chicago Sugar Refining Company and named Behr its superintendent. Behr would later receive the Perkin Medal for distinguished service to applied chemistry from the Society of Chemical Industry; the citation noted that the practicability of the glucose and starch industry depended in large part on his inventions, discoveries, and appliances.[22]

Like many things during the Gilded Age, the Chicago Sugar Refining Company was enormous and expensive. Covering eight acres of land between railroad tracks and the south branch of the Chicago River, it could handle some eight million barrels of corn a year, or about six times as much as the next-largest plant in the country. The concern comprised a sugarhouse of eleven stories plus a basement, a filter house for reprocessing the bone black used to whiten the sugar, a corn house with machinery to empty the railroad cars automatically, a corn-processing building that held five large steam engines and several pumping engines, and a power house with twenty Babcock & Wilcox boilers and machinery for feeding them automatically with coal. All this cost some $1.5 million and was financed, in large part, by Marshall Field, George Pullman, and other local industrialists. Behr, the chemist, served as superintendent.[23]

A reporter noted that the new firm had negotiated a sweet water deal, adding that water "must be had at much less than the ordinary meter rates, as otherwise the cost would absorb the profits." With favorable rates and pumps of its own, however, the firm would receive millions of gallons of water a day for a mere $25 or $30, and the city would make up the reduction of $50 or so by increasing property taxes.[24]

Fires were as much a problem in Chicago as elsewhere. One incident knocked out the walls of one building of the Chicago Sugar Refining Company and wrecked the machinery. After another incident, the firm erected a separate drying house for the starch in order to contain any further explosions.[25] The new structure was built with care, no fire was allowed in the building itself, and the boilers that provided the high-pressure steam for drying the starch were set in a separate building located some 150 feet away. Yet, when a fire broke out in one of the large kilns, the explosion demolished walls and hurled the roof into the air, killing a dozen men and injuring seventeen. Initial reports trumpeted the heartlessness of the firm, which kept production going while the search for bodies was still under way, but management quickly paid off the widows and heirs for $800 to $2,200 apiece.[26]

The Chicago Sugar Refining Company ran continuously from 12 a.m. on Monday mornings until midnight the following Saturday. It had some 450 employees, most of them laborers who earned between $1.50 and $1.75 per day. Each carried a brass check with a number that established his right to draw pay from the cashier, and most were immigrants, unknown to each other except by their first names. Thus, in 1902, following a fire described as "one of the most spectacular seen in Chicago in years," it was difficult to identify the eight men killed and the many others injured.[27]

Since "great skill and a thorough knowledge of chemistry" were needed in order to "prevent the losses that must inevitably result from defective manufacture," several chemists were on company payroll.[28] One verified every figure and weight used in the houses and kept careful records of processes, materials used, prices of apparatus and raw materials, profits, and other important factors.[29]

Horizontal and Vertical Integration

Like so many capitalists, glucose entrepreneurs sought to control the market. Cicero Hamlin and his sons acquired a controlling interest in the American Grape Sugar Company,[30] then joined with the Firmenichs to form the American Glucose Company. A reporter covering this merger noted that the interested persons were "reticent" about the matter; nonetheless, it was understood to have been done "with the object of ending competition, creating a practical monopoly, and putting up prices." Since the new firm was incorporated in New Jersey, it could avoid making an annual report such as the laws of New York would compel and paying a corporation tax or any tax except on real estate. It was capitalized with $15 million, most of which had been made from the business during the previous six or seven years.[31]

A National Glucose and Grape Sugar Association was formed in 1882 but could not hold the line in the face of widespread overproduction.[32] A second voluntary association suffered a similar fate.[33] A third effort, made under cover of "all the secrecy that ingenuity could suggest," failed when the Chicago Sugar Refining Company declared it would enter no combine, amalgamation, trust, or pool.[34] But that situation was soon to change.

The Glucose Sugar Refining Company was formed in 1897, largely through the efforts of Conrad H. Matthiessen, president of the Chicago Sugar Refining Company, and his uncle, Franz O. Matthiessen, who was said to have more money than he knew how to spend. This organization, widely known as the Glucose Trust, was chartered in New Jersey and patterned after the Sugar Trust. With $40 million in hand, it bought the Chicago Sugar Refining Company and several other firms. The Hamlins received $3 million for their firm and agreed not to engage in the manufacture of glucose within one thousand miles of Chicago. Three firms not included in the agreement were said to be "in sympathy with the scheme" to elevate prices.[35]

The Glucose Sugar Refining Company succeeded in raising prices and changing the business climate in several ways: it abolished the cash discount that had been offered to jobbers, which meant that these men could no longer offer a cash discount to retailers, and it abolished its long-standing guarantee on corn products.[36] The Glucose Sugar Refining Company was in turn bought out by the Corn Products Company, an enterprise that again aimed to eliminate the cut-throat policy in the industry, raise prices, and turn every bit of the corn available into such products as oil, rubber, starch for laundry or culinary purposes, glucose, and animal feeds and fodders. Conrad H. Matthiessen served as president, and his $75,000 annual salary was said to be the highest in any industrial concern in the country.[37]

The Corn Products Company was subsequently taken over by the Corn Products Refining Company (CPRC), an organization formed by men who, having worked for Standard Oil, understood the benefits of vertical integration and international markets. Accordingly, as reported by business historian Alfred Chandler, they "built up the enterprise's purchasing and sales organizations, moved aggressively into European and other overseas markets, and instituted new policies of packaging, branding, advertising, volume purchasing, and scale economies. The Corn Products Refining Company, the successor to four failures, quickly became, by the definition of a careful student of the merger movement, an 'outstanding success.'"[38] After abandoning most of the facilities of its constituent firms, the CPRC built a new factory in the cornfields some twelve miles southwest of the Chicago loop. This Argo facility was for many decades the largest wet-milling corn plant in the world. As the *Chicago Defender* reported in 1922, the CPRC sought to hire fifteen men and drew no color line. Mamie Bradley, mother of Emmett Till, lived at Argo while her father worked for the firm.[39]

When the federal government brought suit against the CPRC, Judge Learned Hand decided that the firm was a trust in restraint of trade. Without admitting any wrongdoing, it agreed to divest itself of some properties in the United States as long as it could expand its foreign business, which before World War I accounted for 25 percent of total revenues. The Justice Department sought an injunction against "price-fixing and curtailment of production activities" in 1932, but apparently to no avail.[40]

Glucose and Health

In 1882, when Congress began considering a bill to tax and regulate the manufacture and sale of glucose, one witness affiliated with the National Glucose and Grape Sugar Association explained that glucose contained nothing unhealthful or unnatural, that pure honey was 78 percent glucose, and that cane sugar, when eaten, turned partly into glucose. Also, since corn had become king of American agriculture, glucose manufacturers should be applauded for obtaining the greatest commercial value from that magnificent crop. Another witness argued that the bill must be intended to increase federal revenue (which was patently not needed) or to annihilate the industry. A third explained the process of manufacture and outlined the extent of the industry.[41]

In response to considerations of this sort, the commissioner of internal revenue asked the National Academy of Sciences to examine the composition, nature, and properties of glucose, paying special attention to its saccharine content and "deleterious effects when used as an article of food or drink, or as a constituent element of such articles." Written by three eminent academics—Ira Remsen, professor of chemistry at The Johns Hopkins University; Charles F. Chandler, professor of chemistry at the Columbia School of Mines; and George F. Barker, professor of physics at the University of Pennsylvania—the academy's report stated that the manufacture of sugar from starch was a long-established, scientifically valuable, and commercially important industry; that the processes employed were unobjectionable in their character and left the product uncontaminated; that the starch sugar thus made and sent into commerce was of exceptional purity and uniformity of composition and contained no injurious substances; and that, although starch sugar had at best only about three-fifths the sweetening power of cane sugar, it was in no way inferior to cane sugar in healthfulness and had no deleterious effects on the system, even when taken in large quantities.[42]

The American Glucose Company sent reprints of the academy's report to customers around the country to counter the efforts of those who would prevent the use of glucose in articles of food and drink. The firm went on to observe that, since the publication of the report, its sales had increased to such an unexpected extent that it was confident that the use of corn sugar would soon become "as universal as that of any other acknowledged and established ingredient."[43]

The Consumption of Corn Sweeteners

Describing glucose as a triumph of science over nature, *American Grocer* noted in 1869 that it could be made for half the price of cane sugar, providing a "strong inducement" for its use in candy, jellies, jams, and other forms of preserved fruit.[44] Glucose was also used in place of malt in the manufacture of beers and ales. The subsequent expansion of the packaged food industry led to increased sales for both sugar and glucose.

Starch sugar in France and Germany had long been used to adulterate cane and beet sugar and to manufacture brandy. In the United States, by the early 1870s, mixtures of glucose and sucrose were known as New Process Sugar, Niagara ABC, Harlem B, and Excelsior C.[45] Obviously concerned about this practice, Havemeyers & Elder declared their sugars and syrups to be absolutely unadulterated, announcing that neither glucose nor any other foreign substance "is, or ever has been mixed with them."[46] The firm also provided $500 for the New York Board of Health to make a "full and complete analysis of the refined sugars made by the New-York refiners."[47] Since refiners received a tax refund on sugar brought into the United States, refined, and then exported, adulterating sugars for export was fraud. When this practice came to light, federal agents reported that the scam had cost the government nearly $4.5 million in just two years.[48]

Recognizing that they could not stop the proliferation of glucose, sugar refiners argued that glucose should be sold as glucose.[49] The New York State Assembly decided that mixed sugars and syrups must be marked "adulterated" and provided with a certificate of analysis from a "reputable chemist or Professor in some college or university in the State." Violators risked a fine of not less than $10 and imprisonment of not less than sixty days. The governor agreed that goods should be offered and sold for what they were and believed the state's laws regarding adulteration, like those of several other states, were adequate for the purpose.[50]

Corn Syrup

Following its introduction in the early 1870s, corn-based glucose was used primarily in candy and other processed foods. Vast amounts were also used as table syrup. Corn syrup flavored with small amounts of honey or maple syrup sold under such names as Golden Drip Syrup and Crook's Hyper-Sweet Corn Syrup.[51]

Karo, introduced to the market in 1903, was neither the first canned corn syrup nor the first with a brand, but it was the first with a national reach.[52] An advertising campaign that may have been more extensive, expensive, and well illustrated than any theretofore for any food product—except, perhaps, Coca-Cola—helped ensure Karo's success. So did the fact that the CPRC drove competitors out of business by lowering prices, sometimes to below the cost of production, and by refusing to provide the basic syrup for other "mixers"—or so the government alleged after it had investigated the CPRC with regard to restraint of trade. Whatever the truth of this matter, a whopping 430 million pounds of Karo were produced in 1920.[53]

Besides being wholesome and good, Karo was said to be a "pre-digested food which the weakest stomach of infant or invalid [would] readily assimilate." *The Food of the Infant and the Growing Child* told parents to think of Karo when it was time to wean the baby. The Dionne quintuplets, born in Canada in 1934 and nurtured with a mixture of corn syrup and water, were living proof of this contention and in time became Karo's celebrity endorsers.[54]

For homemakers who wished to make optimum use of Karo, the CPRC offered a host of free or inexpensive product cookbooks written by such well-known experts as Helen Armstrong, Emma Churchman Hewitt, and Ida B. Allen. The recipes in one had been "originated by leading professional cooks and endorsed by Domestic Science Experts." Those in another were endorsed by Oscar of the Waldorf Astoria. Jane Ashley, a spokeswoman perhaps as concocted as Betty Crocker and Aunt Jemima, came on board in the 1940s.[55]

All of these cookbooks explained how to make peanut brittle, a delicacy that came to public attention in the 1890s, and those from the early 1940s had recipes for pecan pie. While pecan pie made with sugar had long been known, recipes were few and far between. In the 1930s, however, the editors of the *Washington Post* described pecan pie made with Karo syrup as a southern dish that should become a "necessary addition to every recipe book."[56]

The CPRC marketed Karo as corn syrup, believing that the public associated the word "glucose" with such unpleasant things as the glue factory. Harvey Wiley, the pure-food crusader at the U.S. Department of Agriculture (USDA), opposed this move, arguing that the term "corn syrup" pertained to the sweet product made from the juice of the corn stalk. In the ensuing brouhaha, a slew of chemists, many of them affiliated with elite universities, testified that the word "syrup" could apply to any thick, sweet liquid. Wiley countered that these men had been offered bribes of up to $50 for their testimony, while the state chemists to whom he had submitted the question were wholeheartedly in his camp. The secretary of agriculture overruled Wiley, and Theodore Roosevelt agreed with that decision. "The President and I thought that as a matter of plain common sense," the secretary reported, "it should be called corn syrup if that was the truth of the matter."[57] Food Inspection Decision 87, issued in February 1908, said that the viscous syrup from corn could be labeled corn syrup, and if there was a small amount of refiner's syrup in the mix—that is, syrup made from the dregs of a sugar refinery—it should be labeled corn syrup with cane flavor.[58]

This decision obviously pleased the corn syrup people, but others saw it as a politically motivated nullification of the Pure Food and Drugs Act of 1906 and a "knock-out blow for honest labeling."[59] When Wiley continued to rail against the decision, a corn syrup lawyer forced him to fess up to his connections with the sugar industry: while holding stock in the American Sugar Refining Company, Wiley had sold several patents to members of that firm, and while drawing a government salary, he had enjoyed the hospitality of another sugar tycoon.[60]

As Food Inspection Decision 87 flew in the face of various state regulations, conflicts were inevitable. A Wisconsin grocer was convicted of having sold Karo Corn Syrup in contravention of a state regulation that it must be labeled glucose. The U.S. Supreme Court overturned his conviction on the basis that this instance involved interstate commerce; therefore, federal regulations trumped those of the states. Wisconsin countered by insisting that the word "glucose" appear on the label along with the phrase "corn syrup."

Minnesota, home to many sorghum growers, insisted that corn syrup flavored with sorghum be labeled "sorghum syrup substitute."[61]

In *McDermott v. Wisconsin*, a case involving information placed on shipping container(s) but not the packages that customers would see, the U.S. Supreme Court again decided that federal regulations prevailed in instances of interstate commerce.[62] Elsewhere, however, the Court sided with the Kansas Board of Heath, which required that information on the percentage of each of the ingredients appear on each can of syrup. This state regulation, the Court said, did not interfere with federal requirements but dealt with an issue that Congress had left untouched.[63]

Wiley returned to the question in 1911, issuing a tentative food-inspection decision stating that corn syrup was not a normal ingredient of such food products as mincemeat and thus, when used, must be declared on the labels of such products. Sugar, however, was a normal ingredient, and so need not be declared. The CPRC disagreed, arguing that such labels suggested that corn syrup was a harmful adulterant rather than a wholesome ingredient. After two hearings and a spate of briefs and brochures, the board of the Food and Drug Administration sided with the corn lobby. Wiley then resigned from the USDA, took a post with *Good Housekeeping*, and convinced that magazine not to accept advertising for Karo or other CPRC products.[64]

The By-products of Corn

Luigi Chiozza, a botanist living in Cervignano, a town near Trieste, devised a process for removing the oily parts of corn in such a way that flour made from the farinaceous parts would not become rancid. After seeing the process in operation, Erhard Matthiessen bought the rights to Chiozza's American patent, and his brother in New Jersey asked Behr to modify the process for larger-scale production. While working on this task, Behr developed two ideas that facilitated the commercial exploitation of all parts of the kernel, yielding by-products that made glucose even more profitable than it had been before.[65] Competitors were soon using these techniques, and when the Chicago Sugar Refining Company sued for patent infringement, the circuit court decided the Behr contrivance "had no right to be wearing a patent, inasmuch as it had not the required novelty which puts an invention under the wing of the patent laws." The circuit court of appeals overturned this decision, but by then Behr's patents were soon to expire.[66]

While Chiozza saw corn oil as a problem to eliminate, the Chicago Sugar Refining Company saw it as a valuable commodity, and the *Chicago Tribune* saw it as one of the "latest of the new products which modern science every now and then throws upon the world." Corn oil came to market in 1889 and was used primarily as a lubricant and to soften leather gloves and harnesses.[67] The Hamlins asked an industrial chemist named Ernest Mas to investigate the possibility of producing a corn oil suitable for use in paint and other industrial products. When his ideas panned out, they built a refinery in Peoria, Illinois, under Mas's plans and supervision, paying him a fifteen-cent royalty on every barrel of oil produced. A later contract with the Glucose Sugar Refining Company (see chapter 7) gave Mas an annual salary of $3,600 plus 1 percent of all proceeds from corn oil turned out under his process.[68] An edible corn oil that could compete with olive oil would soon follow, and Mazola would hit the markets in 1911.[69]

With corn oil out of the way, the Chicago Sugar Refining Company began making fluorine, an inexpensive corn flour that that could be mixed with more costly wheat flour.[70] Along with other glucose firms, it also made a nutritious animal feed from corn gluten. Cicero Hamlin boasted that his valuable show horses ate gluten feed made by the American Glucose Company.[71]

Then there was artificial rubber. A visitor to the Chicago Sugar Refining Company saw "a large chunk of 'rubberish'-looking, offensive-smelling substance" that the company chemists claimed to be "the first practical substitute for rubber yet discovered." The formula was jealously guarded, but the visitor understood that this stuff was the "vulcanized product of corn oil" and that five of the firm's chemists had been working on it for several months. It was not ready for market, but the chemists were sure they were onto something big. By 1898, the corn processors were letting it be known that their rubber substitute would give them control of the bicycle tire market and bring annual profits of $1 million. After much confusion and collusion, however, the process came to naught—much like Goessling's goopy stuff some thirty years before.[72]

chapter 7

CANE SYRUP AND CORN SYRUP

Cane syrup, which is clarified and concentrated cane juice, was made wherever cane was grown but was especially popular where weather conditions forced the cane to be harvested before it was ripe enough to produce substantial amounts of crystallized sugar. In the United States, that meant areas in the Southeast. Cane for syrup was traditionally grown on small farms, processed with simple apparatus, and consumed locally. J. E. Bostick of Albany, Georgia, for instance, advertised 125 barrels of "Choice Georgia Syrup" in 1863. Writing of a local farmer who had made four hundred gallons of syrup and cleared $109 per acre, the *Atlanta Constitution* announced in 1882, "The country seems ripe for a change, and in assailing the waning power of king cotton we hasten to place before the people the merits of king cane."[1]

Cane syrup could yield consistent profits only if its quality was consistently high. Much of the work on this problem was done by scientists paid with federal funds under the congressionally mandated Hatch Act of 1887, which aimed to promote "scientific investigations and experiments respecting the principles and applications of agricultural science." Bennett Battell Ross, professor of chemistry at the Agricultural and Mechanical College at Auburn and state chemist of Alabama, knew that the production of cane syrup was a delicate matter: if removed from the fire too soon, it would be watery, but if cooked too long, it tended to crystallize. While cane syrup was delicious in winter, it tended to ferment when the weather turned warm. Thus, for many months of the year, Alabamans were reduced to eating inferior and often adulterated syrups made elsewhere. To alleviate these problems, Ross recommended that makers use a Baumé hydrometer to determine specific gravity, clarify their syrup by adding sulfur dioxide and lime, and put the syrup into well-stopped containers. Horace Stockbridge, a chemist with a PhD from the University of Göttingen who served as the director of Florida's agricultural experiment station, offered similar advice, but he would filter the cane juice through Spanish moss. Similar publications appeared in Mississippi and Texas.[2]

As director of the Louisiana Sugar Experiment Station, William Carter Stubbs was primarily concerned with the sugar business in the southern part of the state. Nonetheless, he hired a syrup maker to show farmers, especially in northern Louisiana, how to make excellent syrup using locally grown cane and inexpensive equipment. At the behest of Daniel Gurgel Purse, president of the Savannah Board of Trade, Stubbs tested Georgia cane and found it remarkably sweet. He also urged Georgia farmers to expand their acreage and the Georgia General Assembly to invest in scientific agriculture. Led by Purse, a group of Georgia planters, capitalists, and railroad men traveled to New Orleans in 1899, visited Stubbs, and saw up-to-date sugar mills and refineries. Stubbs later spent several weeks travelling across Georgia and Alabama, teaching the local people how to make open-kettle cane syrup.[3]

Purse, who had long worked tirelessly to make Georgia cane products profitable, knew that the cultivation of cane in Florida and Louisiana predated that in Georgia. He was quick to point out, however, that Georgia was the first of the original thirteen states to engage in the business. Cane was planted in Georgia in 1825, a mill was erected on a plantation along the Alatamaha River in 1829, and ribbon cane from Georgia sparked the development of the Louisiana sugar industry.[4]

Aware of the enormous help that the U.S. Department of Agriculture (USDA) had given the cane industry in Louisiana, the sorghum industry in the Midwest, and the beet-sugar industry in the West, Purse visited Washington, DC, in early 1901 and convinced Harvey Wiley to analyze samples of soil and cane from his area. Wiley visited Georgia and Florida in the fall and returned a few months later, accompanied this time by Secretary of Agriculture James Wilson. Wiley and Wilson then convinced Congress to appropriate $11,000 for experiments in the region. Similar appropriations would follow in the next several years, some of them specifying efforts to secure syrup of uniform grade and first-class quality.[5]

Seeing cane syrup as a "wholesome, palatable and nutritious article of diet" in competition with "artificial" corn syrups, Wiley approached the project with gusto. Since many sugarhouses had fairly primitive equipment, the USDA would demonstrate the best manufacturing practices and quality control. After erecting a new facility at Waycross, Georgia, USDA agents managed, through sterilization, to prevent fermentation in barrels and other large containers. Using the faucet devised by Louisiana's agricultural experiment station, they found that syrup could be "drawn off, from time to time, in any desired amount and the empty space filled with sterilized air in such a way that no danger of fermentation is incurred."[6]

The USDA also sponsored a series of agricultural experiments in Georgia. It named Walter B. Roddenbery, a progressive farmer living in the southern part of the state, west of the great Okefenokee Swamp, a special agent for the purpose of educating farmers about fertilizers and other up-to-date farming techniques. Roddenbery was not new to the business. His father had graduated from medical school, built a cabin near the Cairo post office, planted cane, and made syrup that he carried with him as he rode from one patient to another. In 1870, when Cairo was chartered as a town, Roddenbery opened a general store, enticing customers with his syrup. Father and son began canning syrup around 1890.[7]

Interstate Sugar Cane Growers Association

Walter B. Roddenbery went to Macon in May 1903 for the first meeting of the Interstate Sugar Cane Growers Association (ISCGA), an organization spearheaded by Purse and designed for those interested in scientific agriculture, agricultural education, cane cultivation, and the manufacture of sugar and syrup. With eight hundred representatives, this was described as the "greatest gathering of agriculturists ever held in the south." Equipment dealers and manufacturers showed their wares. While wealthy farmers dominated the association, they recognized that centralized mills would serve everyone in the neighborhood, especially those of modest means.[8]

As the featured speaker on the first morning, Roddenbery described syrup as an unmitigated good. Since commercial production had taken off, land prices in the Cairo area had increased from $3 to $15 per acre, and the farms presented "a thoroughly prosperous air," "with better houses, better furnished and filled with happier women and children . . . with better schools, more of them and better attended, less political wrangling and fewer Populists and, best of all, fewer and smaller debts and larger bank accounts." Having kept accurate accounts of his expenditures and receipts, Roddenbery could show that by following the fertilizer regimen recommended by the USDA, he had earned over $5,000 on forty-three acres of cane rather than the $258 he expected from other crops. His wage hands earned $9 per month plus rations, the negro women who did his hoeing earned forty cents a day, and the extra male laborers he hired for the harvest earned between fifty and seventy-five cents a day. Roddenbery's argument reached a wide audience; the text was reprinted in full in the *Louisiana Planter*, the leading cane-sugar publication in the United States, and appeared in a pamphlet issued by the Central Georgia Railway and distributed liberally at the Georgia Farmers' Fair.[9]

The second meeting of the ISCGA was held in Jacksonville, Florida. Rufus Rose, Florida's state chemist, explained that cane could be vastly improved by careful selection and intelligent culture. Stubbs of Louisiana discussed two new varieties of cane developed by the Royal Agricultural Society of British Guiana. Wiley discussed the need for a federal pure foods and drug act.[10] Roddenbery talked about the large amount of low-grade and bogus syrup on the market. To eliminate the former, he would have all syrup makers follow best practices, which included using small, hermetically sealed containers. Barrels, he said, would not do. Bogus syrups, primarily those adulterated with corn syrup, should carry labels indicating the percentages of their constituents so that people who wanted to consume it could do so with adequate knowledge. The Georgia legislature had made it a misdemeanor to adulterate syrups but provided no funds for enforcement. Roddenbery also noted that syrup makers needed to expand their customer base. Since northern manufacturers "covered the country with canvassers and free samples, and in various ways created a demand for the article which before was not known to the market," he was willing to spend $1,000 in advertising. He would also have a booth at the 1904 Louis & Clark Centennial Exposition in St. Louis and distribute free samples. Worried that competitors would benefit from his endeavors, he urged the ISCGA to name a committee to devise ways of introducing syrup into new territories.[11]

Many farmers thought that if the industry were to prosper, the labor base would have to expand. Since Italian immigrants had worked alongside African Americans in the cane fields and mills of Louisiana, the ISCGA asked the land and industrial agent for the Southern Railway to discuss his efforts to attract workers from southern Italy. They heard that the Immigration Committee of the Jacksonville Board of Trade was making unusual efforts for a great immigration movement to Florida and that the immigration and industrial agent of the Louisville and Nashville Railroad was bringing two hundred foreign families a month to work in Alabama and Florida. Many of these people would work on cotton and turpentine plantations, but some would find their way to cane fields and mills.[12]

The secretary of agriculture talked about establishing a standard for grading syrups according to color and consistency. This standard, clearly modeled after the Dutch Color Standard long used for sugars, would consist of twenty small bottles, each filled with a different type of syrup. The colors would range from white to dark red, and the numbers would range from one to twenty. The national standard would be kept in the USDA offices in Washington, DC, just as the national standards of length, weight, and volume were kept at the Bureau of Standards, and copies would be distributed throughout the South. The ISCGA endorsed this idea in 1905 but reversed itself the following year, saying that pure cane syrups "are necessarily of different colors owing to the diverse conditions of cultivation and the process of manufacture." The standards for grades of cane syrup adopted by the USDA in 1951 were more sophisticated, taking into account flavor, clarity, defects (extraneous but harmless matter), and color.[13]

The fourth and final meeting of the ISCGA was held in Mobile, Alabama, in February 1906, and because the proceedings were published, we know a fair amount about the issues on the table. Purse railed against congressional efforts "to modify our sugar tariffs in a pharisaical effort to help the Philippines" at the expense of home cane and beet growers. Professor Edward Daniels presented a paper titled "A Plea for Schools for Agricultural Science and Industrial Art." Others addressed the need for state and federal support.[14]

Growth of the Industry

By 1911, local boosters were describing Cairo as "the second largest cane syrup market of America," noting that the product gave their farmers an "abundant" income. Cairo would later style itself the Syrup City; the male athletes at the Cairo High School would be known as Syrup Makers and the female athletes as Syrup Maids. Leaders of the neighboring Thomas County were equally enthusiastic, noting that "Thomas County soil first produced Georgia Cane Syrup."[15]

Seaborn A. Roddenbery Jr., Walter's brother, was elected to Congress in 1910 and spent much of his time opposing miscegenation and pensions for Civil War veterans. In addition, however, he obtained a $10,000 appropriation so that the USDA could investigate the cause of and remedies for red rot and other diseases affecting sugar cane. Frank Park, who succeeded Roddenbery in Washington, helped convince the USDA to establish an agricultural experiment station in Cairo.[16] Julian Roddenbery, Walter's son, returned to Cairo after seeing service in World War I and expanded the business by erecting new buildings, buying syrup from other farms, and introducing new labels and brands. Many other firms across the South followed a similar trajectory.

Nevertheless, cane syrup in the South remained, in large part, a cottage industry. In 1920, for instance, more than seventeen million gallons of syrup came from farms, while fewer than six million gallons came from factories. To achieve the uniformity that would justify higher prices and expand the market, the USDA promoted canning cooperatives. The Farmers Co-operative Cane Syrup Association of Georgia was formed in 1921 with the aim of marketing the crop and erecting a plant to store, standardize, and classify the syrup. It had about two thousand members and received a $35,000 loan from the War Finance Corporation.[17]

Some of the farmers who made cane syrup were African American. Thus, the *Chicago Defender*, an African American newspaper, noted that M. L. Carrington, a black woman who taught domestic science at the Mobile Training School, had won first prize in canning and second prize in homemade cane syrup at a local fair in 1917. It later told of Rachel Cooper, a widow and "one of the foremost women of our Race in the state of Mississippi," who had made five hundred gallons of syrup and grown substantial amounts of cotton and corn on her 160-acre farm.[18]

Recognizing the rising demand, sugar refiners began producing cane syrup, branding it, and sponsoring national ad campaigns. Thus, there was Domino Golden Syrup, Franklin Golden Syrup, and Federal Honey Dew Table Syrup; from England, there was Lyle's Golden Syrup.[19]

Selling Syrup

"Too much *meat* and too little *sweet* is the trouble in many families," proclaimed the Alabama-Georgia Syrup Company, adding that syrup furnished sugar, "the most important of the three food elements—in its most delicious form." Alaga, the world's "best and healthiest breakfast food," was canned in "sanitary surroundings." Velva Breakfast Syrup on biscuits, waffles, or cakes made "the healthiest, tastiest meal for the children." The Roddenberys' Pure Gold Syrup was said to build "strength and energy."[20]

Since modern equipment captured so much sucrose from the cane juice, little was left over for syrup. Syrup makers faced this situation by being emphatically retro. Alaga was made by the traditional open-kettle plantation process. The Roddenberys showed a primitive mill and an open kettle on the label of their Cross Roads Syrup. Mammy-Ann Brand was "open kettle cane syrup." Nigger in de Cane Patch (shortened to Cane Patch in 1930) was "pure Georgia cane syrup [that] has been the standard of purity, excellence and delicious flavor on the tables of southern folks who believe in good things to eat." Dixie Maid and Old Plantation invoked similar images.

Labels for Lou-Anna brand featured a black mammy. New South brand Pure Sugar Cane Syrup had two labels: one featured steamboats, a locomotive, and two black boys; another featured a black boy saying, "Sho is fine."[21]

Many Alaga advertisements included testimonials from famous Americans, from Scott and Zelda Fitzgerald to Hank and Audrey Williams. Recognizing that African Americans who had fled southern fields for northern cities might be nostalgic for down-home flavor, the firm advertised in such community newspapers as the *Chicago Defender*, the *Amsterdam News*, and the *Pittsburgh Courier*, sometimes including the names and addresses of local grocery stores that carried their syrup; it also bought time on WLW, a high-power, clear-channel radio station in Cincinnati, Ohio. The Roddenberrys showed their syrup at a cooking school and food show sponsored by the *Chicago Defender*.[22]

Following World War II, as many African Americans found better-paying jobs in the North, the Alabama-Georgia Company acknowledged that "Negroes" were their best customers.[23] The firm began advertising in *Ebony* in 1955. Later advertisements featured Hank Aaron, Nat King Cole, and Willy Mays. Indeed, Mays would later recall that Alaga was the first paid endorsement to come his way (Figure 19). In the 1960s, as Alaga became emblematic of down-home cooking, Edna Stewart's diner in Chicago was a gathering place for civil rights activists who spent fifteen cents for a plate of corn bread drenched with the syrup. The *Chicago Tribune* ran a feature on a black family that liked its new home in Oak Park but was disappointed to find that suburban stores did not stock Alaga. And yet, when Martin Luther King Jr. called for a boycott of Alabama products after the city of Selma filed a $100,000 suit to recover money it allegedly spent policing the local civil rights demonstrations, Alaga headed the list.[24]

Scientists and Syrup

Because Louisiana sugarhouses used sulfur dioxide and lime to clarify their cane juice, traces of these substances were invariably found in the finished products, be they sugar, syrup, or molasses. Harvey Wiley saw sulfur dioxide as a harmful adulterant and lobbied to have it listed on labels where appropriate. He resigned his federal position in 1912 after being overruled on this matter.

Georgia sugar makers tended to use traditional mechanical means of clarification and so agreed with Wiley regarding sulfur. Walter B. Roddenbery led a group of Georgia sugar men to Washington, DC, to protest this decision in 1913, arguing that the demand for their product would be greater

Figure 19: Willie Mays shilling for Alaga Syrup in 1966. Collection of the author.

when people could distinguish between pure Georgia cane syrup and the "adulterated article." Georgia syrup, they said, might be murkier than that clarified with sulfur, but it was superior in taste. For its part, Alaga boasted that its syrup "contain[ed] no sulphur dioxide, nor other drug or chemical."[25]

As the United States faced sugar shortages during World War I, the USDA increased its efforts to help the syrup industry. One bulletin described the most suitable varieties of sugar cane for syrup production—Louisiana Purple (or Red), Louisiana Striped (or Ribbon), D74, and D95—and discussed the types of equipment most suitable for use on small farms. Other experiments showed that one could avoid the need for sulfur and lime by using infusorial or diatomaceous earth to clarify the cane juice; the resulting syrup was free from dirt and dregs and retained its "natural, mild, agreeable flavor."[26]

In 1920, the USDA trumpeted its discovery of a new method whereby the cane syrup industry might be "stabilized and placed upon a commercial basis." The secret here was invertase, a yeast enzyme that could split (or invert) ordinary sugar (sucrose) into two simple sugars (fructose and glucose). When the right amount of invertase was added, cane syrup would not crystallize or ferment and thus could be shipped in barrels and held until market conditions were favorable. To make invertase available to all cane syrup producers, the USDA convinced a laboratory in New York to produce it on a commercial basis. Seeking a supplemental appropriation from Congress, the USDA promised to furnish invertase to those desiring to try it, to publish directions for its use, and to pay the costs for experts who would travel through the South to explain and demonstrate the process. Charles F. Walton, an employee of the USDA, patented a method for treating cane syrup with invertase in order to prepare better syrup.[27]

Adulterations and Blends

Since the 1870s, much of the sugar, molasses, maple syrup, and honey on the market had been mixed with glucose, and as cane syrup became commercially important, adulteration became problematic here as well. An early federal food inspection decision (number 75) said that the label on blended syrups must indicate which syrups were in the mix and which predominated. Cane syrup with a touch of maple, for instance, should be labeled "cane syrup with maple flavor" or something of that sort.[28]

The *Atlanta Constitution* reported in 1903 that Wilder & Buchanan had sold five thousand barrels of "pure Georgia cane syrup" that was "better than any other variety" and that "appealed to the taste of the epicure." These latter claims may have been true, but with 15 percent corn syrup added to prevent fermentation, the syrup was hardly pure, so the firm was convicted of misbranding.[29] The Alabama-Georgia Syrup Company boasted that Alaga's old-fashioned taste derived from pure Georgia ribbon cane, but the firm admitted in fine print that Alaga was "a blend of pure ribbon cane syrup, with just enough corn syrup to keep the same from sugaring or souring." When authorities discovered that "just enough" amounted to 28 percent, the firm was convicted of misbranding. In another instance, the courts found that Alaga's Unemo brand syrup—said to be "a blend of pure Georgia cane and high grade Louisiana syrup with corn syrup to keep from sugaring or souring"—was actually 45.6 percent corn syrup.[30] The Roddenberys would take these lessons to heart, acknowledging that Dixie Maid and Cross Roads syrups were mixtures of pure cane and corn syrup and that Sunnyland blended corn syrup and pure sorghum syrup. Nigger in de Cane Patch, however, was "NOT a city compound of glucose and molasses" and "NOT a blend of corn syrup and other cheap products."

The Association of American Dairy, Food and Drug Officials approved various standards for syrups in August 1915, anticipating that the secretary of agriculture would follow their lead. The label, they said, should legibly identify the manufacturer (or brand) and the net contents (weight or measure). Mixed syrups "must not be sold under the name of any one ingredient, but may be sold under the names of all of the ingredients arranged in the order of their proportion, beginning with the name of the ingredient present in the largest proportion"; any syrup present in but small amounts should be listed as a flavor; and the words "compound," "imitation," or "blend" should be used where appropriate.[31]

Cane Syrup as a Hobby

Historian Rachel Maines applies the term "hedonizing" to tasks done for pleasure as well as for necessity.[32] Knitting is a classic example, as are woodworking and gourmet cooking. So, too, is cane syrup manufacture. A few firms, most notably C. S. Steen in Abbeville, Louisiana, and Carson Ann in Georgiana, Alabama, still sell large amounts of homogeneous syrup in packages with tamperproof seals and tracking codes.[33] Enthusiasts, however, enjoy the challenges of crafting a quality product and modifying their techniques to match their microclimates and particular soil conditions. Many belong to the Southern Syrupmakers Association, use antique mills and kettles, which they lovingly keep in repair and display at folk festivals and county fairs, and continue to rely on advice from the USDA.[34]

chapter 8

DEXTROSE, HIGH-FRUCTOSE CORN SYRUP, AND SPECIALTY SUGARS

ndustrial producers accounted for most of the sugar sold in the United States in the late nineteenth century, and most of the sugar that Americans consumed was found in food prepared outside the home. Some of these foods were obviously sweet, but some—such as canned vegetables and catsup—were not. The explanation here is that, besides being a sweetener, sugar served as a preservative and a flavor enhancer.[1]

Corn-based glucose, whether in dry or liquid form, had a large and growing share of this industrial market. By the mid-1930s, Americans were using about a billion pounds of corn syrup annually. About half was used in confectionery, and over 250 million pounds were used in table syrup (much of it mixed with cane syrup, molasses, sorghum, maple, or maple flavoring). Bakers used about fifty million pounds, as did brewers. Smaller amounts were used in other foods (such as ice cream) and for industrial purposes (such as in tobacco, tanning, and rayon—or viscose, as it was then known). Large amounts of dextrose (the new name for very pure corn sugar) were used in bread, confectionery, and carbonated beverages, and since it preserved the red color of hemoglobin, it was also used with pickled, fresh, and canned meats and sausages.[2]

The Development of Dextrose

Scientists began trying to produce dextrose in the 1870s but encountered a host of difficulties. The first to claim success was Arno Behr, a German chemist who came to the United States to work for the New Jersey sugar refinery of Franz O. Matthiessen and William Wiechers and found a way to make "crystallized, anhydrous dextrose of great purity and beauty."[3] Building on this promise, Matthiessen and Wiechers established the Chicago Sugar Refining Company (CSRC) and put Behr in charge of operations. Behr soon found, however, that techniques that worked well in the laboratory were not particularly efficient or cost-effective. Moreover, when mixed with cane sugar from Louisiana, making a product attractive to confectioners, the dextrose absorbed moisture and solidified in the barrel. So, the CSRC turned to more traditional corn sweeteners and other products.[4]

Theodore B. Wagner, an American chemist with a PhD from the University of Würzburg who worked with Behr in Chicago, saw the commercial advantages of anhydrous dextrose. Thus, in 1906, while working for the Corn Products Company, he patented a process in which "the yield of sugar is larger, its quality is purer, the time required for its production is shortened, and the amount of labor required is materially lessened." Still, there were problems.[5] In 1911, the Corn Products Refining Company (CPRC) was making dextrose by pressing the liquid out of the slurry—a technique worked out by Charles Ebert, superintendent of one of the firm's factories—and selling it under the trade name Cerelose (Figure 20).

When the advent of war cut the supply of the very pure sugars made in Germany and used for medical and scientific purposes, Americans were forced to make them on their own. Christian Porst, director of the CPRC research laboratory in Edgewater, New Jersey, managed to make some dextrose that was pure enough for medical purposes. So did Richard F. Jackson, a young chemist at the National Bureau of Standards, the federal agency responsible for standardizing the sugar tests that customs agents used for tax purposes.[6] Porst's visit to the bureau in 1915 led to a long and productive collaboration between government and industry in this regard. Exploiting several key ideas generated by William Newkirk, a sugar chemist who had been detailed to the bureau during the war, the CPRC erected an experimental dextrose plant at Edgewater. Newkirk then joined the staff of the CPRC, continued fine-tuning the methods of producing crystalline dextrose "on a commercial scale and by methods which are economically feasible," and convinced upper management to build an industrial facility at Argo, Illinois.[7]

In the mid-1920s, the CPRC announced that Cerelose could compete with cane and beet sugar on the basis of quality and price, and since it was ready for immediate absorption into the blood, it was "far more healthful" than ordinary sugar. In the 1930s, the CPRC boasted that their commercial success showed what private enterprise was doing to conquer the Great Depression. In a similar vein, enthusiastic journalists saw the industry as

"THAT'S WHAT I CALL

Crust Color!"

How often does that remark slip out of you as you take your first look at the day's bake? How often are you really satisfied and pleased with the external appearance of your bread?

Making bread is a complicated business—many factors, in ingredients and production — all have their effect on quality. Cerelose won't solve all your problems, of course.

But Cerelose will *help* you get richer crust color, more uniformly. For Cerelose is directly fermentable by yeast; and in the dough, it produces a slightly quicker caramelization at the surface — just enough to make sure of that lively golden-brown that bread-buyers like so well.

Grain, texture and volume also show improvement. But the first comment bakers usually make when they change over to Cerelose is the definitely better crust color. Join the parade to Cerelose for all bread and sweet yeast doughs!

CERELOSE

PURE REFINED DEXTROSE SUGAR

CORN PRODUCTS SALES CO.   17 Battery Place, New York City

Figure 20: Cerelose advertisement from the trade publication *Northwestern Miller*. Collection of the author.

"an excellent example of the accomplishment of American research genius and the resourcefulness of private enterprise." The CPRC at that time had three corn-processing plants in the United States. The one in Argo was the largest in the world, with two thousand workers processing seventy thousand bushels of corn a day.[8]

As anhydrous dextrose became commercially important, other firms got into the business. The Clinton Corn Syrup Refining Company manufactured Clintose, and the American Maize Company manufactured Amaizo.[9]

Dextrose and Packaged Foods

When World War I disrupted the production and distribution of cane and beet sugar, soft drink manufacturers began using larger amounts of corn sugar and syrup. The many ads for Cerelose in such trade journals as *National Carbonator and Bottler* suggest how important this practice was in the postwar period.[10] Dextrons, made with dextrose, was "an ideal candy for children." Krystal Rock Ginger Ale was "better for children" because it was "enriched with dextrose the energy sugar." The dextrose in Kre-Mel Dessert was "a food element especially suitable for growing children."[11] The substance was also used in breads, ice cream, and condensed milk.[12]

Acknowledging that moonshiners knew dextrose fermented more quickly than cane or beet sugar, the federal Prohibition commissioner estimated that 95 percent of the illegal whiskey consumed in the United States was made from this source. When accused of having conspired to violate the Prohibition law by selling corn sugar to bootleg distributors, the CPRC entered a plea of nolo contendere (no contest) and paid a fine of $5,000. At the time, some 895 million pounds of corn sugar were produced in the United States, but only 40 percent of it could be accounted for in legitimate manufacture.[13] Dextrose was also used in the near beer that home brewers were allowed to make at that time.[14]

The American Medical Association observed, as early as 1931, that dextrose was not as sweet as sucrose. Thus, more of it would be needed to obtain the same level of taste, which meant a higher consumption of carbohydrates. Obesity was already a problem at that time, and this concern would become ever more important in the postwar period.[15]

In December 1941, just weeks after Pearl Harbor, the CPRC announced a major expansion so that corn syrup could replace the sugar no longer available from the Far East (Figure 21). Ads from this period announced, "[The] first responsibility of all food industries is to see that our armed

Figure 21: Patriotic advertisement touting the importance of dextrose for national defense during World War II. Collection of the author.

forces get all they need—of the best our country affords," and Cerelose was the best way to do this. Home economists, inside government and out, helped the war effort by promoting recipes with sugar substitutes.[16]

Truth in Advertising

The *Wall Street Journal* reported in the mid-1920s that the CPRC hoped to market Cerelose "under the pure food laws in the same class as cane and beet sugar."[17] The problem here was Harvey Wiley, the pure-food crusader who had long fought all efforts to foist dextrose on the American people under the guise of "real sugar."[18] While at the U.S. Department of Agriculture (USDA), Wiley convinced the agency that the word "sugar" should apply only to sucrose, the sweetener derived from cane or beets. Moreover, since sugar was not deemed an adulterant of packaged foods, it did not have to be listed on the label, whereas corn sweeteners, which were considered adulterants, did require listing. Wiley left government service in 1912 but continued his fight in books, speeches, and columns in *Good Housekeeping*.[19]

The CPRC challenged this labeling decision and lost in the U.S. Supreme Court. It then turned to Henry C. Wallace, secretary of agriculture under presidents Warren G. Harding and Calvin Coolidge, whom, as an Iowan, the CPRC expected to be sympathetic to corn interests. Sensing a need for political cover, Wallace asked Congressman Cyrenus Cole and Senator Albert Cummins, both of Iowa, to create the public sentiment needed to remove the "discrimination" against corn.[20] This they did, arguing time and again that a label change allowing corn sweeteners to be listed as sugar would lead to an increased use of corn and so bring relief to farmers suffering from overproduction. They also proposed that the Food and Drugs Act be amended to include a label exemption for corn sugar.[21]

Wallace died before he could rule on this matter, and his two successors ducked the issue. But in the summer of 1930, with the Great Depression under way, Secretary of Agriculture Arthur Hyde held a lengthy meeting on the issue. Wiley was no longer living, but his widow was there to present his arguments: "To the housewife sugar is sucrose, and to sell canned goods sweetened with corn sugar without a label would be deception and a violation of the spirit of the food and drug act, which is intended to protect the buyer."[22] Hyde dithered for months. Then, after a congressional delegation from the Midwest paid him a visit, he ruled that "corn sugar (dextrose) when sold in packages, must be labeled as such; when sold in bulk must be declared as such; but the use of pure refined corn sugar as an ingredient in the packing, preparation or processing of any article of food in which sugar is a recognized element need not be declared upon the label of any such product." With this ruling in hand, the CPRC announced its intention of doubling its production of dextrose.[23]

The Food and Drug Administration (FDA), continuing to believe that consumers had a right to know when a cheaper ingredient was substituted for a more expensive one, saw in the Food, Drug, and Cosmetic Act of 1938 the leverage it needed to overturn Hyde's ruling. Accordingly, it organized public hearings concerning such standard foods as canned peaches and tomato catsup. While consumers testified that they thought sugar meant sucrose and nothing but sucrose, eminent scientists testified that all sugars—including dextrose, levulose, maltose, and sucrose—were interchangeable so far as food value was concerned. The secretary of agriculture at that time, Henry A. Wallace (son of the aforementioned Henry C. Wallace), chose to stand behind Hyde's decision.[24]

The Cane Sugar Refiners' Association and the U.S. Beet Sugar Association challenged this decision in court, arguing that the secretary of agriculture was required to base his decisions on "substantial evidence" presented at public hearings, but to no avail. In 1942, with the country at war and sugar in short supply, the Federal Security Administrator (under whose authority the FDA then operated) adopted amended definitions of canned fruit that included dextrose as a sugar.[25] Not until the summer of 1967, when the Fair Packaging and Labeling Act went into effect, did the labels of processed foods have to list all ingredients in order of their prominence. After that, a distinction was made between sugar (sucrose), dextrose, and corn syrup.[26]

High-Fructose Corn Syrup

Glucose, the sugar made from corn, is less sweet than sucrose and far less sweet than fructose, the sugar found in fruit. But since glucose has always been relatively inexpensive, there has long been a desire to enhance its sweetness. Scientists working for the A. E. Staley Manufacturing Company, a large corn- and soybean-processing firm in Decatur, Illinois, developed the first substance of this sort in the mid-1930s. They began in the traditional way by converting some cornstarch to glucose by means of acid (in scientific terms, by acid hydrolyzation), and then they converted more of the starch to glucose by enzymatic means. The new sweetener was marketed as Sweetose, and its commercial success led other scientists to seek other ways to produce high-glucose corn syrups as well as other enzymes that might be used in the food industry.[27]

Glucose and fructose are isomers. That is, they have the same molecular formula ($C_6H_{12}O_6$) but different structures, hence different properties. In 1957, while working in the CPRC's research laboratory in Summit, Illinois, Richard Marshall and Earl Kooi found a microbial enzyme that would isomerize glucose—that is, turn it into fructose. This enzyme only worked when a poisonous arsenate was added as a cofactor, but it did produce a high-fructose corn syrup (HFCS).[28]

Food scientists in Japan then tackled the problem of isolating an enzyme that did not require the poisonous arsenate. As word spread of their success, managers of the Clinton Corn Processing Company flew to Tokyo, obtained a license from the government agency that controlled the patent issued to Yoshiyuki Takasaki and Osamu Tanabe, and returned home with several test tubes of the new enzyme. The *Washington Post* termed this "corporate gold, panned from the grasp of other American competitors."[29] Clinton introduced its first commercial HFCS in 1967. With 15 percent of its glucose converted to fructose, it was known as HFCS-15. HFCS-42, trade name Isomerose, followed in 1970. It tested well with regard to color, sweetness, shelf life, and microbial stability and was soon used in commercial confectionary, baked goods, still and carbonated beverages, pickles, salad dressings, fountain syrups and toppings, table syrups, ice cream, and catsup. HFCS-55, which was as sweet as sucrose, appeared in 1980.

Like other enzymes used in food processing, the Japanese one could be used only once and thus was relatively costly. Clinton chemists solved this problem by attaching the enzyme to an insoluble support so that it could be recovered and used over and over again. Success was announced in 1972, a patent was granted, and the Clinton Corn Processing Company won the Food Technology Industrial Achievement Award in 1975.[30]

Since Clinton's agreement with the Japanese government included the proviso that the original enzyme be made available to other American firms, a mad scramble soon occurred in the industry. The A. E. Staley Manufacturing Company introduced Isosweet in 1975.[31] Archer-Daniels-Midland acquired a controlling interest in Corn Sweeteners Inc., announced plans to build the world's "newest, largest, most efficient" corn-processing plant at Cedar Rapids, Iowa, and began marketing Cornsweet. The H. J. Heinz Company purchased J. C. Hubinger Brothers of Keokuk, Iowa, for $41.4 million, then invested millions more building facilities to produce HFCS, much of which would go into its catsup and other condiments. Amstar, the nation's largest sugar-refining company, purchased a corn-processing plant in Dimmitt, Texas.[32]

The commercial introduction of HFCS came at a time when the price of sugar was going through the roof, and largely because of federal policies designed to aid American growers and producers, Americans paid much more for sugar than consumers in other countries. When Congress allowed the sugar price-support program to lapse, however, the price of sugar on the world market dropped precipitously, from sixty-five cents per pound in 1974 to less than nine cents per pound in 1976. This situation, so disastrous for American sugar growers who apparently needed a price of around eighteen cents in order to break even, was equally disastrous for the makers of HFCS who no longer enjoyed a healthy price advantage. So, pressured by these different perspectives, Congress agreed to restrict the availability of cheap foreign sugar and reinstate price controls for sugar grown at home.[33]

As the price of sugar began to rise, so did the demand for HFCS. Cargill Inc., which had begun making Isoclear in the late 1970s, built a $100 million plant with an annual capacity of over six hundred million pounds of HFCS. CPC International, as the CPRC had become, began to rethink its "prudent and conservative approach to high-fructose capacity" as HFCS continued to gain market share. Tate & Lyle, the British sweetener conglomerate, paid $1.5 billion for Amstar in 1988.[34]

Beverage makers constituted the largest group of sugar users, accounting for some 30 percent of the annual deliveries, and were the first important users of HFCS. This came to public attention in the summer of 1974 when the Coca-Cola Company announced that Sprite, Mr. Pibb, and Fanta might contain as much as 25 percent HFCS. The Royal Crown Cola Company approved a blend of 25 percent HFCS and 75 percent invert sugar in its soft drinks, and Canada Dry approved a limited amount of HFCS in some of its products.[35]

Coca-Cola resisted using HFCS in Coke, but economic considerations finally forced its hand. By 1980, Coke was 50 percent HFCS. By 1984, Coke in cans contained "only" 75 percent HFCS, while soda fountain Coke was sweetened entirely with HFCS.[36] In 1985, following the return to Classic Coke after the debacle of New Coke, the Sugar Association ran an expensive ad campaign stating that the real Classic Coke had been made with real sugar, which cost the manufacturer only a penny more per bottle than HFCS—but to no avail.[37] PepsiCo too resisted the move to HFCS, but by 1983 Pepsi Cola contained 50 percent HFCS-55.[38]

Consumption statistics offer another view of the market. In 1974, average Americans consumed four pounds of HFCS (along with twenty-three pounds of other corn-derived sweeteners and ninety-seven pounds of sucrose).[39] By 1987, they consumed about 129 pounds of caloric sweeteners, and of this, 52 percent was corn. That is, combinations of glucose and HFCS had surpassed sucrose as the nation's top sweetener. Moreover, the United States was the dominant producer of HFCS, and Americans were the world's top consumers.[40]

As the use of HFCS skyrocketed, some Americans became concerned that this sweetener was contributing to, if not causing, the rising obesity epidemic. The Corn Refiners Association responded in September 2010 with advertisements claiming that corn sweeteners were actually sugars and requested of the FDA that the term "corn sugar" be allowed on food labels in place of "high-fructose corn syrup."

Invert Sugar

Practical sugar makers had long wondered why some sugar did not crystallize as readily or completely as expected. Wanting to understand this problem, a French sugar chemist named Augustin Dubrunfaut treated sugar syrup with acid and heat and obtained similar results. That was in 1830. In 1836, Jean-Baptiste Biot, the French physicist who had discovered the optical properties of sugars, put an acidified sugar solution in his polariscope and found that it rotated the plane of polarization to the left, while untreated sugar rotated this plane the right, and so he named this stuff invert (or inverted) sugar. Other scientists would eventually learn that cane and beet sugar (sucrose) was a large, complex molecule that, by means of inversion, was split into one molecule of glucose and one of fructose.[41]

Inversion, it turned out, was an advantage for candy makers aiming to make a smooth rather than gritty product, and the same held for those who made jelly, jam, and marmalade. While some fruits were sufficiently acidic to invert the sugar by themselves, some (such as figs) needed a bit of lemon juice to do the trick. Cooks also found that invert sugar gave a desirable color to baked goods with somewhat less heat, and being hygroscopic, it kept products moist longer and thus extended their shelf life. It would also serve as a malt substitute in brewing beer. A prominent molasses dealer named Noah Taussig brought Nulomoline to market in 1910, describing it as an "invert sugar of uniform composition" that would enable confectioners "to better control texture and consistency and to preserve the freshness of their candies over a longer period." The former problem was of long standing, while the latter was becoming increasingly important as industrialization increased the lag between production and consumption. Nulomoline was primarily sold in industrial quantities: 50-pound pails, half drums of 350 pounds, and full drums of about 630 pounds, as well as a few 9-pound tins.[42]

According to a government report, Nulomoline quickly found use in "textiles, flexible fibers, patent medicines, soft drinks, tobacco, confectionery, antiphlogistic pastes, tooth pastes, printers' rollers, embalming fluid, oil-proof cement, crown corks, flexible leather, book bindings, flexible glue, coated paper, stamping ink, asbestos packing, puncture-proof compounds, and anti-freeze mixtures."[43] By the 1920s, several products were boasting of their invert sugar. Domino Golden Syrup blended refiners' syrup and invert sugar. Domino Sugar-Honey was "a pleasing combination of fresh honey and invert sugar, pure and of exceptional quality."[44] Cliquot Club ginger ale gave instant energy because the invert sugar it contained was "sugar in its most useful, most digestible form."[45] With sugar again in short supply during World War II, invert sugar was used in carbonated beverages.[46]

As invert sugar became commercially important, chemists at the USDA discovered and promoted a method of artificially producing invertase, an enzyme found in some plants and in the intestines of some animals capable of effecting the inversion of sucrose. USDA chemists also promoted the inversion of sugar by means of tartaric acid.[47] The Nulomoline Company introduced Convertit in 1921, describing it as a purified invertase of standardized activity. By using this product, confectioners could make cream centers firm enough for machine handling, a form of production rapidly coming into use in the United States and other industrialized countries.[48]

Liquid Sugar

Liquid sugar is now "an investment, not an experiment," a major sugar-industry analyst proclaimed in 1945, going on to predict a revolutionary and widespread growth in the production of this saturated solution of sugar in water. Liquid sugar offered several important advantages. It fit easily into a continuous production line, a cheaper and more efficient form of manufacturing that was increasingly replacing batch production. It could contain sucrose or invert sugar or any combination desired by the customer. It saved labor and injuries as workers no longer had to handle heavy bags of dry sugar. It promised greater cleanliness. It was stored in tall and thin tanks (while dry sugar, usually stacked only about six bags high, took up much more floor space). Its controls were simple and fast, and it could be delivered quickly and conveniently through its own port (not tying up the buyer's regular loading dock).[49]

The first liquid sugar in the United States came from Cuba and the Dominican Republic. Refined Sugar & Syrups introduced to market the first liquid sugar made in the United States, Flo Sweet, in 1927. After moving from Brooklyn to a larger facility in Yonkers, the firm acquired a fleet of tank trucks, train cars, and boats to deliver Flo Sweet to customers across the country.[50]

Other sugar refiners entered the business in the postwar period, and in 1955 these firms produced more than nine hundred thousand tons of liquid sugar. Beverages took 32 percent of the total, canned and frozen foods took 23 percent, and confectionery, ice cream, and dairy products took a lot as well. Liquid sugar was also used in pharmaceuticals and blood plasma, tobacco processing, paper sizing, textile finishing, and animal feeds. Canned, frozen, and other convenience foods, however, presented the most promising new use, and this market would grow substantially over the ensuing decade.[51]

chapter 9

THE SORGHUM RAGE
OF THE GILDED AGE

Seeing sorghum as a temperate substitute for tropical cane, Americans promoted it as a profitable crop for farmers facing a host of rapidly changing technological and economic conditions. This tall grass grew well throughout the country, and the sweet sap in its stalk was easily expressed and converted to syrup—but the syrup had a somewhat peculiar taste, and its sugar would not crystallize readily. Scientists, inventors, and entrepreneurs rushed into the breach, often with handsome government support, eventually conceding, however, that sorghum sugar could not compete with cane or beet sugar and that sorghum syrup could not compete with that made from cane or corn.[1]

Sweet sorghum originated in two places, China and South Africa. Western interest in Chinese sorghum began in 1851 when the French consul in Shanghai sent some seeds to the Geographical Society in Paris. The horticulturist at the Marine Gardens at Toulon managed not only to get one of these seeds to germinate but to bring the plant to maturity. Louis Vilmorin, a noted horticulturist in Paris, bought eight hundred seeds from an offspring of this plant, allegedly paying eight hundred francs for the lot, and sold some to Daniel Jay Browne, an agent of the U.S. Patent Office who was seeking botanical materials that might prove profitable in America. At about the same time, Leonard Wray, a British horticulturist, found a similar plant at South Africa's Cape Natal and brought seeds to Europe and the United States.[2]

Seeds of Chinese and African sorghum were planted on government land in Washington, DC, and sent to agricultural societies around the country, promoting an immediate and enthusiastic reaction. Books were published, three alone in 1857.[3] Papers were read at scientific and agricultural meetings.[4] Sorghum associations were established.[5] The American Philosophical Society learned that the rapid spread of sorghum in the middle, northern, and western states had added millions of dollars to agricultural resources, that this crop was "the richest acquisition to our agricultural resources since that of cotton," and that it would probably soon free the United States from the tropical monopoly on sugar. Ever alert to the business of mechanics, *Scientific American* called attention to the many contrivances for processing sorghum adapted for farmers "who raise[d] but little of the cane, and who, of course, require[d] an apparatus entirely different from the great mills on our southern plantations."[6]

Farmers soon found that sorghum syrup was easy to make, especially for home consumption.[7] Others were more skeptical. Augustus Allen Hayes, an industrial chemist in Boston, told the American Association for the Advancement of Science that sorghum did not secrete true sugar, its saccharine matter being "purely glucose in a semi-fluid form." Accepting sorghum for what it was, *Scientific American* reported that many western farmers were buying specialized hydrometers for testing sorghum syrup and opined that this syrup would "take the place, in a great measure, of common molasses, among our rural population."[8]

Interest in sorghum increased during the Civil War after Congress raised the tariff on imported sugar and the embargo on the Mississippi River limited access to sugar coming from and through New Orleans. Isaac Hedges sent some sorghum syrup to President Abraham Lincoln in November 1861 with a note extolling recent improvements in the mode of manufacture, reporting complete success in crystallizing the sugar and looking forward to the day when "our whole North West will realize a full supply of good sugar of their own production." Then, at Lincoln's suggestion, he prepared a report for the agricultural division of the Patent Office.[9] Shortly after it was established in 1862, the U.S. Department of Agriculture (USDA) ordered a fresh supply of sorghum seed from China "to meet a very general demand throughout the western States," where its growth "is attended with the most gratifying success."[10]

The USDA published another report on sorghum in 1864, this one written by William Clough, editor of the *Sorgo Journal* and president of the Ohio State Board of Sorgo Culture, who would in 1867 establish a factory based on a new process of refining and deodorizing the syrup that made it attractive to urban customers.[11] A sugar refinery in Chicago was modified for sorghum in 1861, and the West Side Sorghum Refinery opened in the Windy City in 1863. The *Chicago Tribune* published reports of heavy shipments of sorghum arriving daily from central and southern Iowa, Illinois, and Michigan.[12] Camp Sorghum, a bare-bones facility in Columbia, South Carolina, where Yankee prisoners subsisted on cornmeal and sorghum molasses, suggests the importance of these crops to the Confederacy.

The sorghum enterprise slumped in the immediate postwar period, but then the USDA "took it up, and, by its general inferences from insufficient data, kindled an enthusiasm that amounted to a craze." A vast amount of money was lost, and many planters were ruined. By the late 1880s, however, the technology had so improved that growers could have a "reasonable expectation" of selling their crops and supplying the domestic market with syrup. This analysis came from David Blymer, president of a Cincinnati firm that produced machinery for processing sorghum and cane and attracted customers throughout the Americas and from as far away as Australia and New Zealand. Blymer brought out *The Sorghum Hand Book* in 1868 and reissued it annually for many years.[13]

William LeDuc was a graduate of Kenyon College in Gambier, Ohio, who had served as a quartermaster during the Civil War. Becoming commissioner of agriculture in 1877, he found the American sugar industry to be "greatly depressed" and decided to stake his reputation on achieving American self-sufficiency—largely through sorghum. After visiting the Minnesota State Fair and seeing sugar made from a sorghum variety known as Minnesota Early Amber, which to his eyes appeared identical to the common brown sugar of Louisiana, LeDuc asked USDA chemist William McMurtrie to analyze this and other samples of sorghum sugar from around the country.[14]

This work was continued by Peter Collier, a chemist with a PhD from the Sheffield Scientific School at Yale who taught at the University of Vermont before joining the USDA in 1878.[15] Collier was particularly taken with a process for clarifying and crystallizing saccharine liquids recently patented by a Pennsylvania man named Francis L. Stewart, arguing that it clearly indicated the probability that sugar could be made at a profit. The *Chicago Tribune* agreed, claiming that Stewart's process would revolutionize sugar production and lead to "an industry which promises to add untold millions to the nation's wealth."[16] Few criticisms of the Stewart process have been unearthed, but the scarcity of subsequent references in the public record suggests that it did not live up to early projections.

LeDuc announced in August 1879 that the USDA would promote the manufacture of sugar "from whatever sources within our own country" and asked farmers and refiners to send in reports of their experiments, whether satisfactory or not.[17] His annual report contained gorgeous illustrations of the flowers and seeds of various varieties of sorghum, charts of the percentage of glucose in each, analyses of fertilizers, and a request for a well-equipped laboratory to cost not less than $300,000, an experimental farm of one thousand acres nearby, and five experimental stations around the country. This was not all for sorghum, to be sure, but the sorghum project would be the primary recipient.[18]

In an 1880 report on sorghum prepared for the Senate Committee on Agriculture, LeDuc stated that the USDA chemist had "demonstrated that there is practically but little if any difference in the juice of different varieties; that all varieties produce sugar that can be easily granulated, if the cane be taken at the proper period of growth; and that the only important question yet to be determined is as to the variety that will yield the largest amount in a given soil and climate." Obviously pleased with this report, Congress authorized the printing of fifty thousand copies for distribution to constituents around the country.[19]

LeDuc then requested $15,000 for further experiments, asked the Colwell Iron Company of New York about purchasing a complete sugar mill with vacuum pan and centrifugal similar to those used in Louisiana and Cuba, and wrote a short-term contract with Theodore Kolischer, an "eminent Austrian chemist" who had made beet sugar in Germany and cane sugar in the West Indies. When Congress appropriated a mere $6,500, Colwell offered to make the equipment available at reduced cost so that "an experiment of so much importance to the country" would not be put off to another season. However, the equipment did not work well, and the results were far from satisfactory.[20]

George B. Loring, the Harvard graduate and congressman from Massachusetts who became commissioner of agriculture in 1881, was less enthusiastic about sorghum. Indeed, Collier accused Loring of being insufficiently supportive, and Loring accused Collier of having neglected all duties except those pertaining to sorghum.[21] One point of contention concerned the sorghum grown in the District of Columbia. Some was cut before it was fully ripe, some was left in the field until destroyed by frost, and in one instance the USDA grew fifty-two varieties on 135 acres, apparently heedless of the hybrid varieties that would follow from such close contacts. Another point of contention concerned the USDA refinery. One story, perhaps planted by Collier, noted that this operation, however costly, had proved conclusively that one could manufacture high-quality sugar from common sorghum.[22]

At a meeting of the National Academy of Sciences, Benjamin Silliman Jr. read a paper in which Collier presented "important facts" regarding sorghum and its value as a source of sugar; Silliman urged the academy to investigate the matter. The ensuing report surveyed the history of the enterprise, noted that repeated failures had produced distrust of all attempts to renew an industry, applauded the work of the USDA, and concluded that much remained to be done. The report was published at federal expense: two thousand copies for the Senate, three thousand copies for the House of Representatives, one thousand copies for the USDA, and five hundred copies for the academy.[23]

The report, however, provided no guidance for the federal program and no new information of a scientific or technical nature. Historian J. A. Heitmann has suggested that the academy "may have been less than totally objective in evaluating Collier's work" since three members of the committee were from Yale.[24] One member not from Yale was Charles Anthony Goessmann, an agricultural chemist with a PhD from the University of Göttingen, who had come to the United States in 1857 to work for a sugar refinery in Philadelphia and shown that the saccharine matter in sorghum was the same as that in corn—that is, it was glucose rather than sucrose. Later, while teaching at the Massachusetts State College at Amherst, Goessmann grew Minnesota Early Amber from seeds provided by the USDA and concluded that sorghum sugar might be technically possible, but as it needed costly apparatus and skilled labor, it would not be cost-effective.[25]

Loring eventually concluded that the USDA sorghum experiments were too expensive and unsatisfactory and that the work could be better conducted elsewhere. He asked sorghum farmers for information about their methods of cultivation and manufacture, labor costs, and the merits of different kinds of machinery, promising $1,200 awards to the ten best responses.[26] One award went to the Champaign Sugar Company, a firm established by Henry Adam Weber and Melville Amasa Scovell after the Illinois legislature instituted a bounty for those who succeeded in granulating sorghum syrup. Weber, professor of chemistry at the Illinois Industrial University (now University of Illinois), had studied with the eminent German chemist Justus Liebig. Scovell, professor of agricultural chemistry, had conducted experiments showing, he believed, that some varieties of sorghum could produce sugar. Weber and Collier together held two patents, one of which described a process for producing sorghum sugar and syrup with no objectionable odor and taste. When the Champaign Sugar Company venture failed, Weber became head of the chemical agricultural department at Ohio State University, and Scovell found work with the USDA sugar experiment stations.[27]

Another USDA award went to Magnus Swenson, a Norwegian immigrant who had graduated from the University of Wisconsin with a degree in engineering, then joined the agricultural experiment station in Madison. After the Wisconsin legislature authorized funds for sorghum investigations, Swenson planted a large field and built a refinery on the shores of Lake Mendota equipped with the largest sorghum crusher ever made in this country. He also wrote a graduate thesis on sugar chemistry that received an award of $2,500 from the state.[28]

A sorghum factory was established at Rio Grande, near Cape May, in 1881 after the New Jersey legislature authorized a bounty both to local farmers who raised plants that yielded crystallized sugar and to those who manufactured sugar from these plants. The agricultural experiment station affiliated with the Rutgers Scientific School conducted experiments with sorghum cultivation at this time.[29]

The technical expert at the Lafayette Sorghum Sugar Refinery was Harvey Wiley, professor of chemistry at the Indiana Agricultural College (now Purdue University). A graduate of Hanover College and the Lawrence Scientific School, Wiley had worked with eminent sugar chemists in Europe, conducted research on glucose, and been named chief chemist of the state of Indiana. He was also the scientific advisor of the Mississippi Valley Cane Growers' Association (MVCGA), the most active sorghum organization in the country, if not the world.[30]

Wiley met USDA commissioner Loring in December 1882 at the annual meeting of the MVCGA. Although Loring gave no intimation that he was preparing to sack Collier and replace him with someone he found more congenial, this encounter likely got Wiley thinking that the chief chemist of the USDA had a generous salary, a substantial budget, and a bully pulpit. The one obvious drawback was sorghum. If the project succeeded, he would be a national hero; if it did not, he would be blamed for the fiasco.[31] A month later, at the annual meeting of the Agricultural Association in Indianapolis, Wiley gave a deceptively modest paper on this "democratic and domestic sweet" that we can see as the work of an ambitious, articulate, and ambivalent man. There were many problems associated with sorghum, but chemistry joined with practical skill would surely overcome them. So, it would "be wise to make friends with the coming giant." Wiley also offered such other bromides as "Let me make the sweets of the nation and I don't care who makes the laws" and "The consumption of sugar is a measure of progress of civilization."[32]

Loring fired Collier in May 1883 and hired Wiley in his stead. *Manufacturer and Builder*, a generally optimistic publication, explained that the recent sorghum-sugar experiments were undertaken on the advice of the late chemist of the USDA, who without "mature consideration" had given "by far too favorable an opinion of the possibilities of success" of this material for the production of sugar. The editors went on to say that they believed these enterprises would ultimately fail and that it seemed amazing, "in view of the remarkable success of the beet sugar industry in Europe, that our people should throw away money upon such hazardous experiments as these."[33]

Wiley probably agreed, but he arrived in Washington, DC, announcing that the future of sorghum was full of promise, especially when the diffusion process was employed. This process—in which the stalks were cut into small pieces and placed in hot water rather than crushed in a mill—had been developed for the European sugar beet industry. Diffusion equipment was shown at the Centennial Exhibition of 1876 and discussed at length in the reports of the juries. It was also tried with cane in Louisiana and sorghum in Illinois and New Jersey.[34]

Using a cane cutter and a diffusion battery ordered from the Colwell Iron Company, Wiley was soon making sorghum sugar for "only" $1 per pound. The *New York Times* noted that this achievement must have been "very gratifying to the Agricultural Department, notwithstanding that the commercial value of the sugar is about 6 cents per pound." The *Chicago Tribune* noted that the USDA's sorghum sugar had cost $10 per pound three years before and $5 per pound the previous year. Wiley admitted that the results were far from satisfactory but thought that much could be learned from experience with the new process. After including a complicated and probably useless mathematical formula describing diffusion provided by William Harkness of the U.S. Naval Observatory, he closed by hoping that the "liberality of Congress" would be as large as its recommendations and the needs of this great industry.[35]

Wiley traveled to St. Louis in January 1884 for the annual meeting of the Mississippi Valley Cane Growers' Association and gave two talks, illustrated "on canvas by the camera," that attracted wide and appreciative audiences. In the first he explained the diffusion process. In the second he explained that while sorghum contains some 9 or 10 percent of crystallizable sucrose, it also contains other sugars that will not crystallize in and of themselves and actually prevent the sucrose from doing so. He also argued that sweet sorghum would not grow as far north as "many of his friends ha[d] hoped and many of its devotees prophesied." Wiley's second sorghum report continued in this negative vein. Figures showed, he said, that "the importance of sorghum sugar from a commercial view [was] altogether in its possibilities and not in its actual magnitude." And he did not believe the "prevalent idea that each farmer [would] become his own sugar-maker," especially as many syrups made in small mills were far from palatable.[36]

Though perhaps technically correct, Wiley was no match for Senator Preston Plumb and others who envisioned Kansas as the first northern sugar state. Urged by his constituents, Plumb made a thorough study of the subject and, with what would be remembered as "the foresight of statesmanship," secured an appropriation of $50,000 for the "development of the sugar industry." Saddled with funds he did not want, Wiley ordered an improved diffusion battery and cane cutters that, as it happened, arrived too late for use that season. With little success to show, Wiley warned that the chemical division of the USDA "would be guilty of a great public wrong" if it led investors to expect a higher rate of return than the facts warranted.[37]

Norman Colman, named commissioner of agriculture by President Grover Cleveland, had established *Colman's Rural World* in 1868, published a sorghum column on its front page for many years, and succeeded Isaac Hedges as president of the MVCGA. Not surprisingly, therefore, he was determined to solve the problem of making sorghum sugar. Also not surprisingly, being the first Democratic president since before the Civil War, Cleveland would not discourage the amassing of evidence that would discredit previous Republican administrations. Thus, Colman told reporters that the USDA had wasted thousands of dollars on the sorghum project, spending some of these funds on seed at prices ranging from $1 to $2.95 a bushel. When asked how much seed he would have bought, Wiley, who agreed with Colman on this, responded, "Not a pound." "There was no purpose in view. There was no new variety to be tried." Colman would later describe Loring's $1,200 prizes for sorghum information as "fraudulent" transactions.[38] In the midst of this dispute, Congress appropriated $40,000 for sugar research in 1885, mostly for sorghum experiments at the Franklin Sugar Company in Ottawa, Kansas. Wiley then hired Melville Scovell of Illinois to install the USDA cane cutters and diffusion batteries, as well as new carbonatation equipment. Carbonatation, another process derived from the beet-sugar industry, used large amounts of lime to clarify the juice and carbonic acid to precipitate the debris. The initial results were phenomenal. The diffusion process extracted 98 percent of the juice in the stalk, which was twice the amount obtained in the ordinary way. Carbonatation produced a limpid juice with "a minimum of waste, and a maximum of purity." With the two techniques properly combined, 95 percent of the sugars in the cane could be marketed as either dry sugar or syrup.[39]

While this work was under way, Wiley conducted an extensive review of the American sugar industry, showing that in light of cane, beets, and maple, the sorghum situation did not look so good. The sugar works in Hutchinson, Kansas, was the best-equipped sorghum firm in the country, and its process was supervised by Magnus Swenson, who had enjoyed such success with sorghum in Wisconsin. Yet the cost of production far exceeded the proceeds. When a reorganized venture sustained a heavy loss, even Swenson became discouraged. "Under the present low prices the sorghum sugar industry is barely able to hold its own," he said, but if prices could be raised from .5 cents to 1 cent per pound, or if legislation would provide direct aid, the sorghum-sugar industry might be established on a sound basis and add "very materially to the wealth and prosperity of the country."[40]

Wiley spent the winter of 1885 and 1886 in Europe, studying the equipment and techniques used in the sugar refineries in England, France, Germany, and Spain and becoming thoroughly convinced that beets would beat out sorghum.[41] Congress, however, authorized $94,000 for sorghum and cane experiments in 1886. The Senate offered an amendment stating that all machinery purchased with these funds must be "built in the United States wholly of domestic material," but since Wiley had already ordered European equipment, the final bill stated that "all machinery purchased under the provisions of this act shall be built in the United States, wholly of domestic material, except so much of not exceeding $10,000 in cost, as is now under contract, express or implied, or such parts thereof as cannot be built in the United States within proper time."[42]

With funds in hand, Wiley arranged for the new apparatus to be installed at the Parkinson Sugar Works in Fort Scott, Kansas, a new enterprise managed by a lawyer named William Parkinson. Magnus Swenson was the technical expert and received a government appointment to help with the investigations. The relationship soured, however, when Swenson scorned Wiley's machinery and Wiley refused to try the techniques that Swenson had developed. Wiley's first report telegraphed the complete success of the experiments. Later ones, however, announced that diffusion worked well with sugar cane but was an absolute failure with regard to sorghum. "No known process, save an act of creation," he said, "could have made sugar successfully out of such material. If nothing better than this can be obtained, then it is time to declare the belief in an indigenous sorghum-sugar industry a delusion."[43] There was some hope, however, that experiments with seed selection and cultivation techniques would increase the richness and purity of the sorghum syrup.[44]

In December 1886, in his annual address as president of the Chemical Society in Washington, Wiley described the American sugar situation as bleak. The cultivation of beet sugar was a failure, maple sugar could not last much longer, and sorghum was not yet a profitable business, leaving cane as the main hope of the sugar manufacturers in this country. Commissioner Colman, however, remained upbeat, believing that the reports from Fort Scott indicated that "it [would] be but a few years before this country [was] a great sugar-producing nation."[45]

As it happened, 1887 was a great year for sorghum. The Parkinson Sugar Works made over 16,500 pounds per day using new machinery of American make, "constructed according to the ideas of Judge Parkinson and Prof. Swenson, and peculiarly adapted to the diffusion of the particular cane with which they have to deal." A delighted Colman noted that the experiments at Fort Scott were "equally important to this country as the invention of the cotton gin." "Our nation is paying $100,000,000 annually to the sugar producers of other countries," he said, "and by profitably producing sugar on our own farms, we could keep this money at home and save our people $50 million annually in customs duties."[46]

After Swenson reported that sugar could be produced as cheaply in Kansas as in Louisiana, word of this accomplishment spread quickly.[47] The *Washington Post* reported, "It has been demonstrated that a new source of sugar supply has been found in sorghum, which can be grown in all parts of our country as easily as Indian corn, and that it will yield from a thousand to fifteen hundred pounds of sugar per acre, making its production profitable to the farmer and to the manufacturer, and it is contended that its seed will pay the entire cost of raising the crop, a virtue no other sugar plant possesses." For the *Chicago Tribune*, sorghum sugar promised to rank as one of the "chief wealth-producing crops of the West." According to the *Chicago Journal of Commerce*, "Every dollar judiciously invested in the sorghum sugar business at this time will double itself each year for the next ten years." The General Assembly of Virginia asked Congress to appropriate $100,000 for a sorghum factory in Alexandria and a school at which the manufacturing process could be taught.[48]

The success at Fort Scott stemmed, in large part, from a new carbonatation process developed and patented by Swenson. This used a carbonate of lime rather than caustic lime and apparently prevented the inversion— that is, the splitting of the sucrose into glucose and fructose—of the sugar.[49] Fearing the formation of a sorghum trust similar to that for sugar, Senator Plumb and Rep. John A. Anderson, also of Kansas, argued that Swenson's patent should be canceled, as he had been an employee of the USDA when he made the experiments on which it was based. This issue dragged on with no resolution, only to fade at the end of the sorghum-sugar experiments.[50]

The Kansas congressmen had reason to be worried, for eastern investors were establishing a joint stock company, expecting to control the entire sorghum crop of the country through the use of Swenson's process and to produce sugar from sorghum "cheaper than cane sugars, even if the latter could be admitted free of duty." With this infusion of funds, Parkinson and Swenson remodeled the factory at Fort Scott and doubled its capacity. They also built a $100,000 facility at Topeka, said to be "the most complete and largest factory ever erected in the West." They also profited from the bounty enacted by the Kansas legislature: two cents per pound on sugar manufactured under certain conditions from plants grown in the state.[51]

In the midst of "one of the most exciting and expensive booms ever experienced in this often-boomed State," several more sorghum-sugar mills were built in Kansas the late 1880s. In one instance, promoters convinced the citizens of Minneola to vote for the issuance of bonds worth $65,000 for a mill that would produce first-class sugar; they then built an inexpensive mill with inadequate materials, laced the product with sugar purchased in Dodge City, and skipped town before the fraud could be discovered.[52]

Although the agricultural appropriation bill passed by the House of Representatives in July 1888 said nothing about sorghum sugar, the Senate added a $100,000 line item for these experiments. In the ensuing conference, Congressman William Hatch from Missouri "protested it would be a waste of money, and that the Department of Agriculture did not want it," but Thomas Ryan from Kansas convinced his colleagues to keep the funds in the bill. In his last annual report, Secretary Colman (agriculture having recently been elevated to cabinet status) reiterated the justifications for the sugar experiments and concluded that, since sugar beets offered the most promise, the U.S. government should promote this crop, as the French and German governments had so wisely done. The sorghum experiments had produced valuable information but not solved the problems of crystallization or cost-effectiveness.[53]

Jeremiah Rusk, who became secretary of agriculture in 1889, may have wanted to quash the sorghum experiments, but Congress appropriated $80,000 to subsidize diffusion apparatus in facilities around the country. After inspecting several of these sorghum facilities, Rusk conceded that "we could not have obtained better results for the same expenditure in any other way." Wiley, for his part, remained opposed. The manufacture of sugar from sorghum on a small scale might become commercially successful, he said, and the record of disasters did not prove that the sorghum-sugar industry was impossible. The record did, however, prove that the "conservative and unbiased conclusions" of the USDA were a safer guide for investors than "the representations of irresponsible and interested parties."[54]

Sorghum got a new lease on life in December 1890 when the USDA announced a new process for removing the starch and gummy substances from the juice. One newspaper headline blared, "A Process of Far-Reaching Importance Has Been Perfected." Since the process used lots of expensive and highly regulated alcohol, proponents urged a modification of the revenue laws so that this alcohol could be exempt from the tax. The *Washington Post* asked if Senator Plumb would be "willing to furnish a written guarantee that the Kansas people want free alcohol for purely sorghum sugar purposes." Congress authorized $25,000 so that Wiley could conduct experiments with this technique, but the results were not promising. It is to be regretted, he said, "that certain hallucinations seem to constantly follow the development of the sorghum sugar industry."[55]

The sugar bounty in the McKinley Tariff of 1891 to 1894 provided a temporary boon to sorghum, but the death of Senator Plumb in December

Figure 22: Sorghum boiler, Lancaster County, Nebraska. Photograph by John Vachon for the U.S. Farm Security Administration, 1938. Courtesy Prints & Photographs Division, Library of Congress.

1891 dealt a blow to the industry, as did the drought and insect visita-tions of 1893. J. Sterling Morton, Grover Cleveland's secretary of agricul-ture, believed in strict economy and was adamant that the sorghum work should end promptly. When Congress discontinued the sorghum appro-priations, he was quoted as saying that a "stage is now reached when indi-vidual enterprise can and should take advantage of what the Department has accomplished."[56]

The end of the sorghum-sugar enterprise came quickly. Swenson left the Parkinson Sugar Works, foreseeing that sorghum sugar would not be able to hold its own against American beets and Cuban cane. Parkinson called it quits in 1898, selling the machinery that had cost Nebraska men $104,000 for a mere $9,000 and selling the building to others who converted it into a sorghum syrup plant. By 1908, sorghum was said to be "very largely a product of home manufacture."[57] So it remains to this day (Figure 22).

chapter 10

MAPLE SUGAR AND SYRUP

There is, "in some parts of New England, a kind of tree, so like our Wallnut-tree, that it is there so called; whose juice, that weeps out of its Incision, &c, if it be permitted slowly to exhale away the excess moisture, doth congeal into a sweet and saccharine substance." So wrote Robert Boyle in the early 1660s, after having talked with John Winthrop Jr., the governor of Connecticut, who was then in England seeking a royal charter for his colony and engaging with men of science. Boyle went on to say that sugar was "(at least in these Western Regions) an almost recent discovery" that preserved meats and rendered various substances "exceeding grateful to the Taste"—suggesting that it was not yet an important part of the daily diet.[1]

Maple sugar and syrup (the terms were often used interchangeably) was still a novelty in England in 1684 when John Ray, a prominent botanist, described it as "a thing to me strange and unheard of before" that was made by Indians in Canada who had "practiced this Art, longer than any now living among them, can remember."[2] When this practice began, however, is unclear as no textual or material evidence exists from before Europeans arrived in the Americas bringing metal tools, writing, and a decided taste for sweets.[3]

André Thevet, a French writer whose 1557 book is known to be reliable in some parts and fanciful in others, described an American tree resembling the European walnut in size and form. It was thought, he said, that this tree was good for nothing, but one day someone cut one down "and found that the sap which poured forth from it possessed a fine delicate taste resembling that of one of the good wines of Orleans or Beaune." And the Canadians, "much liking the drink," now care for the trees "in order to make it." Thevet may have gotten this information from Jacques Cartier, the explorer who made three voyages to Canada in the early sixteenth century and claimed the land for France.[4]

Further contact led to further information. Marc Lescarbot, a lawyer who visited the Gaspé region of New France in 1606 and 1607, wrote that if the Indians were "tormented by thirst, they have the skill to suck certain trees, whence trickles a sweet and very pleasant liquor, as I myself have sometimes proved." In his 1672 account of the natural history of North America, a royal official named Nicolas Denys described the method of gashing the trees and collecting the sap, which had "a sweetness which renders it of very good taste." A Catholic missionary to the Micmac Indians, reported in 1691 that when maple water is boiled and reduced to a third, it hardens to something like sugar and takes on a reddish color, adding that it "is formed into little loaves which are sent to France as a curiosity, and which in actual use serve very often as a substitute for French sugar" (Figure 23).[5]

By the early eighteenth century, Native Americans were trading maple sugar for European goods, and colonists were making it for their own consumption. Robert Beverley, author of *The History and Present State of Virginia* (1705), examined some sugar that Indians made from the sap of the "sugar tree" and found it to be bright and moist, with a full grain and a taste like that of good muscovado. Paul Dudley of Massachusetts said that maple sugar was as sweet as that obtained from cane and that some New England physicians looked on it "not only to be as good, for common use as the West India Sugar, but to succeed all other for its Medicinal Value."[6]

Susannah Carter said nothing about maple products in *The Frugal Housewife*, which first appeared in London in 1765 and was often reprinted in England and America, but the anonymous appendix to the 1803 edition of Carter's book published in New York featured "several new receipts adapted to the American Mode of Cooking." And here there are detailed instructions for producing maple sugar, beginning with making an incision in the trees, as well as maple molasses and maple beer.[7]

Maple and the Antislavery Movement

Samuel Hopkins was a student at Yale College when the first great religious awakening swept across the British colonies of North America. After graduation in 1741, he studied with Rev. Jonathan Edwards and there came to see the injustice of enforced servitude. Since sugar was made by slaves and indeed (along with its by-products, molasses and rum) constituted the main article of New England consumption produced by slaves, Hopkins was pleased to learn that the Housatonick Indians obtained their sugar

Figure 23: Native American women collecting sap and cooking maple syrup and other activities. From Joseph Lafitau, *Moers des sauvages ameriquains, comparées aux moeurs des premiers temps* (Paris, 1724), vol. 2, pl. 7. Courtesy Smithsonian Institution Libraries.

from the sap of maple trees. Since these trees grew in profusion in New England, Hopkins imagined they would more than furnish all the British colonies on the American continent with sugar. A contemporary review of Hopkins's book suggested that it might "be prudent for those who have a sufficient number of these Trees on their new Farms, to spare them, and use them from Year to Year, to supply themselves with these Commodities."[8] An antislavery advocate who talked with Pennsylvania Quakers argued that tapping the maple would "drive out the sugar produced by the tears and blood of slaves." An American almanac claimed that sugar made at home "must possess a sweeter flavor to an independent American of the north, than that which is mingled with the groans and tears of slavery."[9]

Benjamin Rush, a prominent Philadelphia physician who had signed the Declaration of Independence, saw in maple "the happy means of rendering the commerce and slavery of our African brethren in the sugar islands as unnecessary, as it has always been inhuman and unjust." To this end, he helped organize the Society for Promoting the Manufacture of Sugar from the Sugar Maple Tree, held a scientific tea party to prove that maple sugar was as sweet as that from cane, and produced a detailed account of the cultivation of the tree and the method of sugar manufacture. He also argued that maple sugar was bug free and the hands of American farmers were probably cleaner than those of West Indian slaves. Maple sugar made a most agreeable molasses that could be used to make a summer beer of which he approved, as well as a spirit of which he did not.[10]

Rush also noted that Thomas Jefferson used "no other sugar in his family, than that which is obtained from the sugar maple tree." That statement was a bit of a stretch, but Jefferson did appreciate the political and economic arguments for maple. Writing to his friend Benjamin Vaughan, he noted that "large countries within our Union are covered with the Sugar Maple as heavily as can be conceived," and "this tree yields a sugar equal to the best from the cane, yields it in great quantity, with no other labor than what the women and girls can bestow, who attend to the drawing off and boiling the liquor, and the trees when skillfully tapped will last a great number of years, yet the ease with which we had formerly got cane sugar had prevented our attending to this resource." Jefferson went on to say, "Late difficulties in the sugar trade have excited attention to our sugar trees, and it seems fully believed by judicious persons, that we can not only supply our own demand, but make for exportation. . . . What a blessing to substitute a sugar which requires only the labour of children, for that which it is said renders the slavery of blacks necessary" (Figure 24). As it happened, however, the sugar maples that Jefferson planted at Monticello did not do well. Nor did those that George Washington planted at Mount Vernon.[11]

Maple and Money

Benjamin Franklin eventually came to oppose slavery, but frugality led him to maple. Recognizing that honest and prudent men would do without those things they could not pay for, he told his readers, in the preface to the 1765 edition of *Poor Richard Improved*, how to supply a syrup in "every Way superior to Melasses." In the text he noted that from the sugar maple great quantities of sugar may be made. "In the frontiers of Connecticut they are now much in the Practices of it. A Friend, who has lately traveled in that Way, assures me, that . . . they make more than they can consume, and sell

Figure 24: "Young Son of Frank H. Shurtleff Gathering Sap from Sugar Trees for Making Maple Syrup." Photograph by Marion Post Wolcott for the U.S. Farm Security Administration, late 1930s. Courtesy Prints & Photographs Division, Library of Congress.

it at Eight Dollars and One Third per Hundred Weight. Some Families last year made Five Hundred Weight."[12]

Also aware of the economic argument, Benjamin Rush noted that hundreds of families in New York and Pennsylvania supplied themselves with maple sugar, with most of the work done by women and children. A French botanist who visited the United States in the late 1780s observed that Americans made some ten million pounds of sugar each year from the tree that the Swedish botanist Carl Linnaeus had named the *Acer saccharinum*. Daniel Jay Browne, the first American botanist to survey the forest trees indigenous to North America, found in the 1830s that a large portion of the sugar used in many eastern parts of the country was derived from the maple.[13]

Maple sugar made financial sense for those with farms and families, but when the price of cane sugar fell, when beet sugar became available, and when Americans became increasingly urban, maple lost its economic advantage.[14] Taste was a factor as well. As long as much of the cane sugar on the market was raw or only partially refined, consumers probably enjoyed the taste of maple and did not mind the color. But as Americans came to prefer refined white sugar and to look on raw sugar with disgust—an attitude promoted by sugar refiners—maple sugar lost its edge. Maple syrup, however, gained in popularity, as did maple flavoring for tobacco products.[15] The agricultural schedules of the federal census show that the production of maple syrup rose slightly from nearly 1.6 million gallons in 1860 to over 2 million gallons in 1900, while the production of maple sugar fell from 40.12 million pounds to nearly 12 million pounds. In this latter year, 62,718 farmers reported making some maple products, suggesting that maple was primarily a domestic endeavor.[16]

New York was the largest maple sugar producer in the country, largely because it was so very large. Vermont was second, but maple accounted for a larger share of the state's economy and identity, and there was a sugar exchange designed to put Vermont's sugar at the head of the market and to create a demand by guaranteeing its purity and quality. The sugar maple became the state tree of Vermont in 1949.[17] Ohio was another important maple state. Indeed, an entire building at the Columbian Exposition of 1893 was devoted to Ohio maple, and Ohio producers, most of them located in the northeastern area known as the Western Reserve, captured most of the medals given for maple goods.[18]

As noted in earlier chapters, Congress removed the tariff on foreign sugar in 1890 and instituted a bounty on sugar produced in the United States. In the three years that the bounty was in effect, only $66,119 was paid for maple products—Vermonters received \$36,225, and New Yorkers received \$11,703—far less than was paid for cane sugar and beet sugar. One observer claimed that the bounty "stimulated men to work the sugar bush as it has not been worked before for years."[19] Another noted that the primary value of maple sugar was its flavor and delicacy, not its sugar content. As the bounty was paid on sugar but not on syrup, less than 7 percent of the maple product of the United States received any encouragement, "and the industry was in no way improved."[20]

More substantial help for the industry would come from the numerous publications concerning the production and analysis of maple sugar and syrup issued by the U.S. Department of Agriculture (USDA) and several state agricultural experiment stations.[21]

Adulteration, Blends, and Artificial Flavorings

Maple syrup is tastier than corn syrup but also more costly, and by the 1880s, reputable syrup dealers were arguing that they were entitled to some protection from "the flood of diluted and doctored glucose" that unprincipled men were pouring out all over the land as pure maple syrup. Analyses done under the leadership of Harvey Wiley at the USDA in the 1890s showed that the commercial articles were largely adulterated with glucose. A 1905 USDA report found that at least seven-eighths of the syrup on the market was "a spurious article, which is only in part maple sugar, or is manufactured entirely from foreign materials." When mixers realized that polariscopes could detect this fraud—sucrose (whether from cane or maple) rotates the polarized light to the left, while glucose rotates it to the right—they began mixing maple syrup with syrups from cane or beet syrup, all of which were chemically identical.[22]

The first efforts to solve the problem were made at the state level. The Dairy and Food Commission of Ohio decided that each container of maple syrup must give the name and address of the packer and the location where it was produced. A Chicago firm advertised "pure maple syrup (is pure)" and offered to give "gold dollars for every drop of adulteration in our pure maple syrup." Wiley reported a can bearing the label, "We guarantee this to be perfectly pure maple molasses, purchased by us of responsible farmers, and hermetically sealed in cans, under our own supervision. It is justly denominated the best of all and is unequaled by any other brands."[23]

The first federal standards for maple products, issued by the secretary of agriculture in 1906, defined maple sugar as the "solid product resulting from the evaporation of maple sap," noting that it contained "not less than sixty-five one-hundredths (0.65) per cent of maple sugar ash" (a substance that contributed to the flavor of the product). Maple syrup contained "not more than thirty-two (32) per cent of water and not less than forty-five hundredths (0.45) per cent of maple sugar ash." A subsequent food inspection decision stipulated that maple syrup "weighs not less than eleven (11) pounds to the gallon (231 cu. in.)."[24]

The Board of Food and Drug Inspection also decreed that when a syrup was a blend, the predominant syrup must appear first on the label; the word "maple" could only be used when maple was present in substantial quantities; and when there was only enough maple syrup or sugar to give a maple flavor, the proper label would be "cane (or corn) syrup flavored with maple." When food inspectors went looking, however, they found numerous instances of fraud. Many cans filled with cane or corn syrup and just a hint of maple were labeled "pure maple" and sold for as much as, or even more than, the real thing (Figure 25).[25]

Log Cabin Maple Syrup

Log Cabin Maple Syrup was introduced in the 1880s by Patrick J. Towle, a grocer in St. Paul, Minnesota, who chose the name to honor Abraham Lincoln, his boyhood hero. After winning top prizes at the international expositions held in Paris in 1900 and St. Louis in 1904, the Towle Maple Syrup Company was on a roll.[26] Towle became a part of Postum (soon to become General Foods) in 1927 and was later acquired by Pinnacle Foods.

The firm began an extensive advertising campaign in 1904, at a time when momentum was building for a national pure-food law, and a careful reading of these texts shows how difficult it might be to know what was in the can. Log Cabin Syrup was said to be wholesome, healthful, and delicious. It had "the true maple flavor." It was "made under conditions that insure its absolute purity and richness of flavor," and it was always "uniform in flavor and quality." The brand on the can was "a guarantee of excellence and a pledge of satisfaction to you." Elsewhere, the syrup was said to be "absolutely pure and full measure," "made from the choicest sugar products in the Maple Orchards of Vermont and Canada," and "used by the best families, clubs and hotels in the United States." Towle's Log Cabin Camp Syrup had "that delicious, wholesome, true maple flavor, because it is syrup made from pure maple sugar right from the maple grove. It is Nature's nectar put up in gallon, half gallon, quart and pint Log Cabin cans." Continuing the

Figure 25: Label reading, "Colonial Brand Pure Maple Sap Syrup. Guaranteed by Rigney & Co. Brooklyn, N.Y. Under the Food & Drugs Act, June 30, 1906. Serial No. 2383." Warshaw Collection of Business Americana—Food. Courtesy Archives Center, National Museum of American History, Smithsonian Institution.

theme in 1905, the Towle Maple Syrup Company claimed that its name had, for twenty years, "stood as a guarantee of purity and deliciousness." In 1906, the firm showed its wares at the National Pure Food Show and advertised in that organization's publication.[27]

Following the passage of the federal Pure Food and Drugs Act in the summer of 1906, the firm modified its claims. While still calling Patrick J. Towle "the pioneer of absolutely pure and full measure maple syrup," it admitted that Towle had discovered that by blending a small amount of cane syrup with the maple, he could greatly improve and balance the color and flavor, producing a deliciously sweet syrup that retained the delicate maple character. The Log Cabin can, moreover, ensured "uniform quality, uniform purity and uniform full measure." By 1911, Towle's Log Cabin Syrup was touted as a "rich blending of pure cane syrup" with "no glucose— no corn syrup, no adulteration." Making a virtue of necessity, Towle would eventually tell customers that the firm had spent years of experiment and patient toil before learning how to combine and blend the ingredients of Log Cabin Syrup and that its process was "one of the really great discoveries in preparing foods."[28]

Cost considerations led to a drastic change in the composition of most maple syrups in the latter decades of the twentieth century. Acknowledging that the maple syrup in Log Cabin was now down to 10.5 percent, General Foods said in 1967 that it "quietly made the cuts only in the states that don't require the percentage to be printed on the label, holding the content in the other states (Ohio, Michigan, North Dakota, Vermont and Wisconsin) at 15%." In the past year, however, it had cut the percentage in these states to 10.5 percent to achieve uniformity. The firm also noted that it had begun adding artificial maple flavoring to its syrup. By 1975, Log Cabin Syrup contained just 3 percent maple, as did Vermont Maid, and so, according to federal regulations, could no longer be termed maple flavored. That designation only applied to foods containing at least 10 percent maple.[29]

As the amount of real maple in syrup declined and the price of the real stuff skyrocketed, maple syrup became a luxury enjoyed by connoisseurs and those who wished to resist, at least in this arena, the dominant commercial culture.[30]

Maple Flavorings

Artificial maple flavoring can be traced back to 1883 when Josiah Dailey of Indiana devised a decoction of hickory bark, or wood that, he said, tasted like maple. Mapleine made according to Dailey's formulation was on the market by 1891.[31] Although the USDA convinced a federal judge that Mapleine was misbranded as it contained no maple juices, the product remains on the market to this day.[32] Mapletone was another artificial maple sold to candy makers, and Bush's Maple Flavor No. 1617 (Purely Vegetable) was said to be "so true that expert candy men have *refused to believe* that syrup flavored with it was not prepared from the *sap of the maple tree*."[33]

Natural maple extract dates from 1927, when USDA chemists devised a way to extract the flavor from the maple sap. These men were federal employees, and they assigned their rights to the government and the people of the United States.[34]

chapter 11

HONEY

Europeans found no cows or honeybees when they landed in North America, but imagining the New World as the land of milk and honey, they set about making it so. The Virginia Company sent bees and hives to Jamestown in 1621, and as colonists elsewhere followed suit, honey became an important part of the American diet. Moreover, as Paul Dudley of Massachusetts noted in 1720, the Indians had no word for bees and thus "called a Bee by the name of English Man's Fly." Thomas Jefferson would later add that bees "generally extended themselves into the country, a little in advance of the white settlers," and so the Indians "call them the white man's fly, and consider their approach as indicating the approach of the settlements of the whites."[1]

Bees were important in symbolic ways as well, as Tammy Horn explains in her delightfully informative *Bees in America*. English writers saw queen and worker bees as industrious and thrifty, drones as idle and slothful, hives as efficient organizations, and hiving off (or swarming) as a proper solution to problems caused by overpopulation. These ideas were easily transferred to the colonies, so it is not surprising that the currency of the Continental Congress featured a beehive with thirteen rings. Nor is it surprising that in his *Letters from an American Farmer* (1781), Jean Crevecoeur described America as a "fruitful hive, sending out industrious swarms and not having useless drones."[2]

Believing that "Nothing but Money is Sweeter than Honey," Benjamin Franklin observed, "Our common Father has divided his Benefits among his Children, with more Equality than is generally apprehended." While the West Indies got sugar cane "from which, by the forced Labour of Slaves, Sugar and Melasses are extracted, for their Masters Profit," the northern colonies got "an Infinity of Flowers, from which, by the voluntary Labour of Bees, Honey is extracted, for our Advantage." But, "This Boon of Providence we do not make the most of. Were people more attentive to the Management of Bees, great Profits would arise to the Owner, as well as to the Country, both from the Honey and the Wax." For those unmoved by economic arguments, Franklin noted that if Americans could "see and know the extreme Slovenliness of the West-India Slaves in making Melasses, and the Filth and Nastiness suffered to enter it, or wantonly thrown into it, their stomachs would turn at the Thought of taking it in either with their Food or Drink." Finally, Franklin wondered how many fine sets of teeth might be saved, and what a quantity of toothache avoided, if Americans ate honey rather than sugar.[3]

Over time, as other forms of sweeteners and wax came to the fore, new technologies made beekeeping safer and more productive and extended its reach. Americans made close to fifteen million pounds of honey in 1850 and over twenty-three million pounds in 1860. With an average value of 22.5 cents per pound, the thirty million pounds produced in 1868 were worth $6.75 million.[4]

The first important new technology, the movable frame hive, permitted an occasional peek at the bees—not so easily accomplished with the traditional basketlike skep. Although this type of hive originated in Europe, it was substantially improved by Lorenzo Langstroth, a Yale-educated minister who, having observed his bees with care, found the optimum spacing to be three-eighths of an inch. If the frames were farther apart, the comb was difficult to crack, but with the correct "bee space," they could easily be removed from the hive. In his patent application, Langstroth claimed that his hive gave bees better protection against the bee moth, as well as against extremes of heat and cold and sudden changes in temperature and dampness. It also let the beekeeper multiply colonies rapidly, obtain surplus honey in the most convenient, beautiful, and salable forms, and perform all necessary operations without injuring the bees. It further enabled "the most timid" to remove the honey without danger to themselves (Figure 26).[5]

The centrifugal extractor threw honey from the comb in the same way that the centrifugal machine threw water from sugar in a refinery, and it enabled beekeepers to offer liquid rather than comb honey. The original form was introduced by F. E. Hruschka, an Austrian beekeeper working in Venice. Langstroth and a colleague patented a hand-powered version with apparently better performance characteristics.[6] The bellows smoker was safer and more efficient than the traditional method of subduing bees by blowing smoke through a simple tube filled with burning tobacco or another organic substance. Moses Quinby, an apiarist in St. Johnsville, New York, introduced the first design in the mid-1870s, and numerous

Figure 26: "George Arnole Exhibits a Super of Honey Raised on His Farm in Chaffee County, Colorado." Photograph by Arthur Rothstein for the U.S. Farm Security Administration, 1939. Courtesy Prints & Photographs Division, Library of Congress.

adaptations followed in quick succession.[7] The comb foundation, the fourth important invention, was introduced by a European beekeeper who found that it made bees more productive. American improvements included shallow, hexagonal cell walls and a roller device, somewhat like a clothes wringer, that pressed the design onto a sheet of wax (Figure 27).

The introduction of the Italian honeybee—a gentle animal that could withstand cold temperatures and had amazingly prolific queens—was another factor in the rise of apiculture in the United States. The first Italian

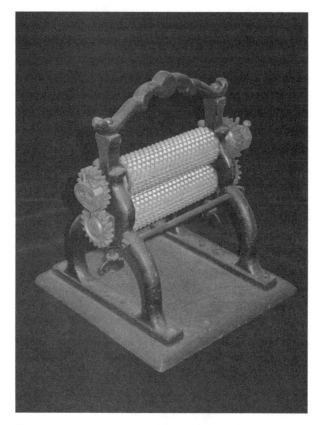

Figure 27: Frances A. Dunham, "Machine for Manufacturing Honey Comb Foundations," U.S. Patent 246,099 (1881). Dunham was a Wisconsin woman active in apiary circles. Patent model now in the collections of the National Museum of American History.

queens sent to this country perished en route. Eventually, however, an American who studied Sicilian horticulture and agriculture at the behest of the federal government succeeded in bringing ten colonies of Italian bees to his nursery on Long Island, keeping them over the winter, and distributing them to beekeepers around the country.[8]

Associations and periodicals helped beekeepers keep abreast of developments across the country and around the world, promote their products, lobby for pure-food laws, and obtain county inspectors who could identify foulbrood and other diseases harmful to hives. *The American Bee Journal* began publication in 1861 and, but for a few years during the Civil War, has appeared every month since. *Gleanings in Bee Culture* began publication in 1873 and, incidentally, published the first accurate reports of the Wright brothers' flights in 1904.[9]

Books were also important. Langstroth's *The Hive and the Honeybee* (1853) offered sound practical advice. Lengthy excerpts appeared in numerous periodicals, and subsequent editions were soon forthcoming, some in such languages as Spanish, Italian, Russian, and Polish, some of which remain in print to this day. Moses Quinby, who believed beekeeping to be "more profitable, with the same capital, than most other kinds of business," brought out the informative and influential *Mysteries of Bee-Keeping Explained* (1853).

Ellen S. Tupper—known in apiary circles as the "Bee Queen of Iowa" and to feminists as "a scientist, a business woman, a lecturer, teacher, neighborhood nurse, citizen and mother, and above all a lover of her kind"—came to public attention in 1865 with a series of articles describing apiculture as a science, an art, and a healthful and remunerative occupation for which women were particularly well suited. She went on to establish the Italian Bee Company in Des Moines, join the North American Beekeepers' Society, give lectures on bees at the State Agricultural College in Ames, Iowa, and edit *The National Bee Journal* and *The American Bee Journal*. Then, having weathered the depression of the early 1870s by passing forged financial papers, she was hauled into court—but acquitted by reason of insanity.[10]

Adulteration

Honey from the hive is a mixture of glucose and fructose, with a sweetness similar to that of sucrose (ordinary sugar). By the 1860s, some honey on the American market was adulterated with glucose made from potatoes and imported from Europe. The introduction of American corn-based glucose in the early 1870s exacerbated the problem of purity, as did the widespread availability of honey extracted from the comb.

Moses Quinby called attention to artificial honey in 1866, noting that it was often made by placing a piece of comb in a jelly cup and filling it with glucose; he added that if this were pure honey, "it would become candied and conceal the comb, yet these are found unchanged upon our grocers' shelves the year round." A few years later, another beekeeper suspected foul play when he sold honey to a Chicago dealer for seventeen cents per pound, then saw it sold, under this dealer's label, for just sixteen cents per pound. The apiary community eventually gathered thousands of signatures from every state in the union and asked Congress to do something—but to no avail.[11]

Harvey Wiley waded into the issue in 1881, calling attention to commercial honey that was "entirely free from bee mediation" in which a paraffin cone was filled with pure glucose by "appropriate machinery." In terms of whiteness and beauty, he said, this honey "rivals the celebrated real white-clover honey of Vermont, but can be sold at an immense profit at one half the price."[12] After joining the U.S. Department of Agriculture (USDA), Wiley examined some forty samples of honey and found that many contained glucose or sugar.[13]

The Pure Food and Drugs Act of 1906 gave Wiley the ammunition he needed to tackle the problem. The first case began in January 1908, when a USDA inspector purchased some "pure strained honey" that a firm in Philadelphia had sent to Detroit. When this batch was shown to contain glucose, the government filed charges and convinced a court that the product was adulterated and misbranded. The government later showed that some "choice pure strained honey" shipped from New York to Philadelphia contained invert sugar—that is, sucrose that had been split into glucose and fructose.[14] The USDA Sugar Laboratory promoted chemical tests for identifying adulterated honey: if the problem was glucose, a bit of potassium iodide would make the sweet stuff red or purple; if the problem was invert sugar, one used a solution of aniline acetate.[15]

Taking Honey to Market

Many beekeepers made just enough honey for their families and friends, but those with numerous hives and sophisticated technology could produce more honey than they could dispose of locally. So, they turned to commercial dealers and packers. H. K. & F. B. Thurber was a wholesale grocer in New York with "a large capital" invested in honey, an extensive department devoted exclusively to it, and some twenty employees who did nothing but cut up honey and repack it properly. The firm offered a gold medal for "the best display of comb honey in the most attractive and marketable shape" shown at the National Bee-Keepers Convention, and its honey won a prize at the international exhibition held in Paris in 1878.[16] It shipped three hundred thousand pounds of American honey to Great Britain and convinced the high steward at Windsor Castle to purchase a case for the queen's table.[17]

The middlemen who stood between producers and commercial users were often seen as sharks and swindlers. Cooperatives, in which neighbors pooled their honey and sent the lot to large and presumably reputable dealers, were known but not popular.[18] In time, however, following the lead of farmers in other fields, some beekeepers formed cooperatives that would pack as well as market their honey. The Sioux Honey Association, formed in 1921 by five beekeepers near Sioux City, Iowa, became the largest of these, and its Sioux Bee Honey (it became Sue Bee Honey in 1948) was the first nationally advertised brand. In 1969, the Sioux Honey Association had more than twelve hundred members in thirty-five states. The numbers have since declined as apiaries have grown, consolidated, and modernized, but the cooperative nature endures.

Federal Scientific and Technical Support

The Patent Office began publishing reports on apiculture in 1850s, and the commissioner of agriculture continued the practice. In-house experiments in apiculture began in the USDA Bureau of Entomology in the 1880s. Frank Benton, the second man in this job, wrote a manual of instruction that went through several editions and sold over twenty thousand copies.[19] Everett Franklin Phillips, an energetic young biologist with a PhD from the University of Pennsylvania, succeeded to the post in 1905. Understanding that from a scientific standpoint, the honeybee was virgin territory, he determined to make apiculture modern, professional, and scientific. To this end he established the Bee Culture Laboratory in the suburbs of Washington, DC, hired additional staff, and began investigating such issues as the causes of foulbrood and other serious bee diseases, the anatomy of the honeybee, and the best ways to keep bees alive over the winter. In 1922, as the Isle of Wight disease devastated bee colonies abroad, Phillips helped convince Congress to prohibit the importation of adult bees except under USDA supervision.[20]

Finding that persuasion could not lift the honey crop above its normal level of about 250 million pounds, Phillips began arguing that the community of skilled commercial beekeepers must be enlarged, that beekeepers should increase the number of their colonies, and that beekeeping must be introduced into new localities. Beekeeping, he said, was an "exacting calling" and good management the "price of success." Phillips also promoted beekeeping as a suitable occupation for veterans returning from war.[21] In 1924, having become disillusioned with the federal bureaucracy, Phillips moved to Ithaca, New York. As professor of entomology at Cornell University, he would teach apiculture to college students and local beekeepers, crusade for federal inspection of apiaries, and promote extension programs in apiculture.

As was the case with other food products, the distribution of honey through distant markets led to calls for standardization. So, in collaboration with A. H. Pfund, an optical physicist at The Johns Hopkins University who worked closely with the Munsell Color Company in Baltimore, scientists at the USDA defined standard grades based on color and developed an instrument for determining the grade of different samples. The resultant Pfund color grader consisted of a wedge of amber-colored glass, a hollow glass wedge that could be filled with honey, and a graduated scale. Though at $50 the Pfund grader was costlier than others on the market, it was widely used by large-scale producers and dealers as well as exporters and importers.[22]

Colony collapse disorder, the most serious and mysterious problem in the early twenty-first century, describes a situation in which honeybees are dying or abandoning their hives. Despite numerous investigations, scientists and apiarists have not managed to identify the cause, be it pathogens, pesticides, diet, or some combination thereof.[23]

Federal Support for the Business

Shortly after the United States entered World War I, the newly formed U.S. Food Administration, in addition to urging Americans to consume less sugar and other foods needed by our armed forces and allies abroad, began prodding American beekeepers to produce more honey. Phillips, for his part, organized a war conference at the USDA, where he urged prominent beekeepers to mobilize the industry to produce "a hundred million pounds of honey extra." In the words of the *Wall Street Journal*, "More than 300,000,000,000 bees are being mobilized by the Government to make up a part of shortage of the sugar crop," and "800,000 beekeepers are to be persuaded to double the number of their swarms." According to the *American Bee Journal*, beekeepers understood that if honey was on every table, apiculture might "rapidly take her rightful place among the agricultural industries of the United States." To help with this endeavor, the Food Administration expedited shipments of apiary supplies and said that beekeepers could feed their bees sugar when necessary. The USDA Extension Service sent men around the country to demonstrate the best beekeeping practices. The USDA Bureau of Markets issued semimonthly reports about honey and touted the advantages of cooperative marketing societies. The U.S. Fuel Administration permitted the chief factories making beekeeping supplies to run on fuelless days. The Post Office allowed bees to be shipped by parcel post.[24]

In 1941, when sugar was again in short supply and beeswax was valued as a protective coating for munitions and canvas goods, Congress added $33,000 to the bee-culture budget.[25] After the United States entered the war, Phillips's book *Beekeeping* was advertised for the hundreds of people who would start keeping bees to supplement the sugar shortage; the makers of Golden Blossom Honey announced it "patriotic to use honey generously"; the American Honey Institute brought out *Old Favorite Honey Recipes*, a pamphlet designed to teach homemakers how to use honey in place of sugar; and the Price Stabilization Board set a maximum retail price for honey.[26]

The end of sugar rationing in 1946 caused a sharp decline in the consumption of honey, and since the weather in 1947 was particularly favorable for honey production, a glut on the market caused prices to fall. In response, the USDA offered to purchase American honey and sell it to the State Department and the U.S. Army for use in aid programs around the world. Congress instituted a formal price-support program in 1949. According to this, the USDA would purchase all honey that met certain standards of quality and cleanliness and could not be sold through regular channels. It would pay the beekeepers 60 to 90 percent of parity and store this honey in warehouses until the Commodity Credit Corporation (CCC) could dispose of it through food banks, school-lunch programs, and prisons. Congress later added a loan program that let beekeepers keep their honey off the market until prices appeared advantageous and to forfeit this honey to the CCC if advantageous prices never materialized.[27]

The support program remained modest for many years; eventually, however, it attracted many beekeepers as inflation made American honey too costly to compete on the world market. Federal funds bought some thirty-seven million pounds of honey in 1982, representing some 20 percent of domestic production. In 1984, the honey program cost taxpayers $94 million. Although this clearly represented a pittance in terms of the overall federal budget, the media repeatedly presented honey as the poster child for federal programs that had outlived their usefulness but would not die. President Ronald Reagan tried to kill these price supports but could only convince Congress to eliminate the parity formula and adopt progressively lower supports.[28]

While campaigning for the presidency in 1992, Bill Clinton promised to eliminate honey price supports (while saying little or nothing about the much larger supports for sugar).[29] Congress debated the issue at length, and although 50 percent of the federal payments went to only 350 individuals, supporters of the program won the day by arguing that honey was produced by many people in every state of the union, that much of it was sold interstate, and that the enterprise comprised small and midsize businesses.[30] Although Congress eliminated the honey program from the 1996 Farm Bill, it soon instituted a recourse loan program that aided the industry in similar ways as before.

Concerned that, in the absence of significant tariff protection, inexpensive foreign honey was pouring into the country, beekeepers asked Congress for relief in 1976, but to no avail.[31] In 1994, having found that China was dumping its excess honey in the United States, the International Trade Commission resurrected a Cold War statute that allowed the president to reduce imports from centrally planned economies that disrupted American markets. This action was never put into effect, but China did agree to limit honey shipments to the United States and to impose a price floor on this product. Following the expiration of this agreement in 2000, the Department of Commerce ruled in favor of levying antidumping duties on China and Argentina.[32] Some Chinese exporters then began laundering their honey by shipping it through other countries. When U.S. agents inspected some of this honey, seeking evidence of country of origin, they found trace amounts of antibiotics banned by the Food and Drug Administration from American foods—and used this argument to protect American consumers and apiarists.[33]

As early as 1968, honey producers began asking Congress to help them compete with sweet spreads that enjoyed multi-million-dollar advertising budgets. "As long as we operate on the free enterprise system," said one honey man, "supply and demand is going to dominate the marketplace." As free enterprise could not compel beekeepers to pay for a service that would benefit the whole community, however, a need existed for "a compulsory collection of funds to carry on promotion and research for the industry."[34] Congress did not respond to this problem until 1984, when, while nibbling away at price supports, it authorized a National Honey Board, funded by an assessment of one cent per pound to be paid by producers and importers, that would conduct programs under the oversight of the USDA.[35]

Killer bees also garnered congressional attention. This story began in 1956 when a Brazilian beekeeper imported some African bees, hoping to develop a strain better adapted to tropical conditions than the European bees then producing honey in South America. The hybrid bees soon flew the coop and headed northward at the rate of two hundred miles per year. In response to rising concern, the National Research Council sent a team of scientists to study the situation in 1971. Even though no one knew how to stop the bees at the border, Congress passed a bill to limit the introduction of "undesirable species" of bees. Senator Strom Thurmond of South Carolina was a cosponsor, and although scientists explained that all honeybees belonged to one species, he described the African bee as a "virulent and aggressive strain" that had "mixed with other breeds" and would soon bring its "Africanized genes" into the United States.[36]

Bees and Pollination

Even as the production of honey and beeswax gained national attention, the importance of honeybees as pollinators of important crops was moving toward center stage. A German botanist named Hermann Müller published a scientific account of the role that bees played in the pollination of plants in the 1870s, and Phillips at the USDA observed that, as a rule, American fruit growers recognized the value of the honeybee to their industry and that "the indirect benefit of the beekeeping industry annually adds to the resources of the country considerably more than the amount received from the sale of honey and wax." By the 1920s, fruit growers in California were paying beekeepers to bring colonies to their orchards when the trees were in bloom. By the 1930s, some members of Congress understood that bees were a modest means to the more important end of pollination. A speech inserted into the 1944 *Congressional Record* argued that the honeybee was "the best possible pollinating agency" for seeds, fruits, and vegetables, and the 1949 discussions that led to honey price supports contained numerous mentions of honeybee-pollinated crops.[37]

By 1971, migratory beekeepers were said to be taking some two million colonies, each containing from twenty-five to sixty thousand bees, from field to field to pollinate crops; California growers alone used 350,000 colonies in their almond groves. Congress later learned that some $3.3 billion worth of agricultural crops depended completely on pollination by insects and that honeybees pollinated some 85 percent of this total.[38]

Honey in Industry

Domestic and industrial bakers used honey to sweeten goods and keep them moist—and in some cases the amounts could be staggering. Nabisco, makers of animal and graham crackers, acknowledged purchasing 1,478 tons of honey in 1903 and reportedly purchased seventy carloads of honey a few years later. In 1977, when asking Congress to maintain a low tariff on imported honey, Nabisco said that rapidly escalating honey prices had forced it to eliminate honey from a number of its products.[39] Large amounts of honey were also used in candy and cough drops, in bacon and in beer, and even in antifreeze and golf balls. The National Honey Board reports that in the early twenty-first century, 37 percent of honey sales have gone to bulk or ingredient use, 13 percent to food service (restaurants and cafeterias), and about 50 percent to retail.[40]

And then there is royal jelly. Apiarists have long known that worker bees create this stuff, using it to nurture larvae and turn an otherwise ordinary larva into a queen. In time some people began to wonder whether it would do equally wonderful things for human health and beauty. Two scientists at the University of Toronto—one was F. G. Bantino, who had figured out how to make insulin—began studying the properties of royal jelly in the late 1920s. A scientist with Hoffman–La Roche told the American Chemical Society in 1947 that royal jelly was especially rich in panthotenic acid, a B-complex vitamin, as well as other related vitamins, and that it increased the longevity of fruit flies. Marie Earle, a costly skin cream enriched with this miraculous substance, was on the market by 1954, and similar products soon followed.[41]

Others, however, were more skeptical. The Federal Trade Commission looked into the legitimacy of royal jelly, or rather the claims made for it. The Food and Drug Administration stated, "Our medical advisers have reviewed carefully the available information on royal jelly and have not seen any convincing evidence that it has any value whatever when used by man either as a food, or as a drug, or as a cosmetic. Hence, we recommend no product that contains it."[42]

SACCHARIN

Constantine Fahlberg, a Russian-born and German-educated sugar chemist, was in Baltimore in connection with a shipment of sugar that federal agents had confiscated. While waiting for the trial to begin, Fahlberg contacted Ira Remsen, professor of chemistry at Johns Hopkins University, and arranged to become a visiting scholar in his laboratory. There, in May 1878, Fahlberg produced benzoic sulfinide, got some in his mouth, and found it to be incredibly sweet. After eating this stuff on a regular basis for several years and suffering no ill effects, Fahlberg applied for patent protection in several countries—not on the compound itself, as he and Remsen had already written that up in the scientific literature, but on his method for making it expeditiously and inexpensively. He then returned to Germany, where the necessary raw materials were readily available and where family funds enabled him to build a factory on the outskirts of Magdeburg.[1]

Saccharin from Fahlberg, List & Company came on the market in 1886 and received immediate acclaim from several quarters. Henry Roscoe, an eminent English chemist, saw it as "the most remarkable of all the marvellous products of the coal-tar industry." William Crookes, an equally eminent Scottish man of science, saw it as one of the "chief triumphs of organic synthesis." It did not decay, mold, or ferment and was a boon for those who were diabetic or obese.[2]

Although Remsen had staked his reputation on the purity of his research, he resented Fahlberg's commercial success and scientific reputation. To resolve the conflict, he embraced the theory, long popular with American scientists, holding that pure research would inevitably lead to practical results. Accordingly, he let it be known that Fahlberg had been a student at Johns Hopkins (rather than a professional with credentials equal to his own) and that he had penned the earliest scientific accounts of the compound (neglecting to mention that he and Fahlberg were coauthors). He succeeded in convincing much of the American chemical community that he deserved credit for the discovery of saccharin.[3]

As saccharin proved profitable, other chemists devised means of making it in ways that did not infringe on Fahlberg's patents. The Chemische Fabrik von Heyden produced Heyden Sugar, Crystallose, and Garantose. Merck announced that its saccharin was especially suitable for those who were stout.[4]

Lutz & Movius, a New York firm that had grown rich importing aniline dyes from Germany, brought the first shipment of saccharin into the United States in 1887. The company saw it as an acid for "medicinal, chemical, or manufacturing purposes" and so expected that it would come into the country duty free. But the collector of customs in New York regarded it as a "chemical compound" that must pay a duty of 25 percent ad valorem.[5] The Tariff of 1894 ratified this decision but placed saccharin with other sugars, rather than in the schedule with other chemicals and coal-tar derivatives. The Dingley Tariff of 1897, enacted by Republicans who had campaigned on the promise to protect American industries, raised the rate to $1.50 per pound plus 10 percent ad valorem. Since saccharin was then worth about $1.50 per pound, this was a steep tax indeed.[6] Since no saccharin makers existed in the United States at that time, this tariff primarily benefitted sugar makers and dealers who saw saccharin as a threat to their industry. Also facing pressure from sugar interests and benefiting from taxes on sugar, the German government limited the production of saccharin and banned its use in foods and beverages. As European firms began manufacturing saccharin in the United States, the amount of saccharin imported into the country dropped precipitously, from 30,256 pounds in 1901 to 3,967 pounds in 1902, then to 1,877 pounds in 1903.[7]

The Monsanto Chemical Works, established in St. Louis in 1901, was in many ways an American firm in name only. The founder was John F. Queeny—Monsanto was his wife's family name—a pharmaceutical salesman who had noticed the growing demand for saccharin. Accordingly, he asked Sandoz, a chemical firm in Basel, to send him the raw materials along with a chemist familiar with the process used to make saccharin in Switzerland. Louis Veillon, a young man with a PhD from the University of Zurich, had the Monsanto operation up and running by 1903. Two Swiss chemical engineers, Gaston DuBois and Jules Bebie, arrived a few years later and helped Monsanto produce vanillin, coumarin, and caffeine. During World War I, when Monsanto could not get the chemicals it had been using to make saccharin, Bebie developed a process that avoided this problem and assigned his patent rights to Monsanto.[8]

The 1950s and 1960 saw the introduction of new preparations and packages. Squibb Pharmaceuticals offered Sweeta, a liquid saccharin that could be used in cooking. Abbott Laboratories offered a reformulated Sucaryl that contained one part saccharin to ten parts cyclamate.[9] Pennex Products in Pittsburgh offered Sacrinpak, cellophane packets containing individual servings of saccharin designed especially for hotels and restaurants. The Cumberland Packing Company in Brooklyn offered Sweet'N Low, a mixture of saccharin, dextrose, and cream of tartar.[10]

The Industrial Uses of Saccharin

Fahlberg always knew that some saccharin would serve as a tabletop sweetener, but most would be used in packaged foods and beverages. This was the main message of the impressive display that his firm mounted at the Columbian Exposition held in Chicago in 1893 and the hefty book issued at that time. Saccharin was healthful, inexpensive, and five hundred times as sweet as sugar. It was excellent in fruit preserves, sauces and syrups, lemonade, mustard, crackers, milk, and delicacies for children and invalids. It could be used to make beer, wines, and liquors, to mask the taste of drugs, and, in connection with glucose, in candy manufacture. It was also a substitute suitable for those afflicted with diabetes, obesity, dyspepsia, and gout.[11]

Saccharin was also used to sweeten and preserve chewing tobacco. Nannie Tilley, historian of the R. J. Reynolds Tobacco Company of Winston, North Carolina, claimed that saccharin was the key factor in the expansion of the firm in the early 1890s. Reynolds was famously reluctant to reveal the secret of how it made and flavored its products, referring only to "recent experiments" that improved the chew and reduced the price by 20 percent. But Tilley found an 1899 contract with the Heyden Sugar Company stating that R. J. Reynolds was to use his influence "as before" to promote the sale of saccharin in Virginia and North Carolina.[12]

As early as 1888, a self-styled technical chemist and practical bottler named Charles Herman Sulz predicted that, because of its price differential, saccharin would be widely used in soda pop, a beverage that Americans favored in the same way that the English drank tea, the French drank wine, and the Germans drank beer. Indeed, such was the case. A test of 121 soft drinks sold in Chicago in 1909 showed that 53 contained saccharin.[13]

The Liquid Carbonic Company, a Chicago firm that made sweet syrups as well as the equipment whereby soda fountains could produce their own carbonated water, provided most of the seed money for Monsanto and offered to purchase all of its saccharin from that firm. Monsanto ledgers show that Liquid Carbonic did purchase vast amounts of Monsanto saccharin, especially in the early years, as did many soda bottlers around the country. The ledgers also show that Monsanto sold saccharin in twenty-five-pound packages, obviously for industrial purposes.[14]

Coca-Cola was the most successful soft drink firm in the country, if not the world, and so providing saccharin for Coke would be a big deal indeed. While the Coca-Cola records are not open to the public, the Monsanto records show this firm was often Monsanto's best customer. In June 1908, for instance, Coca-Cola purchased six large lots of saccharin for a total cost of some $6,779. Coca-Cola did not, however, stick with saccharin, probably because it could not afford to give Harvey Wiley, chief chemist of the U.S. Department of Agriculture (USDA), yet another reason to rail against its product. Wiley's antipathy to iced drinks was well known—he termed them "snares of the devil"—and when testing a large number of fountain drinks, he found all contained caffeine, of which he disapproved, and many contained "a deadly drug" that was probably a coca derivative. So, for Coca-Cola, the cost advantage of saccharin did not outweigh the hassle.[15]

Many smaller firms, however, made a different calculation. In 1920, when Pennsylvania food inspectors tested 1,400 soft drinks, they found saccharin in 325—a higher number than usual but understandable because of the scarcity and high price of sugar due to the recent world war. Meeting in Washington, DC, that same year, the executive board of American Bottlers of Carbonated Beverages went on record as opposing saccharin-sweetened soft drinks. Monsanto, however, continued to advertise saccharin to those who made and bottled soft drinks, focusing on such features as economy, solubility, and the prevention of fermentation.[16]

As the sugar shortage during World War II led to another uptick in the use of saccharin, an editorial in *National Carbonator and Bottler* pointed to a growing tendency among bottlers to substitute saccharin for sugar. A physiologist analyzed twenty bottles of soda and found eighteen of them to contain substantial amounts of saccharine. The director of American Bottlers of Carbonated Beverages reminded his members that the association had "striven for years to firmly establish the fact that soft drinks [were] a food, basing [its] well-founded contentions on the sugar content, a highly energizing food element," and the use of "certain substitutes" would endanger this argument. Another bottler opined that "this nation is sick and tired of the saccharine diet. It wants to get back to sugar."[17]

Saccharin and the Federal Government

While gearing up for the Spanish-American War in 1897, the U.S. Army issued a call for fifty-eight pounds of saccharin in tablets of two grains each, intending to put each tablet into a packet containing two small amounts of coffee or tea and then into kits of emergency rations.[18]

At the same time, saccharin was facing opposition from those who promoted natural sweeteners, pure foods, or both. Foremost among the latter was Harvey Wiley, who went so far as to say, "I do not mean that it is advisable to pass a law solely relating to saccharin, but a law which embodies a principle excluding the use of saccharin or any other substance from food products which tend either to deceive or to injure the consumer is desirable."[19]

Wiley told Congress about the dangers saccharin posed and obtained federal funds to test the substance on the healthy young men who sat at his hygienic table. Although he never published the results of these trials, presumably because they were not to his liking, he continued saying that "the majority of experts" saw saccharin as "harmful to health."[20] When not playing the safety card, Wiley raised the issue of fraud. The name, he said, was chosen for the purpose of deception, leading people to think that saccharin had something to do with sugar. Since saccharin was noncaloric, foods sweetened with saccharin had a lower value than those sweetened with sugar.[21]

Since the Pure Food and Drugs Act of 1906 defined adulteration to include the substitution of any constituent that would reduce the quality of food, Wiley managed to engineer a food inspection decision stating that sugar was both a condiment (as was saccharin) and a wholesome food product (which saccharin clearly was not). A USDA inspector in Memphis, Tennessee, then seized 850 cases of Oriole Brand Sugar Corn, packed by Smith-Yingling Company of Westminster, Maryland, that contained Heyden Sugar, a brand of saccharin advertised as being used by "the most conservative canners."[22]

While Wiley and his men pursued this case in the courts, businessmen took their case to the White House. James S. Sherman, a New York congressman who also served as president of a food-canning company, told President Theodore Roosevelt that saccharin had saved his firm $4,000 in the previous year. Wiley responded, "Everyone who ate that sweet corn was deceived. He thought he was eating sugar, when in point of fact he was eating a coal tar product totally devoid of food value and extremely injurious to health." Roosevelt's final word: "Anybody who says saccharin is injurious to health is an idiot. Dr. Rixey gives it to me every day."[23]

Seeking further support for his opinion, Roosevelt appointed the Referee Board of Consulting Scientific Experts and asked it to investigate the health effects of saccharin and other food additives. Under the leadership of Ira Remsen—who, because of his involvement with saccharin, might not have been totally impartial in this matter—the referee board established sophisticated and well-monitored clinical trials with several groups of healthy young men. Sodium benzoate got a pass, but the saccharin results were somewhat complex. Small amounts of saccharin (0.3 gram per day or less) would not hurt normal adults, but larger amounts (more than 0.3 grams per day and especially over 1 gram daily), if taken for considerable periods, might disturb digestion. Saccharin mixed with food did not alter its quality or strength, but its use as a substitute for cane or beet sugar "must be regarded" as leading to a reduction in the quality of the food product.[24]

In early February 1911, shortly after the referee board submitted its report on saccharin but before anyone in Washington had read it with care, Remsen learned that the Board of Food and Drug Inspection had "practically decided" to prohibit the use of saccharin on the grounds that there was no need for it except to get a "cheaper sweetening agent." He considered objecting on the grounds that the report did "not logically lead to such a decision," but he rejected that idea.[25]

Food Inspection Decision 135, issued on April 29, 1911, and undoubtedly written by Wiley, drew selectively from the referee board's report. Most notably, it said that "the continued use of saccharin, for a long time in quantities over three-tenths of a gram a day, is liable to impair digestion" and that the addition of saccharin as a substitute for sugar "reduces the food value of the sweetened product, and hence lowers its quality." Saccharin manufacturers and their agents raced to Washington to meet with the Board of Food and Drug Inspection to undo the damage. One lawyer argued that the saccharin people had not yet had a chance to read the report and so could not know if it had been quoted correctly. He also said that saccharin was so very sweet that a person would be hard-pressed to ingest more than 0.3 grams per day. The government should not, without good cause, destroy a business in which Americans had invested some $1 million and that employed nearly five hundred American citizens. The cabinet secretaries who sat on the board responded by postponing the effective date of the decision by six months, to January 1, 1912.[26]

When the main points of the report were released and it became clear how cherry-picked they had been, the saccharin manufacturers sought another hearing. In preparation for this event, George McCabe, solicitor for the Department of Agriculture, prepared a confidential report for James Wilson, the secretary of agriculture. The question, he said, rested less on science than on politics. If Wilson permitted saccharin in packaged food and drink, he would come under attack from the "pro-Wiley press bureau." There were already rumors assigning responsibility for the time extension on the saccharin ban to Charles Nagel, the corporate lawyer from St. Louis who, as secretary of commerce and labor, was one of the three signatories of food inspection decisions. Any further leniency toward saccharin would provide fodder for the Democrats, who now controlled Congress, and might even embarrass the president.[27]

The first session with the saccharin makers merely rehashed the old arguments, but the second focused on Casoid biscuits, an English product clearly labeled "for diabetics—sweetened with saccharin." The lawyer for the New York firm that imported the biscuits into the United States got right to the heart of the matter: saccharin was clearly not a food, and since it was not "used for the cure, prevention or mitigation of disease," it was clearly not a drug. Insofar as it made some foods more palatable, however, it benefitted those who could not eat sugar. Moreover, Casoid biscuits were costly, unpalatable, and available only through physicians, thus unlikely to attract a large following.[28]

Following these hearings, the secretaries postponed the effective date of the saccharin ban yet again and asked the referee board to clarify its position. Though clearer than before, the board's response was nuanced nonetheless. Small amounts of saccharin were okay. Large amounts of saccharin were harmful—but then, so were large amounts of almost anything. Saccharin did not affect the quality or strength of food to which it was added, but the substitution of saccharin for sugar might "result in a decided lowering of food value."[29]

The secretaries finally acted in March 1912, issuing yet another food inspection decision concerning saccharin in food. As before, normal foods sweetened with saccharin would be considered adulterated under the law. At the same time, saccharin would be classified as a drug, and those products plainly labeled as intended for "those persons who, on account of disease, must abstain from the use of sugar" would not be considered adulterated.

Food Inspection Decision 146, issued a few months later, restated the main point: foods that "may be required for the mitigation or cure of disease" would be permitted to contain saccharin, "provided they are labeled so as to show their true purpose and the presence of saccharin is plainly declared upon the principal label."[30]

With these decisions in place, Monsanto announced, "The United States government has given permission to use saccharin in food products if the container bears a label as follows: 'This (food product) is sweetened with saccharin for the benefit of those to whom sugar is harmful or deleterious.'" The company also announced that it would support anyone prosecuted for using saccharin in foods if the product labels complied with federal requirements.[31] Several firms—including the Jireh Food Company, the Genesee Pure Food Company, and the Chicago Dietetic Supply House—soon began offering saccharin-sweetened foods for the diabetic and the dyspeptic. Macy's in New York had a special section for foods of this sort.[32]

While the saccharin debates were under way, McCabe at the USDA suggested that the government seize some food sweetened with saccharin and let the courts deal with issue. In October 1912, a Monsanto lawyer told Secretary Wilson about a case of saccharin-sweetened lemon soda shipped from New Jersey to New York, hoping that federal agents would seize it. By this time, however, Wiley had resigned from federal service, and the government was reluctant to take on the issue.[33]

Carl Alsberg, who succeeded Wiley as chief chemist of the USDA in the spring of 1912, was more politic but equally disdainful of saccharin. Time and again, in speeches, publications, and legislative hearings, he argued that saccharin should be used only in the home and only when prescribed by a physician, and he urged state legislatures and regulatory agencies to take appropriate action to keep saccharin out of prepared foods and beverages.[34]

The one federal test began in 1916 when an inspector in Chicago seized a can of saccharin marked "the perfect sweetener," "healthful," and "positively harmless" because he believed these labels to be false and misleading. The trial was put off for a few years until government witnesses could return home from fighting in France. By the time the case got to court, Monsanto had pled nolo contendere (no contest) to charges related to the first two descriptors, and so the only issue had to do with the words "positively harmless." Both sides brought forth a slew of expert witnesses, resulting in a hung jury. A second trial was held in 1924, again with no resolution.[35]

When the United States entered World War I, the federal government shut down the several American saccharin manufacturers that were subsidiaries of German firms, thereby eliminating Monsanto's primary competition.[36] Saccharin, moreover, received a boost as military activities diminished the production of beet sugar in Europe, prevented American sugar beet growers from obtaining European seeds, and disrupted transport of cane sugar from the tropics. Details are impressionistic but undoubtedly accurate. In London, the Ministry of Food announced that saccharin could be served in hotels, restaurants, and clubs as a substitute for sugar as long as the customers were informed. On the Continent and in the United States, it was said to be customary to carry about a package of saccharin tablets for use in coffee or tea.

In the immediate postwar period, a Senate investigating committee found that Americans had made a substantial amount of saccharin for export to Europe. Moreover, while fighting was still under way, the War Industries Board had released a large supply of toluol for saccharin manufacture, even though this chemical was also used to make TNT.[37] Monsanto announced that there was enough saccharin in storage to release one hundred million pounds of sugar from direct family consumption, which could help alleviate the sugar shortage.[38]

The federal Food, Drug, and Cosmetic Act of 1938 enhanced the responsibility and authority of the organization that, by then, was known as the Food and Drug Administration (FDA), and in 1941 this agency issued a regulation concerning labeling statements on foods purporting to be for special dietary purposes. Those containing saccharin—this included saccharin tablets because they consisted of saccharin mixed with some sort of inert substance—must read "a nonnutritive artificial sweetener which should be used only by persons who must restrict their intake of ordinary sweets."[39]

As the problem of obesity gained widespread attention in the postwar years, several firms saw a market for beverages containing saccharin or other nonnutritive sweeteners. When Lo-Cal Dietetic Beverages proposed selling its sodas to "all potential customers" without any restrictions on its system of distribution, the FDA could do nothing but insist that Lo-Cal containers carry the requisite warning label. In a number of instances, said one official, "products have been ostensibly labeled for special dietary use but as a matter of fact they were distributed for indiscriminate use by the public." Another bemoaned the fact that ads for artificially sweetened soft drinks suggested that the beverages were "designed for general family-wide consumption."[40]

Saccharin at the State Level

In the early years of the twentieth century, most packaged food was pre-pared and consumed locally, and so most saccharin-sweetened foods were not the concern of federal drug inspectors. They were, however, the con-cern of those states that prohibited the use of saccharin in prepared foods.[41] The North Dakota ban, the first in the nation, was announced in 1903 by Edwin Freemont Ladd, a friend of Wiley's who served as both professor of chemistry at the North Dakota Agricultural College and the state's dairy and food commissioner.[42]

Since Missouri had a regulation pertaining specifically to soft drinks, the state food commissioner could sue the Empire Bottling Company of St. Louis for selling a bottle of soda sweetened with saccharin. The defense, with Monsanto picking up the tab, argued that one would have to drink thirteen pints of the soda in twenty-four hours—an unimagina-ble amount—to get to the danger point identified by the referee board. Furthermore, if saccharin was harmful in small amounts—which the ref-eree board said it was not—it should not be in any food or drink. The state supreme court then overturned the statute on the basis that it represented an arbitrary discrimination against soda makers who used saccharin in favor of those who used sugar.[43]

Another well-known case pitted the New York Sanitary Code against the Excelsior Bottling Works, which put a slight bit of saccharin in its soda. Ruling in favor of Excelsior, the appellate court decided that saccharin could be regulated but not prohibited and that people drank soda to quench their thirst, not for its food value. This was not a case of deception, as the label said clearly that the soda was "sweetened with sugar and one-hundredth of one per cent of saccharin," and there was more danger of consuming too much sugar rather than saccharin. The Board of Health then decided that a sweetened, nonalcoholic, carbonated beverage would be deemed adulter-ated if it contained less than 7 percent by weight of sugar (or 5 percent in the case of ginger ale) and more than 0.004 percent by weight of saccharin or other synthetic sweetening agent. When this matter went to trial, with Monsanto again financing the defense, the statute was ruled invalid and unconstitutional.[44]

State laws came again to public attention in the 1950s as manufacturers saw a market for diet sodas. Because Pennsylvania had more bottlers than any other state, the Cott Beverage Corporation of Connecticut brought a test case in that state. Pressured by local consumers and local businesses that wanted in on the action, the Pennsylvania supreme court ruled the state ban on saccharin unconstitutional in 1955. The New York ban was overturned in 1963.[45]

Saccharin and the Delaney Clause

As Americans began consuming lots of artificially sweetened prepared foods and drinks, the FDA turned to the Food Protection Committee of the National Academy of Sciences. The ensuing report noted that saccharin had been shown to be lethal in large doses in rabbits but was "essentially without effect when fed at even moderately high levels over long periods" to chickens, dogs, mice, and rats. More to the point, single oral doses of five to ten grams had "repeatedly been taken by men with no ill effect," and oral doses as large as one hundred grams were reported to be "without harmful consequences." Thus, saccharin might safely be used in moderate amounts, but it should "be subject to continuing observation for possible deleterious effects under prolonged and varying conditions of use and should be reappraised whenever indicated by advances in knowledge." Correspondence in the National Archives shows that the FDA did follow this recommendation.[46]

The Dwight D. Eisenhower administration may have been generally unsympathetic to government regulation of industry, as noted by food historian Harvey Levenstein.[47] Yet the president signed the 1958 Food Additives Amendment to the Food, Drug, and Cosmetic Act, intended to prohibit the use of additives that had "not been adequately tested to establish their safety." Recognizing the impossibility of starting from scratch, Congress allowed the FDA to compile a list of substances that, because of prior testing and a long history of use, were generally recognized as safe (GRAS). Additives on the GRAS list—including saccharin and cyclamate—could be used in any foods, in any amounts, and without any sort of warning.[48]

One little-noticed sentence in the Food Additives Amendment that would become important later on stated, "No additive shall be deemed to be safe if it is found to induce cancer when ingested by man or animal, or if it is found, after tests which are appropriate for the evaluation of the safety of food additives, to induce cancer in man or animal." This sentence, known as the Delaney Clause, was introduced by James Delaney, a representative from New York who had long worried about the proliferation of chemicals, pesticides, and insecticides in and with respect to food products. Delaney, for his part, thanked Gloria Swanson, as the famous actress had talked to congressional wives about the dangers of food additives.[49] As "generally recognized" did not mean "absolutely," however, the National Academy of Sciences suggested that nonnutritive sweeteners might be safely used in limited amounts in special-purpose foods.[50]

Following the cyclamate ban of 1969, many Americans who relied on diet foods and soft drinks contacted their senators and representatives, who in turn contacted the FDA and its parent agency. Thus, for instance, Congressman William R. Poage of Texas wrote to Elliot Richardson, secretary of the Department of Health, Education, and Welfare, with regard to constituents who feared "that your Department will take action which will place them at a great disadvantage." Noting that the action on cyclamate had affected many diabetics, Poage went on to say he shared their anxiety because "your Department has awesome power to take action affecting millions of Americans who rely on various medical drugs."[51]

Poage and his constituents were right to worry as the Sugar Research Foundation, the organization that had subsidized the toxicological studies of cyclamate, had set its sights on saccharin. In March 1970, after a scientist at the University of Wisconsin School of Medicine reported that pellets of cholesterol and saccharin implanted in the bladders of rats led to a high incidence of bladder tumors, the FDA asked the National Academy of Sciences to analyze all the studies concerning the effect of saccharin on human health. The report, released in July, found that present and projected use of saccharin posed no hazard to humans, recommended further laboratory experiments, and suggested that the pellet-implantation studies be ignored as they were not applicable to saccharin found in food and drinks.[52]

In 1972, after scientists found bladder cancer in rats that had consumed vast amounts of saccharin, the FDA removed the substance from the GRAS list, gave it a provisional regulatory status enabling it to be used as before, and announced that saccharin consumption should be limited to one gram per day for a 155-pound adult. The response was immediate and widespread. Congressman John Schmitz of California suggested that this decision reached "a new height of bureaucratic absurdity," as a man who suffered as the rats did would have to consume 875 bottles of diet soda in a single day and would probably be dead long before he had time to get cancer. He went on to say, "When we learn that this particular study was conducted for the Sugar Research Foundation, it is hard not to suspect some involvement by saccharin's market competitor." Humorist Art Buchwald described saccharin as "the cocaine of dieters, the mother's milk of weight watchers, [and] the sweet taste of success," and he suggested that scientists breed bionic rats that could stand up to anything shoved down their throats.[53]

At the request of the FDA, the National Academy of Sciences then evaluated the scientific validity of all laboratory findings relating to saccharin. Noticing that tumors might have been caused by an impurity in the saccharin, the academy found that it could not say whether saccharin itself was carcinogenic.[54]

In April 1977, learning that scientists working for the Canadian government had determined that saccharin was indeed the cause of bladder tumors in rats, the commissioner of food and drugs proposed banning the use of the substance and its salts in foods, cosmetics, animal drugs, and animal feeds. At that time, some 6 to 7.6 million pounds of saccharin were used in the United States each year. Of this amount, 74 percent went into soft drinks, 14 percent was found in other foods and beverages, and 12 percent served as a tabletop sweetener.[55]

The response from the public, the diet food and beverage industry, and several medical associations was immediate, widespread, and emotional (Figure 28). After several hearings and impassioned speeches, Congress passed the Saccharin Study and Labeling Act, which called for studies

Figure 28: A protest of the saccharin ban, 1977. A group of protesters from Atlanta arrives at Union Station, Washington, DC, after an overnight train ride. They dubbed the train the "Saccharin Special." © Bettmann/CORBIS.

concerning "the impurities in and toxicity of saccharin, and the health benefits, if any, resulting from the use of nonnutritive sweeteners" and prohibited the secretary of health, education, and welfare from restricting the use of saccharin as a food, drug, or cosmetic. It also required foods containing saccharin to carry a label stating in a conspicuous place, "Use of this product may be hazardous to your health. This product contains saccharin which has been determined to cause cancer in laboratory animals." Further, stores selling these products had to post saccharin warning notices.[56] With biannual updates, this law remained in effect for several years.

And so the scientific studies continued, along with heated discussions involving risk-benefit analysis. Eventually, under pressure from agribusiness, which had become increasingly dependent on pesticides, and from scientists who believed that exposing animals to unnaturally massive amounts of potentially harmful substances was not a good way to judge harm to humans, Congress scrapped the Delaney Clause. Henceforth, residues on fruits and vegetables and additives in processed foods would be permitted if they were present in such small quantities that there could be "reasonable certainty" that they would cause no harm.[57]

In December 2000, following the advice of the National Toxicology Program, the American Cancer Society, the American Medical Association, and other such organizations, Congress passed the Saccharin Warning Elimination via Environmental Testing Employing Science and Technology (SWEETEST) Act, which stated that saccharin-sweetened foods need no longer carry a warning label.[58]

chapter 13

CYCLAMATES

As a chemistry graduate student at the University of Illinois in 1937, Michael Sveda picked up a cigarette lying on his lab bench, put it in his mouth, and found that it tasted surprisingly sweet. Sveda then tasted every substance in sight and traced the sweetness to a compound known as sodium cyclohexylsulfamate. Five years later, Sveda and his professor obtained a patent and assigned the rights to DuPont, Sveda's new employer. DuPont, in turn, licensed the patent to Abbott Laboratories, a pharmaceutical firm in North Chicago that hoped to produce a sweetener suitable for diabetics and other people on sugar-restricted diets. That sweetener would be known as cyclamate, and it would first be commercialized as Sucaryl (Figure 29).[1]

Before Sucaryl went on the market, Abbott and the Food and Drug Administration (FDA) conducted a host of toxicity tests to show that it was safe for human consumption. At a meeting of the two camps in 1949, FDA scientists said that they had fed the stuff to rats, with the only notable symptom being a slight bit of diarrhea. They warned, however, that the experiments were still ongoing, and toxic symptoms might still appear.[2] Dr. Arnold Lehman, director of the FDA's Division of Pharmacology, would later tell a congressional committee that in FDA tests, Sucaryl "came out with flying colors." He went on to say, "Abbott Laboratories did a fairly complete study of it. We also did. Our results came out exactly alike. We have no apprehensions about the safety of Sucaryl." But in a memo found in the FDA files, Lehman described Abbott's test data as "an illustration of how an experiment should not be conducted," adding that approval had been granted on the basis of animal studies conducted by FDA scientists.[3]

Abbott unveiled Sucaryl sodium at a sugar-free luncheon in May 1950, touting it as "the first noncaloric sweetening agent that [could] be used in cooking, baking and canning without loss of sweetness." Each tablet was as

Figure 29: This Abbott Laboratories advertisement for foods sweetened with Sucaryl appeared in the *Saturday Evening Post* magazine, April 19, 1958. Collection of the author.

sweet as a teaspoon of sugar and, unlike saccharin, had no bitter aftertaste.[4] There were, however, some areas of concern. Healthy people should use no more than eight tablets a day, and those with high blood pressure or kidney disease should use it only on the advice of their doctors. Sucaryl calcium, designed for those on a low-sodium diet, came out in 1952.[5]

The success of cyclamates was immediate and profound. By 1954, some 420,000 pounds of Sucaryl had been sold, compared with some 96,200 pounds of saccharin.[6] DuPont began producing calcium cyclamate in 1955, branding it Cylan and selling it in one-hundred-pound drums for commercial use. Merck also produced cyclamates in bulk, as did Squibb, Pfizer, Miles Laboratories, Monsanto, and the Norse Chemical Company.[7]

Regarding Sucaryl in the same light as saccharin, the FDA believed it should be sold in drugstores and labeled "a non-nutritive artificial sweetener which should be used only by persons who must restrict their intake of ordinary sweets." Processed foods sweetened with Sucaryl should carry a similar label and be sold only in health food stores. The FDA also told Abbott time and again that "the use of a nonnutritive artificial sweetener in beverages for general distribution would constitute an adulteration within the meaning of the Federal Food, Drug, and Cosmetic Act."[8]

The Association of Food and Drug Officials agreed that the "indiscriminate use of synthetic sweetened foods is not in the best interest of the public health and welfare." The Food Nutrition Board (organized under the auspices of the National Academy of Sciences and National Research Council) observed that the physiological harmlessness of Sucaryl at levels of "maximum probable intakes" had not been established and that the primary probable hazard would be the use of Sucaryl in soft drinks available for public consumption. While recognizing the usefulness of safe artificially sweetened foods for those who must restrict their intake of sugar, the board found no clear evidence that nonnutritive sweeteners would be effective for weight reduction or control.[9]

Abbott disregarded these suggestions, noting that the Code of Federal Regulations had never been amended to include cyclamate products. When the FDA pushed back, Abbott offered a watered-down warning stating simply that cyclamates were "for persons who desire to restrict their intake of ordinary sweets." When the FDA did not punish this infraction, food manufacturers followed Abbott's lead, and dietetic foods appeared on the shelves of ordinary supermarkets.[10]

However well meaning its intent, federal nagging was powerless in the face of popular opinion. The notion that thin was beautiful—epitomized, for instance, by the flappers of the 1920s—had less purchase during the years of depression and war when many Americans faced the problem of getting enough calories to keep them going, but the return to prosperity in the postwar period led to a renewed concern with figures and fashion. "Perhaps it is a tribute to our well-fed economy—now everybody can afford to get fat," said the president of the firm that made the dietetic soda Trim in 1953, adding that many of these people were on "steady or sporadic weight-reducing diets." That same year, a Gallup poll found that thirty-four million Americans admitted to being overweight, and the American Medical Association deemed obesity the country's number one health problem.[11]

Abbott introduced its most successful formulation of Sucaryl, this one containing ten parts cyclamate to one part saccharin, in September 1955. An unusual and obviously expensive full-color, multipage foldout ad in *Food Engineering* alerted food manufacturers to the fact that Sucaryl would be promoted in *Life* magazine and the Sunday supplements of ninety-one newspapers and that "millions of diet-shoppers" would be looking for Sucaryl on the label of low-calorie foods (Figure 30). This was followed by full-color, often full-page advertisements in consumer and trade publications, food columns in newspapers, and free cookbooks offering recipes for Sucaryl-sweetened dishes. Slenderella, a firm that capitalized on the notion of obesity as "the blight of American civilization," sponsored a cookbook with enticing recipes using Sucaryl in place of sugar.[12]

By the early 1960s, Americans could purchase an astonishing array of dietetic products—including puddings, gelatins, salad dressings, jams and jellies, ice cream, bacon, vitamins, pickles, and dog food. Diet Delight Foods were sweetened with Sucaryl, as was Lucky Pop, "a wonderful new soft drink sensation to make your eyes go pop." This "pop from a pill" was an early example of the many artificially sweetened treats designed for and marketed to children. Cheer Freeze desserts were "artificially sweetened with DuPont Cylan cyclamate."[13]

The list of low-calorie sweeteners designed for use on the tabletop or in the kitchen, many of them combining cyclamate and saccharin, also expanded apace. In addition to Sucaryl itself, there was Calorie Free, Chelten House Sweetener, Crescent Sweetener, Diet Joy, Fasweet, Hance-Sweet, Low-Cal, MCP Jelsweet, Motts Figure Control, My-T-Lo, Slimette, Super-rose, Steda-Sugar, Sweet'n It, Sweet'N Low, Sweetabs, Sweet Tablets, and Sweet-10.

And then there were diet sodas. Kirsch Beverages of Brooklyn introduced No-Cal, a Sucaryl-sweetened ginger ale designed for diabetics in

Figure 30: Diet Delight Foods, established in 1946 and affiliated with fruit growers in California, was an important early industrial user of Sucaryl, and it advertised widely in many women's and other mass-circulation magazines. Collection of the author.

1952 and expected to sell within a year an astounding two million cases of this specialty soda (Figure 31). Diet Rite (from Royal Crown Cola) followed soon thereafter, as did Tab (from Coca-Cola), Patio Diet Cola (from PepsiCo), and perhaps a hundred smaller brands. By 1964, the average American was drinking 227 diet sodas each year, and annual sales of these drinks had reached $2.3 billion. By 1969, Americans were consuming some seventy-five hundred tons of cyclamates and some fifteen hundred tons of saccharin annually, and diet foods had become a $1 billion a year industry.[14]

Alarmed by the explosive growth of diet foods, Sugar Information Inc. organized an advertising campaign informing Americans that one teaspoon of sugar contained but eighteen calories and that "no other food satisfies appetite so fast with so few calories." The Sugar Association promoted the energy story of sugar on Walter Cronkite's evening news program and in advertisements in leading general-circulation magazines, and the Sugar Research Foundation sponsored toxicological and epidemiological studies of cyclamates.[15] As the early work did not turn up anything particularly worrisome, the FDA announced in May 1965 that a review of recent medical literature revealed no evidence showing that cyclamates "at present use levels are a hazard to health."[16] Just five months later, however, the Sugar Research Foundation announced that scientists at the Wisconsin Alumni Research Foundation had found that large amounts of cyclamates stunted the growth of laboratory rats.[17] Over the course of the next several years, the Wisconsin scientists made several visits to the FDA to talk about artificial sweeteners. At one, Drs. Philip Derse and P. H. Nees "pointed out that they [had] been doing a great deal of work for the Sugar Research Foundation, and normally their results [were] reported directly to their clients without any publicity." They now felt they had seen sufficient scientific results "to lead them to the point where they should speak out in the public interest." Cyclamates, they said, should be removed from the federal list of additives generally regarded as safe (GRAS), and pregnant women should be warned about potential harm to their fetuses.[18]

The FDA responded by commissioning a study of thirty-two prisoners to be financed by Abbott Laboratories, Squibb Pharmaceuticals, the Sugar Research Foundation, and the FDA itself. The scientists affiliated with the Albany Medical College who monitored these men found that some prisoners who ingested large amounts of cyclamate on a regular basis experienced a change in the character of their blood (namely, an increased level of protein-bound iodine, which could lead to a mistaken diagnosis of hyperthyroidism), some experienced severe and persistent diarrhea, and some

Figure 31: Advertisements associating dietetic drinks and beauty proliferated in the 1950s. Julie Adams starred in the World War II film *Away All Boats* in 1956. Collection of the author.

converted the cyclamates into cyclohexylamine, a substance known to raise blood pressure.[19]

While the sugar proponents were marshalling their arguments, the cyclamate proponents were strengthening theirs. George Cain, Abbott chairman and chief executive, noted that sugar interests had spent about $500,000 on cyclamate research, meaning that an economic question was "masquerading as science." In collaboration with the Calorie Control Council, Abbott published *Cyclamate Sweeteners in the Human Diet* (1968), summarizing "the most comprehensive reviews of cyclamate published today." Scientists affiliated with Abbott or Pillsbury conducted most of the studies in this book, which, not surprisingly, confirmed cyclamates as safe for human consumption. In fact, the book went on to say, millions of people throughout the world have used cyclamates safely for more than fifteen years—"a remarkable record."[20]

An ad hoc committee of the National Academy of Sciences, however, warned against the unrestricted use of cyclamates and suggested that people should limit their daily intake to 70 milligrams per kilogram of body weight, or between about 3.5 and 5 grams for an average adult. (The UN Food and Agriculture Organization and the World Health Organization were a bit more cautious, recommending a daily limit of fifty milligrams per kilogram of body weight.) While releasing the gist of this report in December 1968, the FDA noted that it said nothing "that can yet justify any serious suspicion of adverse effects" at reasonable use levels; the only problem was a "moderate softening of the stools" that occasionally occurs at consumption levels above five grams a day.[21] Three months later, acknowledging that current regulations and current labeling practices were "not regarded as sufficient to provide consumers with enough information to enable them to safely use the cyclamate-containing products," the FDA proposed that food labels indicate the number of milligrams of cyclamate "supplied by the amount of the product normally consumed as a serving." For beverages, this declaration would be made in terms of the entire contents of the container.[22] After publishing this proposal, the FDA took no action on the matter.

Others, however, argued the proposed warning label did not go far enough. *Consumer Reports* said that more research was urgently needed before even limited amounts of cyclamates could be pronounced safe for "all people of all ages and in all conditions of health." Investigative reporter Jack Anderson unearthed FDA documents indicating that staff doctors and scientists had grave doubts about the safety of cyclamates, and those who read the fine print of the aforementioned ad hoc committee report realized that the FDA did not mention the parts that raised questions about the effect of cyclamates on the human kidney, liver, and chromosomes or in which the scientists observed that cancer studies of cyclamates in species other than rats were "either incomplete, or of relatively short duration, or involved too few animals."[23]

Gaylord Nelson, a senator from Wisconsin who was concerned with the health of the American public and of the planet, spoke for many who thought that the federal regulators had not gone far enough. In a blistering letter to the FDA, he noted that cyclamates were being used by some 125 million Americans "who have no real need for a non-nutritive sweetener and probably are unaware that they are consuming the chemical." Thus, the FDA should restrict cyclamate-sweetened products for distribution on a "prescription-only basis to the public." The Soviet Union had banned cyclamate-based foods, he said, and France had restricted the chemical's use to a prescription basis; yet American mothers were "finding it increasingly difficult to find flavored beverages without cyclamates." Simple labels were clearly insufficient. A better way must be found to inform the public.[24]

A congressional subcommittee concerned about consumer protection focused its attention on the FDA in September and October 1969. When the discussion turned to cyclamates, the testimony of an FDA administrator was classic obfuscation: cyclamate use should be limited; the proposed warning label had not been finalized; cyclamates might be removed from the GRAS list, but the details had not been worked out yet. FDA scientists, on the other hand, were more specific: "Consumption of cyclamate has now reached such a level that 13 million pounds were produced in 1967 and a production of 20 million pounds is projected for 1970. At the time of its introduction, it was believed to be almost totally excreted and to have no metabolic products. Several investigators have since found that cyclamate is metabolized to cyclohexylamine (CHA) in dogs and man. The increase in the consumption of the cyclamates and the realization that metabolic products exist necessitates a reexamination of the entire subject of cyclamate safety."[25]

On October 7, 1969, hoping to distance himself from a problem he did not understand and could not control, Secretary of Health, Education, and Welfare Robert Finch issued a scathing indictment of the FDA and its varying assessments of possible health hazards associated with cyclamates. Finch named no names but clearly referred to Commissioner of Food and Drugs Herbert Ley Jr. and to Jacqueline Jarrett, an FDA biochemist who, when interviewed by NBC, explained that she had found deformations in chicken embryos injected with cyclamates and warned pregnant women to exercise care until all doubts about this sweetener had been removed. Observing that the cyclamate evidence "may not be final or conclusive," Finch thought "we still ought to run up the flag and sound the bugle."[26]

The following day, while Abbott was holding a seminar telling food writers and manufacturers once again that cyclamates were perfectly safe for human consumption, the independent Food and Drug Research Laboratories in Maspeth, New York, called to say that in a study commissioned by Abbott, cancerous tumors had been found in the urinary bladders of rats that had ingested large amounts cyclamates. Although an earlier study had found a significant incidence of similar tumors when cyclamate pellets were implanted into the urinary bladders of the rats, the National Cancer Institute and FDA had decided that tests of this sort "were not suitable for evaluating the hazard of orally ingested compounds."[27] But the Delaney Clause made this latest study a serious matter indeed. As mentioned with respect to saccharin, the clause stated, "No additive shall be deemed to be safe if it is found to induce cancer when ingested by man or animal, or if it is found, after tests which are appropriate for the evaluation of the safety of food additives, to induce cancer in man or animal."[28]

Recognizing the urgency of the situation, Abbott scientists flew to Long Island on Friday, confirmed the results of the animal autopsies, and alerted the FDA on Monday, October 13. The Health, Education, and Welfare deputy secretary for health and scientific affairs sought the opinion of the National Cancer Institute on Tuesday. An ad hoc committee of the National Academy of Sciences began reviewing the cyclamate data on Thursday, and on Friday it unanimously recommended that cyclamates be removed from the GRAS list. Later that day, FDA commissioner Ley informed the *Federal Register* that cyclamates could no longer be used as a food additive. Beverages for general use sweetened with cyclamates would have to be withdrawn from the market by January 1, 1970, while cyclamate-sweetened foods for general use must be withdrawn by February 1. Secretary Finch went public with this decision the following day, Saturday, October 18.[29]

Abbott complied with the ban but argued repeatedly that their products should never have been removed from the market. *Barron's*, a probusiness publication, deemed the ban a "triumph of pseudo-science" and called attention to a class-action suit filed on behalf of millions of persons who "enjoy, and wish to enjoy, the consumption of cyclamates without unwarranted and dictatorial Government interference."[30] Ley and Finch understood, of course, that the amount of cyclamates ingested by the laboratory rats was, in proportion to their size, vastly greater than the average person would likely consume.

In his special message to congress on consumer protection, President Richard M. Nixon asserted that the American buyer "has the right to make an intelligent choice among products and services." After proposing several initiatives—including establishing an Office of Consumer Affairs in the White House and a Department of Consumer Protection in the Department of Justice—the president noted that he had asked the secretary of health, education, and welfare to undertake a thorough reexamination of the FDA and to initiate a full review of food additives because recent findings concerning the effects of cyclamate on rats underscored the importance of "continued vigilance in this field." Then, to soften the blow somewhat, he acknowledged, "The major suppliers and users of cyclamates have shown a sense of public responsibility during the recent difficulties and I am confident that such cooperation from industry will continue to facilitate this investigation."[31] Nixon's statement probably had the support of Jean Mayer, a noted nutritionist then serving as a special consultant to the president who said, in a speech to the Women's National Press Club, "If there is the slightest presumption of guilt" with regard to a food additive, "I just would not put it in the food."[32]

There was clearly no intention of banning cyclamates all together, however. Indeed, in his October notice to the *Federal Register*, FDA commissioner Ley had stated, "Cyclamates and artificially-sweetened products intended for use in the dietary management of disease in man, including the management of such diseases as diabetes and obesity, should be relabeled promptly to comply with the drug provisions of the law if they are to continue on the market." By December, a medical advisory group that endorsed the prohibition of cyclamates in foods and beverages for general use concluded unanimously that for some people, under appropriate medical management, cyclamates "provide medical benefits which outweigh their hazards." Accordingly, it recommended that cyclamates be made available "for such patients on medical advice and on a non-prescription, drug-labeled basis." In compliance with these recommendations, the FDA pushed back the ban on cyclamate-containing food products from February 1 to September 1. It also instituted an abbreviated new drug application for substances, such as cyclamates, that had once been approved as a drug. It further proposed that packages of artificial sweeteners containing cyclamates list active ingredients and safe dosages and carry the statements "For use only with calorie-controlled diets by diabetics or obese patients under medical supervision" and "Caution: medical supervision is essential for safe use." The FDA would later extend this sort of labeling to containers of foods and drinks containing cyclamates.[33]

That might have been the end of the story but for the publication of *The Chemical Feast* (1970) written by James S. Turner, one of "Nader's Raiders," a team of consumer advocates led by Ralph Nader. The team had spent the previous summer investigating the food-protection aspects of the FDA and uncovering evidence showing that the FDA had long known much more about the problems relating to cyclamates than it had shared with the public. The center of action then moved to Capitol Hill, where FDA officials were again put on the hot seat, and FDA documents pertaining to cyclamates were entered into the public record.[34] A subsequent report by the full Committee on Government Operations concluded, among other things, that the "FDA failed for several years to protect the public against possible health hazards associated with cyclamates despite a clear legal obligation to do so"; that the "FDA aggravated the consequences of its inaction by permitting the use of cyclamates in food to reach massive proportions"; and that the "FDA attempted to permit the continued marketing of cyclamate-containing products through illegal regulations and procedures."[35]

In mid-August 1970, the FDA's Medical Advisory Group on cyclamates slashed the "prudent" daily limit to 168 milligrams for a person weighing 154 pounds. Equivalent to a mere twenty-one calories of sugar, this amount was hardly enough to make a difference to a dieter. Accordingly, the FDA did another about-face, ordering a complete ban on cyclamates in foods and beverages to go into effect on September 1.[36] Shortly thereafter, Nevis Cook, the deputy regional food and drug director in New Orleans, called the FDA in Washington, DC, to say that Abbott planned to dispose of fourteen freight carloads of cyclamate products in a landfill in Jefferson and Orleans parishes. The Louisiana Health Department did not want to permit the dumping because the area involved was swampy, occasionally flooded, below sea level, and home to crayfish and other animals. The FDA, however, had no "definitive information" on the biodegradability of cyclamates or their toxicity to fish and seafood, and thus had no recommendation.[37]

Still there were problems. In May 1971, the *Washington Post* reported that drugs containing cyclamates had been on the market "for as long as 9 months after such formulations became illegal."[38] Under pressure from growers, manufacturers, packers, and distributors, Congress considered indemnifying those who thought themselves unfairly harmed by the cyclamate ban. No indemnity bill was ever enacted into law, but the hearings and floor debate revealed yet more details of the cyclamate story. Some of the discussion concerned questions of ethics and responsibility. Proponents of indemnity argued that Abbott had promoted cyclamates long after it knew that serious charges had been made against the product. Opponents argued that concerns about cyclamates had long been widespread, and smart entrepreneurs were ready with new formulations when the ban was finally enacted. In the words of Congresswoman Leonor Sullivan, "The artificial sweetener turned sour. Why should the taxpayers pick up the tab—no pun intended."[39]

Loathe to lose its lucrative product, Abbott kept asking the FDA to reinstate its approval. In November 1973, the firm submitted a sixteen-volume petition citing twenty recent independent studies that showed no link between cancer and cyclamates. The FDA acknowledged "that it was currently 'the apparent opinion' of world cancer experts that 'cyclamates when tested in accordance with appropriate protocols (procedures) are not carcinogenic.'" Because, however, some staff scientists thought the carcino-genicity of cyclamates remained an open question, the FDA deferred to a panel organized by the National Cancer Institute.[40] Most members of this panel voted to approve cyclamates. The one holdout, I. Bernard Weinstein of Columbia University, agreed that cyclamate was not a "potent carcino-gen" but worried that it might be a weak one. So, the FDA refused to lift the ban.[41] Abbott then brought suit against the FDA but lost when an adminis-trative law judge decided that the firm had "failed to prove cyclamates don't cause cancer."[42]

With the cyclamate question still hanging fire, the FDA asked the National Research Council to evaluate all the evidence pertaining to the potential carcinogenicity of cyclamates. The report of this evaluation, pub-lished in 1985, stated clearly that "the weight of experimental and epidemi-ological evidence does not indicate that cyclamate by itself is carcinogenic. However, there is suggestive evidence from in vivo and in vitro studies in animals that cyclamate has cancer-promoting or cocarcinogenic activity, and from epidemiological studies in humans that the use of cyclamate-saccharin mixtures may be associated with a small increase in risk of bladder cancer." Although the council had not been asked about this, the report also noted evidence of other adverse effects, such as testicular atrophy in laboratory animals.[43] With this report in hand, the FDA had the support it needed to maintain the ban on cyclamates—which it does to this day. Abbott, how-ever, could and did still produce cyclamates for use in Canada and other countries.[44]

chapter 14

ASPARTAME AND SUCRALOSE

Aspartame

In the 1960s, James Schlatter was a young chemist working for G. D. Searle & Company, a pharmaceutical company best known for Metamucil, Dramamine, and Enovid birth control pills. In December 1965, while investigating peptides for use in treating ulcers, he licked his finger and found it to be intensely sweet—an action that Daniel Searle, president of the firm, would call "a great discovery but lousy lab technique."[1] Then, working with other scientists in the firm, Schlatter ran taste tests to determine which peptide was the most promising and toxicity tests to determine its safety for human consumption. Finally, in April 1969, after submission of the first patent applications, Searle announced that a staff chemist had synthesized a compound potentially "suitable for use as an artificial sweetener." By 1971 Searle said that the new sugar substitute "could be a big item for us," especially as it would be the only serious competitor to saccharin. Annual sales of cyclamate before the ban, he noted, had totaled some $16 million. Also, because the new substance comprised two amino acids that occur naturally in the body, it should be safe for human consumption. Indeed, he had downed several small bottles of it without any apparent ill effects.[2]

Searle soon reported that aspartame, as the substance was now known, was some 150 times sweeter than sugar. Unlike saccharin, it left no unpleasant aftertaste, and it had been tested extensively for safety. Some would be sold as a tabletop sweetener, and some would go into such products as cereals, gum, powdered drink mixes, and diet sodas. Searle also announced plans for a new plant, to be built in the United States and operated jointly with the Japanese firm Ajinomoto, the world leader in the manufacture of amino acids.[3]

In the summer of 1974, the Food and Drug Administration (FDA) agreed that aspartame could be used in tablet or granulated form. If it was

209

an ingredient in processed foods, the label had to say, "Phenylketonurics: contains phenylalanine." This latter warning came at the urging of John Olney, a physician in the Department of Psychiatry at Washington University Medical School, who worried that aspartame, when combined with MSG, might cause mental retardation in children, especially those suffering from the disorder known as phenylketonuria, or PKU.[4]

When Senator William Proxmire asked that aspartame be held off the market until detailed hearings could be held and additional scientific tests made, the FDA agreed to what was deemed an "unprecedented" public hearing. In May 1975, having learned that aspartame had harmed the livers of rats, General Foods stopped all market tests of products containing the sweetener. In October, FDA representatives noticed numerous errors in Searle's records pertaining to aspartame. In December, citing concern over the general reliability of Searle's test data, the FDA rescinded the company's authority to sell the new sweetener altogether.[5] Prospects for aspartame dimmed even further when James Turner, a public interest lawyer and author of *The Chemical Feast* (1970), urged the FDA to stay the approval of aspartame pending further investigation. The FDA then sought a federal grand jury investigation of Searle's practices, alleging that the firm had submitted phony data and withheld test results in its applications for aspartame and three prescription drugs.

Faced with this situation, Searle turned to Donald Rumsfeld, hiring him as a consultant just days after he left his post at the Department of Defense following President Jimmy Carter's inauguration. Rumsfeld became president and chief executive officer of Searle a few months later, earning the then princely salary of $200,000 plus, presumably, handsome stock options. His primary tasks would be restructuring the business and obtaining FDA approval for aspartame and other products in the pipeline.[6] During Rumsfeld's first year at Searle, aspartame accounted for more than 40 percent of the firm's sales and a much larger percentage of its profits. Since aspartame could not be sold in the United States at that time, these profits came from abroad. Under the brand name Canderel, aspartame was approved for sale in Mexico, France, Luxembourg, and Belgium, followed by Brazil, the Philippines, Tunisia, Switzerland, and Canada.[7]

In April 1979, having become impatient with the pace of the federal bureaucracy, Searle sent a strongly worded letter urging the FDA to lift the suspension on aspartame. The FDA, however, opted for a three-man public board of inquiry that would consider the various issues relating to safety.[8] This board found no evidence of toxicity, but since the data at hand did not rule out the possibility that aspartame might cause brain tumors in rats, it called for more research. The FDA Bureau of Foods was willing to recommend approval, but the commissioner remained unconvinced. So, Searle filed suit in federal court to force a decision.[9]

Rumsfeld campaigned for Ronald Reagan in 1980, in part because the Gipper was determined to free business from the straitjacket of what he deemed excessive government regulation. Then, in January 1981, shortly before Reagan's inauguration, Rumsfeld let it be known that he intended to call in his markers. Arthur Hull Hayes, the new FDA commissioner said to dislike regulatory delay, announced in July that aspartame was approved for use in dry foods and beverage mixes.[10]

With this approval in hand, Searle began marketing aspartame as Equal (the substance in blue packets designed for use at the table) and as Nutra-Sweet (for use as an ingredient in prepared foods). The firm also launched a massive campaign promoting such goodies as Kool-Aid, Lipton's Country Time Lemonade mix, and Quaker Oats' Halfsies (with half the sugar of most sugarcoated cereals). Approvals for aspartame in other applications soon followed: in soft drinks (July 1983); in a broad range of juice drinks, frozen desserts, canned and instant teas, and breath mints (December 1986); and in six more areas, including flavored milks and various frozen desserts (1988).[11] In 1993, after Monsanto scientists developed a version of aspartame that remained stable at high temperatures, the FDA approved it for use in baked goods. The agency lifted all restrictions in 1996.[12]

Soft drink industry analysts hailed the FDA approval as a major development, for it would now be possible to make a tasty diet soft drink, something unknown since the cyclamate ban in 1969. Coca-Cola, the first major manufacturer to sign an agreement with Searle, announced that Tab and Diet Coke would contain a combination of aspartame and saccharin, so designed to overcome "aspartame's inability to stay sweet after long periods of time and saccharin's metallic aftertaste." Just two years after the pertinent FDA approval, some one hundred million Americans were drinking some twenty billion cans of diet soft drinks, most of them containing NutraSweet.[13]

Searle worried that, due to the protracted approval process for aspartame, its patents would expire before it could profit from the product. Relief came late one evening in October 1982 when Senator Howell Heflin, a Democrat from Alabama who had received substantial campaign contributions from members of the Searle family, proposed an amendment to the Orphan Drug Act. Aspartame and Searle were not mentioned per se, but as Andrew Cockburn explains in his muckraking biography of Rumsfeld, the amendment had "no other effect than to extend the aspartame patent into 1992."[14] Rumsfeld's final task at Searle entailed selling the company. After a few false starts, Monsanto offered a whopping $2.7 billion in 1985. After a decade in the private sector, Rumsfeld had a net worth said to be around $10 million.[15]

Proponents of aspartame noted that, prior to its approval as a food additive, the substance had been tested in more than one hundred scientific studies, "making it one of the most thoroughly tested ingredients in our food supply." Moreover, the regulatory bodies of more than fifty nations had approved it for use, and it was used in some forty countries.[16]

Many people around the world, however, including Senator Howard Metzenbaum, remained concerned about the safety of aspartame. In the fall of 1987, after his staff conducted a lengthy and exhaustive investigation into aspartame and Searle, this Ohio Democrat sponsored a hearing, which he opened by proclaiming, "It has been hailed as the most successful food additive in history. One hundred million Americans use it. It commands over $700 million in sales. Over 20 billion cans of diet soft drinks containing it are sold each year. And it seems to be in everything—cereal, kids' vitamins, cocoa, puddings, even over-the-counter drugs. There is no doubt about it. NutraSweet has captured the hearts and the tastebuds of the American consumer." But for some consumers and scientists, he went on to say, "the dream may be too good to be true." "The FDA has received close to 4,000 consumer complaints, ranging from seizures to headaches to mood alterations. Studies and letters in the medical journals have warned of possible neurological and behavioral effects in humans, particularly in children and susceptible individuals." Only animal tests were required of NutraSweet, and now some FDA scientists have suggested it should have undergone human tests as well. "I am frank to say that the NutraSweet Company, the food and beverage industry, and their various institutes exert tremendous influence over scientific research and investigation. I want to make sure such work is genuinely independent." In the end, however, no actions came of all this sound and fury.[17]

Aspartame was the leading nonnutritive sweetener in the United States in the latter years of the twentieth century, commanding some 37 percent of the market, but that situation was soon to change. By 2008, Splenda had grabbed 60 percent of the market, saccharin (in the form of Sweet'N Low) came in second, and aspartame followed with a mere 11 percent. During the credit crunch of early 2009, Merisant, the Boston-based investment company that had bought NutraSweet from Monsanto, filed for bankruptcy.[18]

Splenda

Each yellow packet of Splenda contains a tiny bit of sucralose (the sweetener) and a lot of dextrose and maltodextrin (the filler). Since sucralose is about six hundred times sweeter than ordinary sugar (and about three times sweeter than aspartame), a little goes a very long way. Dextrose and maltodextrin are large molecules, similar in size to sucrose itself; thus, a spoonful of Splenda has the same sweetening power and heft as one of sugar. Sucralose is also more stable at high temperatures than aspartame and so can be used in baked goods.[19]

Sucralose (technically trichlorogalactosucrose, or $C_{12}H_{19}C_{13}O_8$) is derived from sugar by a multistep process in which three hydrogen-oxygen groups are replaced with three chlorine molecules. Its story begins with Leslie Hough, a chemist at Queen Elizabeth College (now a part of King's College London), who was not looking for new sweeteners but studying chlorinated sugars for other purposes. Hough asked Shashikant Phadnis, a young Indian chemist, to test this molecule, and Phadnis, who had understood him to say "taste," found it incredibly sweet. Hough, Phadnis, and other chemists in the group obtained British patents in 1976 and assigned the rights to Tate & Lyle, the leading British sugar company. Tate & Lyle then joined with Johnson & Johnson, the world's largest health-care company, to develop sucralose and build a factory in McIntosh, Alabama.[20]

In 1987, having completed toxicity studies in several animal species and clinical tests in human volunteers, Johnson & Johnson sought FDA approval.[21] Eleven years later, after reviewing 110 studies in animals and humans, the FDA decided that sucralose could be used in fifteen major food categories ranging from beverages to syrups. The estimated daily intake of sucralose, according to the FDA, was 98 milligrams per person per day (or approximately 1.6 milligrams per kilogram of body weight per day), which was well below the acceptable daily intake established from the toxicity data. Sucralose received FDA approval for all uses in 1999.[22] The Joint UN Food and Agriculture Organization and World Health Organization Expert Committee on Food Additives had endorsed the safety of sucralose in 1990.

Commercialized by McNeil Specialty Products Company, a Johnson & Johnson division formed for this purpose, Splenda was on the market by 1998 and originally used as an ingredient in Diet Rite cola.[23] It was also promoted by the Atkins low-carbohydrate diet. An advertising campaign projected to cost $40 million began in September 2000 with television spots featuring soccer star Mia Hamm.[24] By 2004, Splenda had become the number one branded sweetener in the United States, and nearly every major food company incorporated sucralose into one or more of its products. In 2005, Splenda introduced a brown sugar blend, and Tate & Lyle announced that it could not keep up with demand.[25]

The original Splenda slogan—"Made from sugar so it tastes like sugar"—tested well with consumers but irked the competition. The Sugar Association (which calls Splenda a "chlorinated artificial sweetener" and tells consumers that they "are actually eating chlorine") and Merisant (the maker of NutraSweet) brought suit, contending that the marketing misled consumers. McNeil lost the case, settled out of court, and agreed to a new slogan: "Starts with sugar, tastes like sugar, but is not sugar." The company sustained no serious harm, as Splenda had 60 percent of the consumer artificial sweetener market in 2007.[26]

NOTES

Introduction

1. Robert Boyle, *Some Considerations Touching the Usefulnesse of Experimental Naturall Philosophy* (Oxford: Printed by Hen. Hall for Ric. Davis, 1663), 2:112.

2. "Increase in the Consumption of Sugar," *New York Times* (November 9, 1858), 2. "Sugar and Salt," *New York Times* (December 17, 1858), 4. R. R. Bowker, "A Lump of Sugar," *Harper's New Monthly Magazine* 73 (1886): 72–95, on 94. Murphy Foster in *Congressional Record* 44 (1909): 2354–79, on 2379.

3. Sidney Mintz, *Sweetness and Power: The Place of Sugar in Modern History* (New York: Viking, 1985).

Chapter 1: Sugar Refining in New York City

1. *Martha Washington's Booke of Cookery and Booke of Sweetmeats*, transcribed by Karen Hess (New York: Columbia University Press, 1985), 225. Eliza Leslie, *Seventy-Five Recepts for Pastry, Cakes and Sweetmeats* (Boston, Munroe & Francis, 1832), 7. Eleanor Parkinson, *The Complete Confectioner* (Philadelphia: J. B. Lippincott & Co., 1864), 13.

2. Ads of James Smith in *Boston News-Letter* (September 2 to 9, 1717), 2, and of Peter Delage in *Pennsylvania Gazette* (March 30 to April 6, 1738), 4. "Will of Nicholas Bayard of New Amsterdam," in *Abstract of Wills* 6 (1760–1766), 192–95. An excerpt is available online at http://apps.olin.wustl.edu/faculty/mcleanparks/bearswamp/Bayard-Will.htm (accessed December 13, 2010).

3. "London January 4," (Philadelphia) *American Weekly Mercury* (April 14 to 21, 1737), 2.

4. "The Sugar Refineries of New York," *New York Times* (August 11, 1856), 6. *Journal of Commerce* cited in "The Refineries of New York," *Merchants Magazine* 35 (1856): 500–2. Paul L. Vogt, *The Sugar Refining Industry in the United States* (Philadelphia: University of Pennsylvania Press, 1908), 6.

5. R. N. Burnett, "Henry Osborne Havemeyer," *Cosmopolitan* 34 (1903): 701–4. "T. A. Havemeyer's Story," *Brooklyn Eagle* (February 2, 1896), 14. "Theodore A. Havemeyer," *Brooklyn Eagle* (April 26, 1897), 14. Jack S. Mullins, "The Sugar Trust: Henry O. Havemeyer and the American Sugar Refining Company" (PhD diss., University of South Carolina, 1964).

6. Paul Casamajor letter in *Journal of the Franklin Institute* 87 (1869): 398–99. "The Williamsburgh Sugar Refineries," *American Grocer* 3 (1870): 368. "Havemeyers & Elder" in *Industrial America* (New York: Atlantic Publishing & Engraving Company, 1876). Hamilton Wicks, "Refining Sugar," *Scientific American* 40 (January 25, 1879), 47, 49, and front cover. "Local Manufactures: The Interesting Process of Sugar Making," *Brooklyn Eagle* (August 17, 1884), 10.

7. "Frederick C. Havemeyer," *New York Times* (July 29, 1891), 5. "Frederick C. Havemeyer," *Louisiana Planter* 7 (1891): 17 (from *American Grocer*). "Increased Terminal Facilities," *New York Times* (January 8, 1876), 2.

8. Noël Deerr, *The History of Sugar* (London: Chapman and Hall, 1950), 559–62. Frederick Kurzer, "The Life and Work of Edward Charles Howard FRS," *Annals of Science* 56 (1999): 113–41.

9. "R. L. and A. Stuart," in *Encyclopaedia of Contemporary Biography of New York* (New York: Atlantic Publishing and Engraving Co., 1882), 2:402–6. "For an Improvement in Manufacturing Confectionary of All Kinds," *Journal of the Franklin Institute* 12 (1833): 167. "Sugar," *Literary Inquirer* (March 12, 1834), 78. "Seventh Annual Fair of the American Institute," *New York Farmer* (1834): 81, and *Niles Weekly Register* 47 (October 25, 1834): 126–29. "The Great Sugar Candy Establishment," (Charleston) *Southern Patriot* (April 5, 1836), 2. "Alexander Stuart," *Scientific American* 42 (1880): 17. "Alexander Stuart," *New York Times* (December 24, 1879), 5. "The Will of a Millionaire," *New York Times* (December 30, 1879), 3. "An Old Merchant Gone: The Death Yesterday of Robert L. Stuart," *New York Times* (December 13, 1882), 5. "Reminiscences in the Sugar Trade," *American Grocer* 22 (1879): 668–69. "The Sugar Refineries of New York."

10. Henry O. Havemeyer Jr., *Biographical Record of the Havemeyer Family* (New York, 1944). T. A. Havemeyer and Henry Schnitzpan, "Improved Carriage for Sugar-Molds," U.S. Patent 34,686 (1862; reissued in 1863). T. A. Havemeyer, I. L. Elder, and C. F. Loosey, assignees of Carl Kronig, "Improvement in the Manufacture of Sugar-Molds and Other Articles," U.S. Patent 43,376 (1864; reissued in 1868). "Applications of Paper," *Scientific American* 41 (1867): 396. Henry Medlock, "Revivification of Animal Charcoal," *Scientific American* 12 (1865): 220 and 224–25.

11. Deerr, *History of Sugar*, 2:573–74. "Reminiscences in the Sugar Trade." "An Improvement on the Water Front," *Brooklyn Eagle* (June 1, 1866), 2.

12. "Sales at the Stock Exchange" *New York Times* (June 21, 1854), 6. "Business Notices," *New York Times* (September 1, 1852), 2. Charles Hersey, "Improvement in Driers," U.S. Patent 114,137 (1871). Hersey Manufacturing Co. ad in *Facts about Sugar* 15 (1922): 484.

13. David O. Whitten and Bessie E. Whitten, *The Birth of Big Business in the United States, 1860–1914* (Westport, CT: Praeger, 2006).

14. Sugar Trade Monopoly, *New York Times* (May 24, 1878), 2. "Freezing Out Refiners: The Bankrupt Condition of the Trade," *New York Times* (February 23, 1880), 2.

15. "Regulating Prices of Sugar," *New York Times* (June 13, 1880), 10.

16. "The Sugar Market," *New York Times* (January 11, 1882), 8. Woodcut in "A Disastrous Fire," *Harpers Weekly* 26 (1882): 36. "Destruction of Havemeyers & Elder's Refinery," *American Grocer* 27 (1882): 67–68. Harry W. Havemeyer, *Merchants of Williamsburgh* (Brooklyn, NY: H. W. Havemeyer, 1989).

17. "The New Sugar Trust," *New York Times* (October 19, 1887), 9. Vogt, *The Sugar Refining Industry*. Alfred S. Eichner, *The Emergence of Oligopoly: Sugar Refining as a Case Study* (Baltimore: The Johns Hopkins University Press, 1969).

18. William H. Doyle, "Capital Structure and the Financial Development of the U.S. Sugar-Refining Industry, 1875–1905," *Journal of Economic History* 60 (2000): 190–215.

19. "World's Greatest Refinery of Sugar for New Orleans," *Christian Science Monitor* (January 20, 1909), 4. "American Sugar Will Soon Build Cuban Central," *Wall Street Journal* (May 1, 1920), 10. "American Sugar to Spend $8,000,000 on Baltimore Plant," *Wall Street Journal* (December 23, 1919), 11. "Open Baltimore Sugar Refinery," *Wall Street Journal* (May 18, 1922), 9.

20. U.S. Bureau of the Census, *Historical Statistics of the United States, Millennial Edition Online*. Richard Peters, ed., *Public Statutes at Large of the United States of America* (Boston: Charles C. Little and James Brown, 1845), 1:24–27.

21. *Report of the Secretary of the Treasury*, 23rd Cong., 1st sess., May 2, 1832, S. doc 139. Benjamin Silliman, *Manual on the Cultivation of the Sugar Cane and the Fabrication and Refinement of Sugar* (Washington, DC: F. P. Blair, 1833).

22. "The New Tariff Bill," *New York Times* (May 28, 1864), 4. "The Sugar Tariff," *New York Times* (December 23, 1870), 5. "The Sugar Tariff Ring," *New York Times* (April 14, 1880), 10.

23. F. W. Taussig, "The Tariff Act of 1894," *Political Science Quarterly* 9 (1894): 585–609.

24. "Our Customs Department," *The American Sugar Family* 1 (February 1920): 24.

25. Eichner, *The Emergence of Oligopoly*, 213–28.

26. Eichner, *The Emergence of Oligopoly*, 229–63.

27. Eichner, *The Emergence of Oligopoly*, 172–80 and 184–86. Charles McCurdy, "The *Knight* Sugar Decision of 1895 and the Modernization of American Corporation Law, 1869–1903," *Business History Review* 53 (1979): 304–42.

28. Eichner, *The Emergence of Oligopoly*, 273–82.

29. "Freezing Out Refiners." Harold J. Howland, "The Case of the Seventeen Holes," *Outlook* (May 1, 1909), 25–38. Harold J. Howland, "The Men Higher Up," *Outlook* (August 6, 1910), 771–84.

30. "Getting Higher in Sugar Fraud," *New York Times* (November 13, 1909), 1. "Employees to Turn on the Sugar Trust," *New York Times* (December 1, 1909), 3. "Sugar Checkers Are Found Guilty," *New York Times* (December 18, 1909), 1. "Sugar-Fraud Gang to Jail for a Year," *New York Times* (January 11, 1910, 4). "Indicts Secretary of the Sugar Trust," *New York Times* (January 15, 1910), 1. "Heike, Gebracht, Both Found Guilty," *New York Times* (June 11, 1910), 1. Theodore Roosevelt, *Autobiography* (New York: Macmillan, 1913), 484–87.

31. "Sugar Weighers Said to Be Indicted," *New York Times* (November 18, 1909), 1. "Loeb to Clean Up Customs Service," *New York Times* (November 19, 1909), 3.

32. House Special Committee on the Investigations of the American Sugar Refining Co. and Others, *Hearings*, 62nd Cong., 1st and 2nd sess., 1911–1912, and *Report under House Resolution 157*, 62nd Cong., 2nd sess., 1912. "Sugar Trust Found by House Searchers," *New York Times* (February 18, 1912), 14.

33. Eichner, *The Emergence of Oligopoly*, 302–25.

34. "The Labor Issue," *New York Times* (June 24, 1872), 8. "Sugar Refining," *Brooklyn Eagle* (December 6, 1896), 21. "Serious Accident at Stuart's Steam Sugar Refinery," *New York Times* (August 10, 1852), 4. "Sad and Fatal Accidents," *New York Times* (March 2, 1853), 3. "Sad Death of a Son of Ex-Mayor Havemeyer," *New York Times* (November 28, 1861), 8.

35. "The Strike Movement," *New York Times* (April 27, 1853), 1. "The Strike Trouble," *New York Times* (June 16, 1872), 1. "Sugar Workers May Strike," *New York Times* (July 29, 1900), 6.

36. "Strike of Sugar Refiners," *New York Times* (April 22, 1886), 5. "End of the Sugar House Strike," *New York Times* (May 12, 1886), 8. "Sugar Workers May Strike." "Large Wage Advances by Sugar Refiners," *Wall Street Journal* (April 7, 1923), 9. "The Coal Troubles," *New York Times* (January 13, 1887), 8.

37. American Sugar Refining Co., *Annual Report for 1915*, n.p. "American Sugar Refining Plans Cooperage Plant," *Wall Street Journal* (January 24, 1920), 11.

38. "Coopers on Strike," *Brooklyn Eagle* (June 17, 1873), 4. "The Coopers' Strike," *Brooklyn Eagle* (June 24, 1873), 4. "What the Cooper's Say," *Brooklyn Eagle* (August 6, 1873), 4. "Labor and Capital," *Brooklyn Eagle* (August 4, 1873), 4. "The Labor Issue," *New York Times* (June 24, 1872), 8. "Brooklyn's Fatal Fire," *New York Times* (June 14, 1900), 14.

39. "Sugar in Bags," *Louisiana Planter* 5 (1890): 448. "Granulated Sugar in Bags," *Louisiana Planter* 6 (1891): 14. "Bags Instead of Barrels for Sugar," *Louisiana Planter* 9 (1892): 330.

40. "American Sugar," *Fortune* 7 (February 1933): 59–65 and 115–16. "Automated Sugar Refinery to Open Near Boston's Bunker Hill," *Christian Science Monitor* (April 14, 1961), 10. "New Sugar Refinery 'Most Automated,'" *Christian Science Monitor* (September 20, 1960), 13.

41. "Sugar Stocks Ebb in Philadelphia," *New York Times* (July 10, 1938), 9.

42. "2,500 Go on Strike at Sugar Refineries," *New York Times* (November 15, 1941), 8.

43. "Amstar Closing Refinery 200 Jobs Lost in Charlestown Decision," *Boston Globe* (February 18, 1988), 57. "At Sugar Refinery, a Melting-Pot Strike," *New York Times* (February 15, 2000), B1. "Striking Domino Workers Feel Bitterness and Resolve," *New York Times* (November 25, 2000), B2. "Bitter Strike at Refinery Finally Ends," *New York Times* (February 27, 2001), B1. "The Last Grain Falls at Sugar Refinery," *New York Times* (January 31, 2004), B3.

44. "Alexander Stuart," *Scientific American* 42 (1880): 17.

45. "Paul Casamajor," *Journal of the American Chemical Society* 9 (1887): 206–8. "Sudden Death of Mr. Casamajor," *Chicago Tribune* (November 13, 1887), 13. "Poison Caused His Death," *New York Times* (November 13, 1887), 2.

46. Marsden Taylor Bogert, "Charles Frederick Chandler," in *Biographical Memoirs, National Academy of Sciences* 14 (1931): 125–81.

47. Marcus Benjamin, *Sugar Analysis* (New York, 1880). John Henry Tucker, *A Manual of Sugar Analysis* (New York: Van Nostrand, 1881, 1883, 1893, and 1912). Ferdinand G. Wiechmann, *Sugar Analysis for Cane-Sugar and Beet-Sugar Houses, Refineries and Experimental Stations, and as a Handbook of Instruction in Schools of Chemical Technology* (New York: John Wiley & Sons, 1890 and 1893; third edition, extensively revised and expanded, 1914).

48. E. Hendrick, "Charles Frederick Chandler, His Life and Works," *Columbia Alumni News* 17 (January 15, 1926): 301–8.

49. "The Adulteration of Sugar," *Scientific American* 6 (1850): 88. Arthur Hill Hassall, "On the General Occurrence of the Sugar Acarus," in Great Britain, Parliament, Select Committee on Adulteration of Food, Drinks and Drugs, House of Commons,

Reports on the Adulteration of Food, Drink and Drugs (London: David Bryce, 1855). "The Little Monster in Your Sugar," *Ohio Farmer* (July 26, 1856), 120, from *News of the World* (English paper); this also appeared in *Saturday Evening Post* (August 16, 1856), 4. Robert Nichol, *An Essay on Sugar and General Treatise on Sugar Refining* (Greenock: A. Mackenzie & Co., 1864), appendix, 62 and plate. Robert Nichol, "The Sugar Insect *Acarus sacchari* Found in Raw Sugar," *Journal of the Franklin Institute* 56 (1868): 327–33. Robert Niccol, "What We Eat in Unrefined Sugar: The Sugar Insect," *DeBow's Review* (March 1869): 239–42. "The Obligations of Sugar-Refiners and Consumers to Physics and Chemistry," *Manufacturer and Builder* 3 (1871): 78. "Sugar Active," *American Grocer* 8 (1873): 362. "What Are Raw Sugars?" *Willett & Gray's Weekly Statistical Sugar Trade Journal* (April 23, 1891), 6, and several subsequent issues.

50. "Sugar Adulteration!" *New York Times* (December 4, 1878), 8, and several subsequent issues. Havemeyers & Elder repeated this text in weekly, full-page advertisements in *American Grocer* 21 (March 27, 1879): 796, and many later issues. "The Alleged Adulteration of Sugar," *New York Times* (January 7, 1879), 5. "The Refined Sugar Question," *New York Times* (January 22, 1879), 8. "The Sugar-Dealers' Quarrel," *New York Times* (May 3, 1879), 3. Pierre de Peyster Ricketts, "A Method for the Detection of Artificial or Dextro-Glucose in Cane Sugar, and the Exact Determination of Cane Sugar by the Polariscope," *Journal of the American Chemical Society* 1 (1879): 2–6.

51. Charles A Browne, "A History of the New York Sugar Trade Laboratory," typescript, Browne papers, Box 33, Library of Congress.

52. J.-B. Biot, *Instructions pratiques sur l'observation et la mesure des propriétés optiques appelées rotatoires, avec l'exposé succinct de leur application à la chimie medicale, scientifique et industrielle* (Paris, 1845).

53. "Description du nouveau saccharimètre de M. Soleil, opticien, rue de l'Odéon, 35," *Bulletin de la Société d'Encouragement pour l'Industrie Nationale* 45 (1846): 543–49, with illustration.

54. R. S. McCulloh, *Reports from the Secretary of the Treasury of Scientific Investigations in Relation to Sugar and Hydrometers* (Washington, DC, 1848), 48. "Science and Its Uses," *New York Times* (November 15, 1852), 2. "How Sweet Is Sugar?" *American Grocer* 11 (1874): 228. "Gifts for the University," *New York Times* (January 7, 1896), 10.

55. Elmore & Richards, *Price List of Chemical Apparatus* (New York: Kelly & Bartholomew, 1880), 115–16.

56. Deborah J. Warner, "How Sweet It Is: Sugar, Science, and the State," *Annals of Science* 64 (2007): 147–70.

57. Franklin ad in *Grocery World* 45 (January 13, 1908), 39.

58. "Refining Sugar," *Scientific American* 40 (January 25, 1879), 48–49 and illustration on front cover. William Möller, "Improved Sugar-Cutting Machine," U.S. Patent 33,260 (1861). Ads in *New York Times* (June 12, 1902), 5; *Atlanta Constitution* (March 30, 1902), 21; *The Youth's Companion* 76 (April 24, 1902), 220; *The Ladies' Home Journal* 19 (June 1902), 35; *Cosmopolitan* 33 (October 1902), 768; *Town & Country* (October 3, 1903), 30; *Christian Advocate* 41 (October 8, 1903), 1644; *Harper's Bazaar* 38 (January 1904), A27; *American Grocer* 77 (January 2, 1907), 2. *The Independent* 71 (November 30, 1911), 2.

59. "New Director Elected to American Sugar Refining Co.," *Wall Street Journal* (March 13, 1915), 5. "Sweet Squawk," *Time* (March 22, 1937), 98. "Earl Babst, Industrialist, Dead; Led American Sugar Refining," *New York Times* (April 25, 1967), 43.

60. "American Sugar in Syrup," *Wall Street Journal* (May 5, 1919), 2. Ads in *Christian Science Monitor* (August 30, 1920), 4, *New York Times* (July 7, 1929), 84, and *Chicago Tribune* (October 9, 1964), B10.

61. "Sugar Institute Formed at New York," *Christian Science Monitor* (December 13, 1927), 3. "Sugar Institute Is Organized Here," *New York Times* (January 8, 1928), 43.

62. "Too Much Sugar for the World to Eat," *New York Times* (April 8, 1928), SM7–SM8. Notice in *Science* 72 (1930): 651.

63. "*U.S. v. Sugar Institute*," *Time* (March 19, 1934), 63. David Genesove and W. Mullin, "Rules, Communication, and Collusion: Narrative Evidence from the Sugar Institute Case," *American Economic Review* 91 (2001): 379–98.

64. Robert C. Hockett, "The Progress of Sugar Research," *Scientific Monthly* 65 (1947): 269–82. "Research Awards Set," *New York Times* (December 10, 1945), 24. Sugar Research Foundation, *Ten Years of Research, 1943–1953* (New York, 1953).

65. "Cuba's the Island We Want," *New York Times* (January 31, 1893), 2.

66. "American Sugar Will Soon Build Cuban Central." See also *Louisiana Planter* (1919). "American Sugar," *New York Times* (July 23, 1959), 39. "U.S. Investment in Cuba Shrinks," *New York Times* (August 21, 1960), E1. "American Sugar," *New York Times* (March 6, 1961), 50.

67. American Sugar Co., *Annual Report* (1963). "American Sugar Co. Becomes Amstar," *New York Times* (October 28, 1970), 96, and full-page ad on 86. American Sugar Co., *Annual Report* (1970). "Amstar to Increase Corn Syrup Output," *New York Times* (August 20, 1974), 54.

68. "Amstar-Essex Deal Sparks Criticism of Merrill Lynch," *Wall Street Journal* (July 21, 1989), 1.

69. "If All Else Fails, Try Plain English," *New York Times* (August 30, 1991), D1. "Tate & Lyle Sets Deal to Buy Amstar Sugar," *New York Times* (September 27, 1988), D2. P. Chalmin, *The Making of a Sugar Giant: Tate & Lyle, 1859–1989* (New York: Harwood, 1990). "Domino Sugar Is Sold and Sign Stays," *Baltimore Sun* (July 27, 2001), 1.

70. "Alexander Stuart," *Scientific American* 42 (1880): 17. Paul S. Sternberger, "Wealth Judiciously Expended: Robert Leighton Stuart as Collector and Patron," *Journal of the History of Collections* 15 (2003): 229–40.

71. Alice Frelinghuysen et al., *Splendid Legacy: The Havemeyer Collection* (New York: Metropolitan Museum of Art, 1993). Frances Weitzenhoffer, *The Havemeyers: Impressionism Comes to America* (New York: H. N. Abrams, 1986). Cynthia Saltzman, *Old Masters, New World: America's Raid on Europe's Great Pictures, 1880–World War I* (New York: Viking, 2008).

Chapter 2: Molasses

1. "Richard Vines to John Winthrop, July 19, 1647," in *Winthrop Papers* (Boston: Massachusetts Historical Society, 1947), 5:171–72. *The Journal of John Winthrop, 1630–1649* (Cambridge, MA: Harvard University Press, 1996), 692–93. Russell Menard, *Sweet Negotiations: Sugar, Slavery, and Plantation Agriculture in Early Barbados* (Charlottesville, VA: University of Virginia Press, 2006).

2. Gilman Ostrander, "The Colonial Molasses Trade," *Agricultural History* 30 (1956): 77–84. John C. Fitzpatrick, ed., *The Writings of George Washington* (Washington,

DC: U.S. Government Printing Office, 1931), 2:437, 2:522, and 3:98–99. Rita Gottesman, comp., *The Arts and Crafts in New York, 1726–1776* (New York: New York Historical Society, 1938), 188, 210, 214.

3. *Story of New England's Traditional Baked Beans* (Portland, ME: Burnham & Morrill Co., 1944). Lura Woodside Watkins, "Beans and Bean Pots," *Antiques* 46 (November 1944): 276–77. "Baked Beans," *Haverhill Gazette* (November 26, 1825), 1. "Boston Baked Beans," *Southern Planter* 7 (1847): 73, and *Prairie Farmer* 7 (August 1847): 256, both allegedly from *Massachusetts Ploughman*.

4. Carl Bridenbaugh, "The High Cost of Living in Boston, 1728," *New England Quarterly* 5 (1932): 800–11. "Proceedings of Congress," *Gazette of the United States* (April 29 to May 2, 1789), 21–22. "Debate upon the Impost of 6 cents on Molasses," *Massachusetts Centinel* (May 9, 1789), 62.

5. "Visiting the Sick Poor," *New York Times* (July 16, 1872), 8. "The Shirt Makers of New York," *Home Magazine* 2 (August 1853): 131–35. "Out Among the Newsboys," *New York Times* (August 17, 1879), 5.

6. Benjamin Rumford, *Essays, Political, Economical and Philosophical* (London, 1796), 1:266. Thomas Eddy, *An Account of the State Prison or Penitentiary House in the City of New York* (New York: Isaac Collins and Son, 1801), reviewed in *American Review* 2 (January–March, 1802): 19–23. "Report of the Committee Appointed to Investigate the Local Causes of Cholera in the Arch Street Prison," *Hazard's Register of Pennsylvania* (March 23, 1833), 177–84.

7. Lucia C. Stanton, *Slavery at Monticello* (Charlottesville, VA: Thomas Jefferson Memorial Foundation, 1996), 27. Herbert Covey and Dwight Eisnach, *What the Slaves Ate: Recollections of African American Foods and Foodways* (Santa Barbara, CA: Greenwood Press/ABC-CLIO, 2009), esp. 13. W. O. Atwater and C. D. Woods, *Dietary Studies with Reference to the Food of the Negro in Alabama in 1895 and 1896* (Washington, DC: U.S. Government Printing Office, 1897). Alan M. Kraut, *Goldberger's War: The Life and Work of a Public Health Crusader* (New York: Hill and Wang, 2003).

8. Giles Silvester, *A Breife Discription of the Ilande of Barbados* (1651), quoted in Frederick H. Smith, *Caribbean Rum: A Social and Economic History* (Gainesville: University Press of Florida, 2005), 16. John J. McCusker, *Rum and the American Revolution* (New York: Garland, 1989), 476.

9. Quoted in Hugh Barty-King and Anton Massel, *Rum Yesterday and Today* (London: Heinemann, 1983), 11.

10. Increase Mather, *A Sermon Occasioned by the Execution of a Man Found Guilty of Murder Preached at Boston in N. E. March 16th 1685 [1686]*, 2nd ed. (Boston, 1687), 25. Cotton Mather, *Sober Considerations, on a Growing Flood of Iniquity, or, an Essay, to Dry Up a Fountain of Confusion and Every Evil Work; and to Warn People, Particularly of the Wonderful Consequences Which the Prevailing Abuse of Rum, Will Be Attended Withal* (Boston, 1708), 2–3. W. J. Rorabaugh, *The Alcoholic Republic: An American Tradition* (Oxford: Oxford University Press, 1979), 29.

11. Ned Ward, *Trip to New England 1698* (London, 1699). Howard W. Troyer, *Ned Ward of Grubstreet: A Study of Sub-Literary London in the Eighteenth Century* (Cambridge, MA: Harvard University Press, 1946). John Joselyn, *An Account of Two Voyages to New-England* (1674; rpt. Hanover, NH: P. J. Lindhodt, 1988), 99. Mather, *A Sermon*, 24. For Wheelock, see the Dartmouth drinking song quoted in

Marie Kimball, "Some Genial Old Drinking Customs," *William and Mary Quarterly* 2 (1945): 349–58.

12. Transcription of a manuscript copy of "Rhode Island Remonstrance against the Sugar Act" (1764) in *Silas Downer: Forgotten Patriot*, ed. Carl Bridenbaugh (Providence: Rhode Island Bicentennial Foundation, 1974), 59–68. *Observations of the Merchants at Boston in New-England on Several Acts of Parliament* (London, 1770). Jay Coughtry, *The Notorious Triangle: Rhode Island and the African Slave Trade, 1707–1807* (Philadelphia: Temple University Press, 1981). McCusker, *Rum and the American Revolution*, 492–93.

13. McCusker, *Rum and the American Revolution*, 434–43.

14. Bernard Bailyn, *New England Merchants in the Seventeenth Century* (Cambridge, MA: Harvard University Press, 1955), 129. Samuel Eliot Morison and Henry Steele Commanger, *The Growth of the American Republic* (New York: Oxford University Press, 1962), 1:96 and 1:111. Rorabaugh, *The Alcoholic Republic*, 61 and 64. William Burke, *An Account of the European Settlements in America* (London, 1758), 174.

15. Charles S. Sydnor, *American Revolutionaries in the Making* (New York: Free Press, 1965), esp. ch. 4, "Swilling the Planters with Bumbo." (Bumbo is a punch made of rum, water, sugar, and nutmeg.) Peter Thompson, *Rum Punch and Revolution: Tavern Going and Public Life in Eighteenth-Century Philadelphia* (Philadelphia: University of Pennsylvania Press, 1999). David W. Conroy, *In Public Houses: Drink & the Revolution of Authority in Colonial Massachusetts* (Chapel Hill: University of North Carolina Press, 1995).

16. John Adams to William Tudor, August 11, 1818, in *The Works of John Adams* (Boston: Little, Brown & Company, 1856), 10:345. Arthur Schlesinger, *Colonial Merchants and the American Revolution* (New York: F. Ungar, 1918, 1957).

17. *The Importance of the Sugar Colonies to Great Britain Stated, and Some Objections against the Sugar Colony Bill answer'd* (London, 1731). *Observations on the Case of the Northern Colonies* (London, 1731). *Some Considerations Humbly Offr'd upon the Bill Depending in the House of Lords Relating to the Trade between the Northern Colonies and the Sugar Islands* (London, 1732). *A Comparison between the British Sugar Colonies and New England as They Relate to the Interest of Great Britain* (London, 1732).

18. John W. Tyler, *Smugglers & Patriots: Boston Merchants and the Advent of the American Revolution* (Boston: Northeastern University Press, 1986). Albert Southwick, "The Molasses Act: Source of Precedents," *William and Mary Quarterly* 8 (1951): 389–405.

19. James Otis, *The Rights of the British Colonies Asserted and Proved* (Boston, 1764), in *Pamphlets of the American Revolution, 1750–1776*, ed. Bernard Bailyn (Cambridge, MA: Belknap Press of Harvard University Press, 1965), 1:408–82. Emily Hickman, "Colonial Writs of Assistance," *New England Quarterly* 5 (1932): 83–104.

20. *Considerations upon the Act of Parliament: Whereby a Duty Is Laid of 6d. Sterling per Gallon on Molasses and Sugar Imported into Any of the British Colonies Shewing Some of the Many Inconveniences Necessarily Resulting from the Operation of the Said Act* (Boston, 1764), in *Pamphlets of the American Revolution, 1750–1776*, ed. Bernard Bailyn (Cambridge, MA: Belknap Press of Harvard University Press, 1965), 1:355–77.

21. Transcription of a manuscript copy of "Rhode Island Remonstrance against the Sugar Act" (1764) in *Silas Downer: Forgotten Patriot*, ed. Carl Bridenbaugh (Providence: Rhode Island Bicentennial Foundation, 1974), 59–68.

22. Allen Johnson, "The Passage of the Sugar Act," *William and Mary Quarterly* 16 (1959): 50–514. Ostrander, "The Colonial Molasses Trade."

23. Samuel Hopkins, *The Rights of Colonies Examined* (Providence, 1764). Frederick B. Wiener, "The Rhode Island Merchants and the Sugar Act," *New England Quarterly* 3 (1930): 464–500, on 465, states, "Trade was the principal source of Rhode Island's livelihood, and molasses was the staple of the trade."

24. Worthington C. Ford, ed., *Journals of the Continental Congress* (Washington, DC: Government Printing Office, 1904–1937), 1:43 and 1:51–52. Nathaniel Greene, *The Evidence Delivered on the Petition Presented by the West-India Planters and Merchants to the Hon. House of Commons, As It Was Introduc'd at the Bar, and Summed Up by Mr. Glover* (London, 1775). Andrew Jackson O'Shaughnessy, *An Empire Divided: The American Revolution and the British Caribbean* (Philadelphia: University of Pennsylvania Press, 2000), esp. ch. 6, "The Crisis of American Independence."

25. *Journals of the Continental Congress*, 16:196–97. *State of Massachusetts-Bay in the House of Representatives, February 5, 1777. Whereas the Rum, Molasses . . . Are All Needed for the Supply of the Army and the Inhabitants of This State* (Boston: Printed by Benjamin Edes, 1777). *State of New Hampshire, an Act to Prevent the Transportation of Live Cattle Beef, Rum and Molasses, Out of This State* (Exeter, NH: Printed by Zechariah Fowle, 1780). *State of New Hampshire, Act for Supplying the Continental Army with Ten Thousand Gallons of West India Rum* (Exeter, NH: Printed by Zechariah Fowle, 1781). Fitzpatrick, *The Writings of George Washington*, 2:228, 5:406–7, and 5:436–407.

26. *House of Representatives Journal*, 1st Cong., 1st sess., May 13, 1789.

27. U.S. Congress, 1st Cong., 1st sess., 1789, "Statutes at Large." Alexander Hamilton, *Report on Manufactures* (Washington, DC, 1913), 52. "More Rum Discovered," *Boston Globe* (February 10, 1895), 7.

28. Donald Canney, *Rum War: The U.S. Coast Guard and Prohibition* (Washington, DC: U.S. Coast Guard, 1989). Malcolm Willoughby, *Rum War at Sea* (Washington, DC: U.S. Government Printing Office, 1964).

29. Daniel Drawbaugh, "Improvement in Faucets," U.S. Patent 59,792 (1866). Enterprise Manufacturing Co., *Illustrated Catalogue and Price-List* (Philadelphia, 1884), 38. Enterprise Manufacturing Co., *Descriptive Catalogue of Patented Hardware Specialties* (Philadelphia, 1911), 92–97. Advertisements in *American Grocer* 77 (January 2, 1907): 5, and 79 (January 1, 1908), 5.

30. "Something New: A Boston Molasses Ship," *Chicago Tribune* (January 27, 1869), 2. "A Novel Experiment," *New York Herald* (January 22, 1869), 7. "Molasses and Petroleum," *Maine Farmer* (April 26, 1894), 62. "Launching Next Tuesday," *Boston Globe* (December 15, 1915), 13.

31. "An Interesting and Satisfactory Experiment—Pumping Molasses," *Journal of the Franklin Institute* 89 (1885): 234.

32. Harriet Beecher Stowe, *Uncle Tom's Cabin* (Boston: John P. Jewett & Co., 1852), 27. Advertisements in *Richmond Times-Dispatch* (April 12, 1907), 4, and (September 3, 1907), 5. W. Hearn, "High Grade Syrup and Molasses," *Sugar Planters' Journal* 39 (1908): 93.

33. Penick & Ford advertisement in *Sugar Planters' Journal* 39 (November 14, 1908), outside back cover. But see Herbert M. Shilstone's comments in D. D. Colcock, ed. *Resumé of the Hearing at Atlanta, Ga. . . . on Sulphur Dioxide in Syrups and Molasses* (New Orleans, 1908), 12, also published in *American Food Journal* 4 (January 1909): 8–11.

34. Examples of Trixy advertising buttons that probably date from around 1900 can often be found on eBay. Penick & Ford ads in (Richmond) *Times* (April 23, 1907), 7, and in *Richmond Times-Dispatch* (April 28, 1907), 4. Brer Rabbit Brand Molasses ad in *Boston Globe* (October 25, 1916), 14.

35. Pierre Lacour, *The Manufacture of Liquors* (New York: Dick and Fitzgerald, 1863), 265. Charles Harrington, *A Manual of Practical Hygiene for Students, Physicians and Health Officers* (Philadelphia and New York: Lea Brothers, 1902), 203.

36. William K. Lewis, "Improvement in Canning Beans and Pork for Food," U.S. Patent 193,880 (1877). "The Navy Bean," *Massachusetts Ploughman* 40 (November 13, 1880): 1. William D. Howells, "Indian Summer," *Harper's New Monthly Magazine* 71 (1885): 616–34, on 617. "Canned Beans Tabooed," *Boston Globe* (January 5, 1891), 8. Jelks & Tappan advertised five three-pound cans of Boston baked beans for $1 in the *Atlanta Constitution* (March 18, 1883), 10. *Canned Vegetables*, pt. 8 of Harvey W. Wiley, ed., *Foods and Food Adulterants* (Washington, DC: Government Printing Office, 1893).

37. Penick & Ford ads in *Dallas Morning News* (November 25, 1899), 5, and in *Richmond Times-Dispatch* (April 28, 1907), 4.

38. Arthur P. Greeley, *The Food and Drugs Act* (Washington, DC: J. Byrne & Company, 1907).

39. Oscar E. Anderson Jr., "The Pure-Food Issue: A Republican Dilemma, 1906–1912," *American Historical Review* 61 (1956): 550–73. James Harvey Young, "Food and Drug Regulation under the USDA, 1906–1940," *Agricultural History* 64 (1990): 134–42. Clayton Coppin, "James Wilson and Harvey Wiley: The Dilemma of Bureaucratic Entrepreneurship," *Agricultural History* 64 (1990): 167–81. Clayton Coppin and Jack High, *The Politics of Purity: Harvey Washington Wiley and the Origins of Federal Food Policy* (Ann Arbor: University of Michigan Press, 1999).

40. Harvey W. Wiley et al., *Influence of Food Preservatives and Artificial Colors on Digestion and Health. III: Sulphurous Acid and the Sulphites* (Washington, DC: Government Printing Office, 1907). "Dr. Wiley's Basis for Promulgating Food Inspection Decision Number 76," *Sugar Planters' Journal* 38 (1908): 254–56. Harvey W. Wiley, *Experimental Work in the Production of Table Sirup at Waycross, Ga., 1905* (Washington, DC: Government Printing Office, 1906). Harvey W. Wiley, "Table Sirups," in *Yearbook of the United States Department of Agriculture for 1904* (1905): 241–48.

41. "The Sulphur Test," *Sugar Planters' Journal* 38 (1907–1908): 17, 65, and 81. "An Unjust Reflection on the Use of Sulphur," *Sugar Planters' Journal* 38 (1907–1908): 97. "Unsulphured Molasses," *Sugar Planters' Journal* 38 (1907–1908): 113.

42. R. E. Blouin, P. E. Archinard, and J. A. Hall Jr., *The Effects on the Human System of Louisiana Manufactured Syrups and Molasses* (Baton Rouge: Agricultural Experiment Station of Louisiana State University and A. and M. College, 1907). Notice of "dried fruit" before the Pure Food Committee, *American Food Journal* 1 (October 1906): 708. "Another 'Poison Squad,'" *American Food Journal* 2 (August 1907): 21. Colcock, *Resumé of the Hearings at Atlanta, Ga.*

43. "Food Inspection Decision 75 and 76," *American Food Journal* 2 (September 1907): 20–23. "The Pure Food Law and Decision No. 76 Concerning Sulphur," *Louisiana Planter* 39 (1907): 81.

44. "Amending F. I. D. No. 76," *Sugar Planters' Journal* 38 (1908): 353. Russell Chittenden, "The Influence of Sulphurous Acid and Sodium Sulphite on the Nutrition and

Health of Man." This is not dated, but a telegram from Ira Remsen to Agriculture Secretary Wilson, dated March 1, 1913, states that the summary of the sulfur report was in the mail and that a big box of supporting papers had been sent the month before. Both are in Remsen papers, RG 39, Referee Board Box 5, Special Collections, Milton Eisenhower Library, The Johns Hopkins University.

45. Boston Molasses Co. ads in *Boston Globe* (September 23, 1923), 27, and (September 30, 1923), 30. Fritz Zerban, *The Clarification of Cane Juice without Chemical Treatment* (Baton Rouge: Agricultural Experiment Station of the Louisiana State University and A. and M. College, 1920). This also appeared in *Louisiana Planter and Sugar Manufacturer* 65 (1920): 204–6.

46. "Penick & Ford Molasses Seized by Government," *Grocery World* 46 (July 6, 1908), 9–10.

47. "Federal Food and Drug Board Will Look into New Orleans Molasses," *Grocery World* 48 (October 8, 1909), 22. "Federal Board May Rule That Only Best Grade Louisiana Molasses Can Be Called 'New Orleans Molasses,'" *Grocery World* 48 (October 15, 1909), 6. "The Federal Board Debates New Orleans Molasses with Leading Molasses Men," *Grocery World* 48 (December 6, 1909), 12. For the survey, see Box 208, Folder Sugar, in Harvey Wiley Papers, Library of Congress.

48. "Molasses Makers Forming a $10,000,000 Trust," *Chicago Tribune* (July 18, 1899), 2. "Charles William Taussig," in *National Cyclopaedia of American Biography* (Clifton, NJ: J. T. White), 36:78–79. For Tugwell, see "Molasses Man," *Time* (November 30, 1936), 12–13. Ads in *Food Engineering* 49 (February 1977): 38, and 49 (March 1977): 165–66.

49. "$4,000,000 Penick & Ford, Ltd., Inc.," *New York Times* (January 16, 1924), 27. "Hey Popeye! Molasses Has More Iron Than Spinach," *Wall Street Journal* (March 27, 1940), 2. Penick & Ford ad in *Woman's Day* (April 1, 1943), 74.

50. "A Sweet and Sticky Problem," *American Farmer* 73 (September 1893): 4. "A Furnace Wanted to Burn Molasses," *Scientific American* 78 (1895): 90.

51. "Good Results from Molasses Feeds," *Massachusetts Ploughman* (December 23, 1905), 5. "The Story of the Miracle Molasses Process," in Anglo-American Mill Co., *The "Miracle Ace" Hammer Mill* (Owensboro, KY, 1927), 43–47.

52. "Molasses for Fuel," *New York Times* (July 27, 1891), 4. "What Alcohol Is Used For," *Chicago Tribune* (August 4, 1889), 26. "Making Alcohol from Molasses," *Chicago Tribune* (July 17, 1892), 13.

53. "Owned by Distillers' Co.," *New York Times* (February 10, 1907), 15. "Distillers Management in Control Industrial Alcohol," *Wall Street Journal* (March 12, 1915), 7. U.S. Industrial Alcohol Co., *Alcohol for Industrial Purposes* (New York: Shoen Printing Co., 1919).

54. Stephen Puleo, *Dark Tide: The Great Boston Flood of 1919* (Boston: Beacon Press, 2003).

55. Benjamin Gayelord Hauser, *Look Younger, Live Longer* (New York: Farrar, Strauss, 1950), 29.

Chapter 3: Cane Sugar in Louisiana

1. Louis McLane to Professor Silliman, August 31, 1832; C. U. Shepard to McLane, September 3, 1832; Silliman to McLane, October 2, 1832; McLane to Silliman, October 18,

1832; Silliman to McLane, October 26, 1832; McLane to Silliman, October 30, 1832; Silliman to McLane, November 7, 1832; and Silliman to McLane, December 25, 1832, in RG 56, Entry 446, National Archives and Records Administration (hereinafter NARA).
2. John A. Heitmann, *The Modernization of the Louisiana Sugar Industry, 1830–1910* (Baton Rouge: Louisiana State University Press, 1987). J. Carlyle Sitterson, *Sugar Country: The Cane Sugar Industry in the South, 1753–1950* (Lexington: University of Kentucky Press, 1953). Glenn R. Conrad and Ray F. Lucas, *White Gold: A Brief History of the Louisiana Sugar Industry, 1795–1995* (Lafayette: University of Southern Louisiana, 1995).
3. Thomas B. Thorpe, "Sugar and the Sugar Region of Louisiana," *Harper's New Monthly Magazine* 9 (1853): 764–67.
4. Bennett H. Wall, "Leon Godchaux and the Godchaux Business Enterprises," *American Jewish Historical Quarterly* 66 (1976): 50–67. Paul L. Godchaux, *The Godchaux Family of New Orleans* (New Orleans, LA: Godchaux, 1971). Alma Godchaux, "Godchaux Began Humbly, Rose to Financial Peaks," and other articles in the Godchaux centennial supplement in the *Times-Picayune* (March 1, 1940). "Death of Louisiana's Largest Sugar Planter," *Sugar Planters' Journal* 29 (1899): 481.
5. "Paul L. Godchaux Dead," *New York Times* (September 30, 1924), 23. "C. Godchaux Dies; Leader in Sugar," *New York Times* (October 24, 1954), 89, and *National Cyclopaedia of American Biography* (Clifton, NJ: J. T. White), 41:538–39. "Jules Godchaux," *New York Times* (July 6, 1951), 18.
6. "Leon Godchaux Dies Suddenly," *Daily Picayune* (May 19, 1899), 3. "Leon Godchaux Dead," *Times Democrat* (May 19, 1889), 2.
7. Advertisements in *New Orleans Times* (December 26, 1876), 4, and *Louisiana Sugar Bowl* (January 18, 1877), 1. Notice in *Louisiana Sugar Bowl* (September 6, 1877), 2. In July 1871, the paper carried an advertisement for "Leon Godchaux, Manufacturer of and Dealer in Clothing & Furnishing Goods."
8. "Organization of the Louisiana Sugar Planters' Association," *Louisiana Sugar Bowl* (December 13, 1877), 1. Notice in *Louisiana Planter* 3 (1889): 243.
9. "Sugar and Sentiment," *Louisiana Planter* 1 (1888): 252.
10. "Plantation Bookkeeping," *Louisiana Planter* 1 (1888): 164–65. "The Cost of the Sugar Harvest," *Louisiana Planter* 1 (1888): 286. "Leon Godchaux's Report," *Louisiana Planter* 3 (1889): 333. "Lafourche Crops," *Sugar Planters' Journal* 39 (1908–1909): 104 and 108. "Sugar Refinery Burns," (Boise) *Idaho Daily Statesman* (September 5, 1922), 10.
11. "Elm Hall Plantation," *Louisiana Planter* 5 (1890): 238; "Trade Notes" and "Assumption," *Louisiana Planter* 17 (1896): 90 and 293; "Elm Hall Sugar Refinery Burns," *Louisiana Planter* 68 (1922): 173. "Echoes of the Storm," *Sugar Planters' Journal* 39 (1908–1909): 821–24. "Lafourche Crop Notes," *Modern Sugar Planter* 41 (September 30, 1911): 7–9.
12. "Beauties of Government," *Liberty* 8 (June 13, 1891): 6. "Louisiana Sugar Crop," *Sugar* 7 (1893): 106. "The Sugar Bounty," *Louisiana Planter* 16 (1896): 337.
13. "Lafourche," *Louisiana Planter* 15 (1895): 133–34 and 149; "Assumption," *Louisiana Planter* 16 (1896): 324. For the 1853 sale of Madewood, see Sitterson, *Sugar Country*, 164.
14. "Lafourche Letter," *Louisiana Planter* 3 (1889): 101. "Sugar Making Items," *Sugar Planters' Journal* 39 (1908–1909): 88.

15. "Sweet Prosperity," *Idaho Statesman* (December 8, 1893), 4. For Reserve, see William E. Butler, *Down among the Sugar Cane: The Story of Louisiana Sugar Plantations and Their Railroads* (Baton Rouge: Moran Publishing Corp., 1980), 88–108. For Elm Hall, see "Assumption," *Louisiana Planter* 35 (1905): 132–33.

16. "Crop and Plantation News," *Sugar Planters' Journal* 40 (November 6, 1909): 69. For the bridge across the Bayou Lafourche, see "Assumption Sugar News," *Sugar Planters' Journal* 34 (1903–1904): 732–33. For the Southern Pacific Railroad, see "Lafourche Letter," *Louisiana Planter* 2 (1889): 68. "Elm Hall Gets the Horns," *Sugar Planters' Journal* 40 (January 15, 1910): 232. "'The Horns' to Go to Sterling," *Modern Sugar Planter* 42 (February 3, 1912): 1.

17. "Closing the Crevasse at Reserve," *Daily Picayune* (June 22, 1893), 1, 12. "The Grit of Godchaux," *Daily Picayune* (June 23, 1893), 1, 6. "The Levee Breaks Near Raceland," *Age Herald* (April 25, 1899), 3. "Lafourche," *Sugar Planters' Journal* 29 (1898): 724.

18. Benjamin Silliman, *Manual on the Cultivation of the Sugar Cane and the Fabrication and Refinement of Sugar* (Washington, DC: F. P. Blair, 1833), 36–37.

19. J. B. Benjamin, "Soleil's Saccharometer [*sic*]," *DeBow's Review* 5 (1848): 347–64. Deborah J. Warner, "How Sweet It Is: Sugar, Science and the State," *Annals of Science* 64 (2007): 147–70.

20. Figerio ad in *Louisiana Sugar Bowl* (January 27, 1876), 1. Duhamel ad in Louis J. Bright & Co., *New Orleans Price Current Yearly Report of the Sugar and Rice Crops of Louisiana: Crop Year 1875–1876* (New Orleans: New Orleans Price Current, 1876), 117. "What Sugar Plantations Need," *Louisiana Sugar Bowl* (December 27, 1877), 1. Henry Studniczka, *The Sugar Manufacturers' and Cane Growers' Handbook* (New Orleans, LA, 1885).

21. "The Sugar Chemists," *Louisiana Planter* 2 (1889): 283. L. A. Becnel, "Need for Chemical Control in Sugar Houses," *Louisiana Planter* 2 (1889): 286–87. Louisiana Sugar Chemists's Association, *Report of the Committee on Methods of Sugar Analyses* (Baton Rouge, 1889).

22. M. Trubek, "On the Estimation of the Extraction in Sugar Houses," *Journal of the American Chemical Society* 17 (1895): 920–23. "Moses Trubek" in *National Cyclopaedia of American Biography*, 36:43–44. "Lafourche Crops," *Sugar Planters' Journal* 39 (1908–1909): 4, and "St. John Parish Crop and Plantation Notes," 712. "Personal," *Louisiana Planter* 37 (1906): 346.

23. John C. Rodrigue, *Reconstruction in the Cane Fields* (Baton Rouge: Louisiana State University Press, 2001), 10. Michael Tadman, "The Demographic Cost of Sugar: Debates on Slave Societies and Natural Increase in the Americas," *American Historical Review* 105 (2000): 1534–75.

24. L. Bouchereau, *Statement of the Sugar and Rice Crops Made in Louisiana in 1868–1869* (New Orleans, LA: M. F. Dunn & Bro., 1869), introduction.

25. Rodrigue, *Reconstruction in the Cane Fields*, epilogue, "The Sugar War of 1887."

26. "Lafourche," *Louisiana Planter* 14 (1895): 53. Frank C. Carpenter, "In the Land of Sugar," *Weekly Indiana State Journal* (December 2, 1896), 2. "Sugar Planters Raise Wages," *New York Times* (August 1, 1897), 3.

27. "Elm Hall," *The Sugar Bowl and Farm Journal* 21 (1891): 728–29. "Elm Hall Plantation" and "Assumption," *Louisiana Planter* 17 (1896): 118–19 and 340. F. J. Sheridan, "Tenant Systems, Costs and Contracts," *Modern Sugar Planter* 43 (June 21, 1913): 12–13.

"An Impending Revolution in Sugar Cane Raising in Louisiana," *Modern Sugar Planter* 41 (December 24, 1910): 10. "The Cane Tenantry System," *Modern Sugar Planter* 42 (February 24, 1912): 12–15.

28. "Louisiana Sugar News," *Louisiana Planter* 65 (1920): 152. "Labor Outlook Worries Louisiana," *Sugar* 41 (June 1946): 44.

29. Bouchereau, *Statement*. "Sugar Manufacture," *Louisiana Sugar Bowl* (June 14, 1877), 2. "Central Sugar Factories," *Louisiana Sugar Bowl* (July 26, 1877), 2. "Improved Sugar Machinery," *Louisiana Sugar Bowl* (March 21, 1878), 2. "Central Sugar Factories," *Louisiana Sugar Bowl* (April 4, 1878), 2.

30. "Elm Hall Plantation," *Louisiana Planter* 17 (1896): 118–19. "Elm Hall Gets the Horns." "'The Horns' to Go to Sterling."

31. "Filterpresses," *Louisiana Planter* 2 (1889): 107. "Lafourche Crops," *Sugar Planters' Journal* 39 (1908–1909): 104, 108. J. G. Brewer, "Mechanical Filter," U.S. Patent 920,909 (1909).

32. Sitterson, *Sugar Country*, 147–50. "Assumption," *Louisiana Planter* 6 (1891): 581.

33. "St. John," *Louisiana Planter* 17 (1896): 69. E. W. Deming, "Let Us Manufacture High Grade Sugars," *Louisiana Planter* 16 (1896): 59–60. Hubert Edson, *Sugar: From Scarcity to Surplus* (New York: Chemical Publishing Co., 1958), 78–79.

34. *Year Book of the Louisiana Sugar Planters' Association for the Crop Year 1910* (New Orleans, 1910), 17, 32, and 33.

35. A Godchaux family album holds a photograph of the assembly on the steps of the Reserve plantation house.

36. "St. John Parish Crop and Plantation Notes," *Sugar Planters' Journal* 39 (1908–1909): 712–13. N/t, *Modern Sugar Planter* 41 (June 10, 1911): 13; B. Sandmann, "The Reserve Factory Process of Making Granulated Sugar," *Modern Sugar Planter* 43 (April 19, 1913): 3–7.

37. "American Cane Growers' Association of the United States," *Louisiana Planter* 17 (1896): 33–34 and 44–45.

38. *Convention of the Representatives of the Louisiana Protected Industries* (New Orleans, LA: City Item Publishing Co., 1884). "The American Cane Growers' Association," *Louisiana Planter* 17 (1896): 33–34. "The American Sugar Growers' Society," *Louisiana Planter* 18 (1897): 55.

39. Deming, "Let Us Manufacture High Grade Sugars." "Louisiana License Bill," *New York Times* (July 10, 1900), 1. "A Sugar Trust Plant Closes," *New York Times* (March 27, 1901), 1.

40. Circular for the proposed Louisiana Planters' Sugar Refinery, in *Louisiana Sugar Bowl* 21 (1890–1891): 536. G. P. Anderson, "How Can the Sugar Planters Make a White Granulated Sugar Economically? Answer: By Building and Operating a 'Planters' Co-Operative Sugar Refinery," *Louisiana Planter* 13 (1894): 25–26. "Plans a Planters' Trust," *New York Times* (August 10, 1902), 3. "Sugar Growers to Combine," *Wall Street Journal* (December 7, 1906), 6.

41. "Planters Frightened," *Boston Globe* (December 5, 1912), 20.

42. "$60,000,000 Sugar Company," *New York Times* (March 6, 1913), 5. "Consolidation of Louisiana Sugar Interests," *Louisiana Planter* 50 (1913): 70. "Big Sugar Corporation," *Louisiana Planter* 50 (1913): 156–57. "Legislative Lynch Law," *New York Times*

(May 1, 1916), 10. Supreme Court of Louisiana #21743, *State of Louisiana v. American Sugar Refining Company* (1915–1916).

43. "Organize Godchaux Sugars, Inc.," *New York Times* (June 24, 1919), 25. Display ads, *New York Times* (July 1, 1919), 27, and *Washington Post* (July 1, 1919), 11. "Godchaux Interests Consolidate," *Sugar* 21 (1919): 362–63. Notices in *Louisiana Planter* 62 (1919): 312 and 409; 69 (1922): 120.

44. "Outlook for Godchaux Sugars," *Barron's* (September 30, 1929), 14. "Godchaux Sugar Makes Progress," *Wall Street Journal* (September 24, 1929), 6. Godchaux ads in *National Carbonator and Bottler* 65 (December 1937): 58; 82 (March 1946): A30.

45. Notice in *Louisiana Planter* 62 (1919): 359. "Sugar Company Employes [*sic*] Insured," *New York Times* (November 13, 1927), N12. Quincy Montz, *My Recollection of the Reserve Community Club Complex and Life, the Way It Was: 1920s–1950s* (Reserve, LA, 2004).

46. "Activist Advanced Racial Equality," *Times-Picayune* (June 18, 1995), 1. "Edward Hall, 96, Civil Rights Activist," *Times-Picayune* (May 18, 2001), 4.

47. *Godchaux Sugars, Inc. and Sugar Workers' Local Union No. 21934, Affiliated with the American Federation of Labor* (Washington, DC: National Labor Relations Board, 1940). *Godchaux Sugars, Inc. and the United Sugar Workers, Local Industrial Union CIO* (Washington, DC: National Labor Relations Board, 1942). Virginia Seits, "Legal, Legislative, and Managerial Responses to the Organization of Supervisory Employees in the 1940s," *American Journal of Legal History* 28 (1984): 199–243.

48. *The Louisiana Sugar Cane Plantation Worker vs. The Sugar Corporation, U.S. Department of Agriculture et al.: An Account of Human Relations on Corporation-Owned Sugar Cane Plantations in Louisiana under the Operation of the U.S. Sugar Program, 1937–1953* (Washington, DC: Inter-American Educational Association, 1954).

49. "Negro Labor Group Backed in Red Fight," *New York Times* (October 25, 1953), 36. "The Cane Mutiny," *Time* (November 2, 1953), 26. Thomas Becnel, *Labor, Church, and the Sugar Establishment: Louisiana, 1887–1976* (Baton Rouge: Louisiana State University Press, 1980), ch. 7, "The Cane Mutiny."

50. Norman Thomas, "'Right-to-Work' Laws," *New York Times* (September 16, 1956), E12.

51. "Light on Godchaux Clan," *Pittsburgh Courier* (September 10, 1955), 24. "The South at the Crossroads," *Chicago Defender* (May 3, 1956), 8. "Godchaux Says It Is Refining Sugar Again," *Wall Street Journal* (May 19, 1955), 10. "Sugar Walkout Settled," *New York Times* (December 15, 1955), 37. "Reminiscing: Sugar Cane, Trains, Important to Family," *Times-Picayune* (April 7, 2002), 5. "Ex–Refinery Worker Recalls Sugar Strike," *Times-Picayune* (September 19, 1993), 4.

52. Correspondence with Quincy Montz of Reserve, Louisiana.

53. "Webb & Knapp Agrees Conditionally to Buy Control of Godchaux," *Wall Street Journal* (January 26, 1956), 22. "Webb & Knapp Confirms Unit's Offer to Buy Godchaux Shares," *Wall Street Journal* (February 14, 1956), 3. "35,000 Acres in Louisiana Sold to Webb & Knapp for Plant Sites," *New York Times* (May 22, 1956), 43 and 46. Godchaux Sugars Inc., *Annual Statement, Fiscal Year Ending January 31st. 1938.*

54. "Southern Industries Holders Vote Sugar Firm Acquisition," *Wall Street Journal* (June 13, 1966), 14.

Chapter 4: Cane Sugar in Florida

1. Gail M. Hollender, *Raising Cane in the "Glades": The Global Sugar Trade and the Transformation of Florida* (Chicago: University of Chicago Press, 2008).

2. Senate Committee on Public Lands, *Report . . . to Authorize the Draining of the Everglades*, 30th Cong., 1st sess., 1848, S. Rep. 242.

3. Pat Dodson, "Hamilton Disston's St. Cloud Sugar Plantation, 1887–1901," *Florida Historical Quarterly* 49 (1921): 356–69. Florida Land and Improvement Co., *Descriptive List Catalogue of the Disston Lands in Florida* (Philadelphia: Edward Stern & Co., 1885), 23.

4. N/t, *Washington Post* (July 24, 1891), 4. "Drowned Lands of Florida," *Chicago Tribune* (January 12, 1891), 7 (from *Baltimore Sun*). "The Sugar Industry," *Atlanta Constitution* (March 15, 1891), 10. "Sugar Making in Florida," *New York Times* (June 12, 1892), 20. "To Grow Sugar in Florida," *New York Times* (December 7, 1892), 1.

5. William Lloyd Fox, "Harvey W. Wiley's Search for American Sugar Self-Sufficiency," *Agricultural History* 54 (1980): 516–26. Harvey W. Wiley, *An Autobiography* (Indianapolis: Bobbs-Merrill, 1930), 183. "Report by Dr. H. W. Wiley, of the Bureau of Chemistry, United States Department of Agriculture, in 1891, on the Muck Lands of the Florida Peninsula," USDA, *Annual Report* (1891), 163–71, quoted in Duncan Upshaw Fletcher, ed., *Everglades of Florida* (Washington, DC: Government Printing Office, 1911), 73–81.

6. Rufus E. Rose, *The Possibilities of Sugar Production in Florida* (Tallahassee: Appleyard, 1910). Rufus E. Rose, *The Swamp and Overflow Lands of Florida: The Disston Drainage Company and the Disston Purchase, a Reminiscence* (Tallahassee, 1916). *Everglades of Florida*, 62nd Cong., 1st sess., 1911, S. Doc. 89. James Oliver Wright, *The Everglades of Florida: Their Adaptability for the Growth of Sugar Cane* (Tallahassee, 1912). *Florida Everglades*, 63rd Cong., 2nd sess., 1914, S. Doc. 379. A. P. Spencer, *Sugar Cane* (Gainesville: University of Florida, Division of Agricultural Extension, 1919).

7. "Message of Gov. W. S. Jennings to the Legislature Relative to the Reclamation of the Everglades, 1903," in *Everglades of Florida*, 84–90. "To Produce Sugar from Everglades," *Atlanta Constitution* (January 12, 1902), 8. "Vast Reclamation Schemes," *New York Times* (August 17, 1902), 24.

8. "Florida May Raise Abundance of Sugar on Swamp Lands Reclaimed by Drainage," *Boston Globe* (March 6, 1910), SM3. "Broward Dead," *Boston Globe* (October 2, 1910), 40. N. Broward, "Reclaiming the World-Famed Florida Everglades," *Christian Science Monitor* (November 23, 1910), SEC35.

9. "Gov. Gilchrist Tells of Great Progress and Wonderful Promise of Florida," *Atlanta Constitution* (November 27, 1910), C6. "Canals Finished in Everglades," *Atlanta Constitution* (April 22, 1912), 10.

10. "Florida Plans Draining of Everglades," *Christian Science Monitor* (January 10, 1917), 3. "Tamiami Trail Being Constructed through Heart of Everglades," *Atlanta Constitution* (February 11, 1917), 18. "Severe Floods in Everglades Test New Drainage System," *Wall Street Journal* (December 22, 1922), 11. "Conquering the Everglades with Canals and Ditches," *Christian Science Monitor* (January 11, 1924), 12.

11. Alfred Hanna and Kathryn Hanna, *Lake Okeechobee: Wellspring of the Everglades* (Indianapolis: Bobbs-Merrill Co., 1948), ch. 25, "The Agro-Industrial Empire." John A.

Heitmann, "The Beginnings of Big Sugar in Florida, 1920–1945," *Florida Historical Quarterly* 77 (1998): 39–61.

12. "The Man behind the Plant," *Facts about Sugar* 25 (April 1930): 409–10.

13. Joseph T. Elvove, "The Florida Everglades: A Region of New Settlement," *Journal of Land & Public Utility Economics* 19 (1943): 464–69. "The Everglades: America's Youngest Sugar Cane Area," *Facts about Sugar* 13 (November 1940): 22–29.

14. Southern Sugar Co., *Operations of the Southern Sugar Company* (Chicago: Southern Sugar, 1928). Southern Sugar Co., *100,000 Shares, the Southern Sugar Company Common Stock* (Chicago: Southern Sugar, 1929). Southern Sugar Co., *The Story of the Southern Sugar Company* (Chicago: Southern Sugar, 1929). "Florida to Resume Cane Sugar Industry," *New York Times* (June 22, 1927), 39. "New Sugar Mill in Florida," *New York Times* (December 7, 1927), 51. "Celotex Expanding in Florida Field," *Wall Street Journal* (December 7, 1927), 18. "Florida Sugar Cane Richer in Content," *Wall Street Journal* (March 23, 1928), 15. "Florida Entering Sugar Production," *Wall Street Journal* (November 8, 1928), 19. "The Alchemy of Muck and Sunshine," ad in *Wall Street Journal* (September 19, 1928), 6.

15. "Putting Everglades to Work," *Chicago Tribune* (January 13, 1929), B2. "Carlton Starts First Sugar Mill in Florida," *Atlanta Constitution* (January 15, 1929), 18. "Will Extend Plantation," *New York Times* (April 14, 1929), 49. "Southern Sugar Expansion Large," *Wall Street Journal* (December 16, 1929), 16. "Florida's Great Sugar Factory," *Sugar* 25 (April 1930): 401–4. "The Man behind the Plant."

16. "Company Maintains Research Department," *Facts about Sugar* 25 (April 26, 1930), 411.

17. "Florida Entering Sugar Production." "Thundering Night Plowers," *Chicago Tribune* (January 13, 1929), B2. "A Motorized Sugar Plantation," *Facts about Sugar* 25 (April 1930): 405–7.

18. "Allis-Chalmers Gets Orders," *Christian Science Monitor* (November 21, 1930), 10. "Southern Sugar Active," *Wall Street Journal* (December 15, 1930), 19. "Allis-Chalmers Gets Machine Order," *Wall Street Journal* (January 5, 1931), 16. "Southern Sugar Co. to Use Harvesters," *Wall Street Journal* (January 19, 1931), 3. See also Falkiner's several subsequent patents.

19. "Australia: Where Cane Sugar Is Made by White Labor," *Facts about Sugar* 28 (1934): 225–26. Geoff Burrows and Ralph Shlomowitz, "The Lag in the Mechanization of the Sugarcane Harvest: Some Comparative Perspectives," *Agricultural History* 66 (1992): 61–75, on 67–69.

20. Dahlberg Corporation of America, *The Dahlberg Sugar Cane Industries* (Chicago: Dahlberg Corporation, 1929).

21. "We Should Build Up Our Domestic Sugar Industry," ad in *Washington Post* (March 4, 1929), IE19. "Need of Sugar Duty Stressed," *Los Angeles Times* (April 21, 1929), B5.

22. Heitmann, "The Beginnings of Big Sugar in Florida."

23. "U.S. Sugar Corp. Starts Grinding at New Florida Mill," *Wall Street Journal* (December 14, 1936), 14. Harry T. Vaughn, "Clewiston: Largest Raw Sugar House in the United States," *Facts about Sugar* 35 (December 1940): 28–32. "32 Farms Got Big AAA Aid," *New York Times* (September 10, 1937), 15. Clarence Bitting, *More Talks on Sugar* (New York: Bitting Inc., 1937). "U.S. Sugar Corp. Starts $500,000 Building Program," *Wall Street Journal* (January 29, 1938), 5.

24. "U.S. Sugar Corp.'s Industrial Farm Located in Florida Everglades Provides Year-Round Work by Integrating Field and Factory Operations," *Wall Street Journal* (June 19, 1945), 5. "U.S. Sugar Industry a Domestic Affair," *New York Times* (June 21, 1959), F1.

25. Clifford L. James, "International Control of Raw Sugar Supplies," *American Economic Review* (1931): 481–97. "Fable in Sugar," *Time* (January 25, 1932), 48. "Sweet Satisfaction," *Time* (May 10, 1937), 78.

26. "Text of Compromise Sugar Bill," *Wall Street Journal* (April 2, 1934), 11. "Sugar Bill Goes to the President," *New York Times* (August 26, 1934), 33. "U.S. Sugar Quota: How It Works," *New York Times* (February 21, 1960), E4.

27. Clarence R. Bitting, *Some Talks on Sugar* (New York: Bitting Inc., 1938). Bitting, *More Talks on Sugar* (New York: Bitting Inc., 1938). Bitting, *Sugar in the Everglades* (New York: Press of Benj. H. Tyrrel, 1938). "50 Congressmen Off Tonight on Junket to Sugar Plants," *Washington Post* (December 1, 1938), X1. "Florida to Take 200 Officials of U.S. on Vacation," *Chicago Tribune* (December 1, 1938), 8. "Congress Goes A-Junketing and Florida Pays the Bill," *Christian Science Monitor* (December 2, 1938), 1.

28. Federal Workers' Project of Florida, *Florida's Sugar Bowl* (Tallahassee: State of Florida, Department of Agriculture, 1939), 15. "Florida Sugar Growing Costs at Low Level," *Wall Street Journal* (January 8, 1940), 11.

29. The FBI report is quoted in Alec Wilkinson, *Big Sugar: Seasons in the Cane Fields of Florida* (New York: Vintage Books, 1989), ch. 21. "Peonage Charged to Sugar Grower," *New York Times* (November 5, 1942), 11. "Florida Sugar Firm Charged with Peonage," *Chicago Tribune* (November 5, 1942), 15.

30. "Florida Seeks West Indian Help," *Christian Science Monitor* (January 30, 1943), 3. "Florida Sugar Workers Shifted," *Wall Street Journal* (June 22, 1943), 13. Cindy Hahamovich, "'In America Life Is Given Away': Jamaican Farmworkers and the Making of Agricultural Immigration Policy," in *The Countryside in the Age of the Modern State: Political Histories of Rural America*, ed. C. McNicol Stock and R. Johnston (Ithaca, NY: Cornell University Press, 2001), 134–60.

31. "U.S. Sugar, Savannah Sugar to Build 2 Plants in Florida," *Wall Street Journal* (December 14, 1960), 28. "First Sugar Mill Cooperative Organized in Florida," *New York Times* (March 25, 1963), 13.

32. "New Sugar Industry Is Flourishing in South Florida," *New York Times* (November 1, 1961), 53. "Cuban Exiles Growing Sugar Cane in Florida," *New York Times* (January 19, 1962), 54 and 51.

33. Phyllis Berman, "The Fanjuls of Palm Beach: The Family with a Sweet Tooth," *Forbes* (May 14, 1990), 56–60, 64, 69. "Kingdom of Cane," *Sun Sentinel* (October 20, 1996), 1.

34. "Chavez's Farm Union Seeking to Organize Florida Citrus Workers," *New York Times* (February 7, 1972), 25. "Chavez Union Tackles Sugarcane Planters," *Washington Post* (February 22, 1972), A4. Wilkinson, *Big Sugar*, ch. 44.

35. "Judge to Rule on Legality of Importing Cane Cutters," *New York Times* (September 24, 1972), 52. "Clash Likely between Sugar Industry and Farm Union," *New York Times* (October 10, 1972), 16. "Bar on Migrants Denied in Florida," *New York Times* (October 17, 1972), 54. "Florida Cane Cutters: Alien, Poor, Afraid," *New York Times* (March 12, 1973), 24.

36. Cesar Chavez, "Inhuman Treatment of Farm Workers Must End," *Los Angeles Times* (February 11, 1974), A7. "Jamaicans Harvest Florida Sugar Cane," *Los Angeles Times* (December 19, 1975), K3.

37. House Committee on Education and Labor, Subcommittee on Labor Standards, *Job Rights of Domestic Workers: The Florida Sugar Cane Industry*, 98th Cong., 1st sess., 1983, Committee Print. "Machines Stalk the Cane Cutters. Move by Sugar Cooperative Could Hasten Demise of Foreign Workers," *Fort Lauderdale Sun Sentinel* (June 21, 1992), 1D.

38. Wilkinson, *Big Sugar*, ch. 43. Colman McCarthy, "The Bitter Lot of the Sugar Cane Worker," *Washington Post* (September 17, 1989), F2.

39. Burrows and Shlomowitz, "The Lag in the Mechanization of the Sugarcane Harvest." Emile Maier, *A Story of Sugar Cane Machinery* (New Orleans, LA: Sugar Journal, 1952).

40. House Committee on Education and Labor, *Report on the Use of Temporary Foreign Workers in the Florida Sugar Cane Industry*, 102nd Cong., 1st sess., 1991. "Florida Sugar Cane Growers Seek Cleaner Image," *New York Times* (December 3, 1991), D1. "Machines Displacing Foreign Sugar Cane Cutters in Florida," *Washington Post* (December 30, 1991), A1 and A4. "Cane Cutters Face Job Loss to Machines," *Orlando Sentinel* (July 12, 1992), D1.

41. "Sugar Subsidies Called Boon for Big Companies," *Washington Post* (May 20, 1977), A18. "Battle Looming on Bill to Boost Sugar Supports," *Washington Post* (March 22, 1978), A14. "In Washington, Sugar Lobby Works on Sweet Deals," *Chicago Tribune* (October 24, 1982), 1.

42. "Sugar's Sweetest Deal," *Time* (April 8, 1996), 34. "Foreign G.O.P. Donor Raised Dole Funds," *New York Times* (October 21, 1996), B8. Alfonso Fanjul and J. Pepe Fanjul, "Sugar Industry Isn't the Villain in Everglades," *New York Times* (May 14, 1997), A20.

43. Robert Mykle, *Killer "Cane": The Deadly Hurricane of 1928* (New York: Cooper Square Press, 2002). "Hoover as Engineer Views Flood Area," *New York Times* (February 16, 1929), 4. "Creating a New Empire of Sugar," *Facts about Sugar* 25 (April 1930): 397–400. "W. Palm Beach Plans Ambitious," *Wall Street Journal* (December 15, 1930), 17. "Hoover Honored at Dike in Florida," *New York Times* (January 13, 1961), 11.

44. "The Proposed Everglades National Park," *Science* 77 (1933): 185.

45. Michael Grunwald, *The Swamp: The Everglades, Florida, and the Politics of Paradise* (New York: Simon & Schuster, 2006). David McCally, *The Everglades: An Environmental History* (Gainesville: University Press of Florida, 1999).

46. "Florida Area Pushes Plan to Drain Swamps," *Chicago Tribune* (February 8, 1949), 8. "The 50-Year War on the Everglades," *New York Times* (April 20, 1997), E14.

47. "National Parks in Danger," *Christian Science Monitor* (November 11, 1963), 14. "Florida Wildlife Hurt by Progress," *New York Times* (July 28, 1982), 17. "Putting Bends Back in a River to Help the Everglades," *Christian Science Monitor* (January 26, 1987), 3.

48. "Trouble in Florida's 'River of Grass,'" *Christian Science Monitor* (July 8, 1993), 3. "Argentine Toads 'Hop' to Task of Defending Florida's Sugar," *Christian Science Monitor* (March 14, 1945), 9. "Imported Toad Turns Out to Be Big Problem," *New York Times* (September 22, 1996), 29.

49. "U.S. and Florida Growers Reach Pact on Everglades," *New York Times* (July 14, 1993), A12. "Growers Pushed to Help Everglades," *New York Times* (January 16, 1994),

18. "Sugar Firms Asked to Pay for Everglades Damage," *Christian Science Monitor* (January 20, 1994), 3. "Florida Sugar Cane Growers Seek Cleaner Image."

50. "Tax Plan Spurs New Sugar Battle in Florida," *New York Times* (February 17, 1996), 6. "The 50-Year War on the Everglades."

51. "Sugar's Latest Everglades Threat," *New York Times* (April 29, 1998), A24. "Sugar Companies Play a Pivotal Role in Effort to Restore Everglades," *New York Times* (April 16, 1999), A20.

Chapter 5: Beet Sugar: Profitable and Patriotic

1. Noël Deerr, *The History of Sugar* (London: Chapman and Hall, 1950), vol. 2, ch. 29, "The Beet and Beet Sugar." An English translation of Napoleon's decree of March 25, 1811, appears in *The Annual Register, or a View of the History, Politics, and Literature for the Year 1811* (London, 1825), 317–18.

2. Joel Barlow to Mrs. Madison, December 21, 1811, in *The Selected Letters of Dolley Payne Madison*, ed. D. B. Mattern and H. C. Shulman (Charlottesville, VA: University of Virginia Press, 2003), 152–53.

3. *The Works of Daniel Webster* (Boston: Charles C. Little and James Brown, 1851), 3:132.

4. "Cultivation of the Sugar Cane, &c," *Gale & Seaton's Register* (January 25, 1830), 554–55.

5. *Memorial of Charles Louis Fleischmann, in Relation to the Smithsonian Legacy*, 26th Cong., 1st sess., 1838, H. Doc. 128. *Memorial of Charles Louis Fleischmann, on the Manufacture of Beet-Sugar*, 25th Cong., 3rd sess., 1839, H. Doc. 62. *Memorial of Charles L. Fleischmann, in Relation to the Manufacture of Beet Sugar*, 25th Cong., 3rd sess., 1839, S. Doc. 147. *Charles Lewis Fleischmann: Sugar Beet*, 25th Cong., 3rd sess., 1839, H. Rep. 319. Paul W. Gates, "Charles Lewis Fleischmann: German-American Agricultural Authority," *Agricultural History* 35 (1961): 13–23.

6. *Mulberry and Sugar Beet*, 25th Cong., 2nd sess., 1838, H. Rep. 815. *House Journal*, 25th Cong., 2nd sess., 280, 796, 869, 1153, 1204, and 1244.

7. *Agreement Relative to the Manufacture of Sugar from the Beet* (Philadelphia, 1836). Extracts of James Pedder's *Report . . . on the Culture, in France, of the Beet Root* (Philadelphia: The Beet Sugar Society of Philadelphia, 1836) appeared in several local papers. See, for instance, "Beet Root Sugar," *Northampton Courier* (July 20, 1836), copied from the *Worcester Palladium*, "Beet Culture," *New York Farmer* 10 (April 1837): 98–100.

8. Henry Clay to Jacob Snider Jr., December 27, 1836, in *The Papers of Henry Clay*, ed. R. Seager II (Lexington : University of Kentucky Press, 1984), 8:874–75. "Mr. Clay's Opinions on Beet Sugar," *Northampton Courier* (April 19, 1837).

9. American Beet Sugar Refining and Manufacturing Co., *Proposals for the Introduction, Manufacture, and Refining of Sugar from the Beet, in Maryland* (Baltimore, 1840). "Beet Sugar in Michigan," *The Genesee Farmer* 8 (1838): 276.

10. "On the Manufacture of Beet Sugar in the U.S.," *The Cultivator* (July 1836): 65.

11. "Sugar Beet," *New England Farmer* 3 (1825): 302.

12. "Premium List of the Massachusetts Society for Promoting Agriculture, for 1831," *Massachusetts Agricultural Repository and Journal* 10 (January 1831): 314.

13. M. Isnard to Thomas L. Winthrop, April 25 and May 16, 1836, and undated, in papers of the Massachusetts Society for Promoting Agriculture, Box 12, Folder 26, in Massachusetts Historical Society. Max Isnard, "Beet Root Sugar," *The Farmer & Gardener, and Live-Stock Manager* 3 (May 24, 1836), 27–28, and *New York Farmer* 9 (August 1836): 225, both from the *Boston Advertiser*. "Beet Root Sugar," *Christian Watchman* 17 (1836): 68.

14. "Beet Root Sugar," *Northampton Courier* (November 30, 1836), copied in *The Farmer & Gardener, and Live-Stock Manager* (December 6, 1836), 254.

15. "New Work on Beet Sugar," *Northampton Courier* (January 18, 1837), 2.

16. Notice in *Northampton Courier* (March 29, 1837) and "Northampton Sugar Beet Company," *Northampton Courier* (May 10, 1837). "Farmers' Work" and "Sugar Beet," *Northampton Courier* (January 25, 1837), 2. "Northampton Farmer and Sugar Beet Culturist," *Maine Farmer* 5 (1837): 1.

17. Commonwealth of Massachusetts, *Report and Bill to Encourage the Manufacture of Indigenous Sugar* (1837). C. A. H., "Sugar Beet," *Northampton Courier* (March 14, 1838), reprinted in *Tennessee Farmer* 3 (1838): 272. C. A. H., "Manufacturing Beet Sugar, &c.," *Northampton Courier* (March 28, 1838).

18. "Beet Sugar," *The Farmer & Gardener, and Live-Stock Breeder and Manager* (May 23, 1837), 28. Benjamin Godfrey and Winthrop S. Gilman, leaders of the Alton company, financed a number of ventures in the area. Notice of Child in *Northampton Courier* (March 28, 1838).

19. "The Massachusetts Anti-Slavery Fair," *Liberator* 9 (1839): 179. "Beet Sugar," *Liberator* 10 (1840): 36. Carol Faulkner, "The Root of the Evil: Free Produce and Radical Antislavery," *Journal of the Early Republic* 27 (2007): 377–405.

20. Henry Braconnot, "Chemical Constituents of Sugar Beet," *Journal of the Franklin Institute* 9 (1845): 357, from *Annales de Chemie* (December 1839). M. Payen, "Manufacture of Sugar from Beets," *Journal of the Franklin Institute* 11 (1846): 68–69, from *Bulletin de la Société d'Encouragement pour l'Industrie Nationale* (1845). "Sugar," *Scientific American* 3 (1848): 178.

21. Fred G. Taylor, *A Saga of Sugar* (Salt Lake City: Utah-Idaho Sugar Co., 1944). Mary Jane Woodger, "Bitter Sweet: John Taylor's Introduction of the Sugar Beet Industry in Deseret," *Utah Historical Quarterly* 69 (2001): 247–63.

22. Charles L. Fleischmann, *Trade, Manufacture, and Commerce in the United States of America* (Jerusalem: Israel Program for Scientific Translation, 1970), an English translation of *Erwerbszweige, Fabrikwesen und Handel der Vereinigten Staaten von Nordamerika* (Stuttgart: F. Kohler, 1852).

23. *Report of the Commissioner of Agriculture for the Year 1867* (Washington, DC: Government Printing Office, 1868), 3, 8–9, 31–57.

24. *Letter from the Commissioner of Agriculture, Relative to the Manufacture of Beet-Root Sugar*, 40th Cong., 2nd sess., 1868, H. Mis. Doc. 84. "Progress of the Beet Sugar Manufacture in Europe," in *Report of the Commissioner of Agriculture for the Year 1869* (Washington, DC: Government Printing Office, 1870), 334–52. "The Beet-Sugar Industry," in *Report of the Commissioner of Agriculture for the Year 1870* (Washington, DC: Government Printing Office, 1871), 210–17.

25. Charles A. Goessmann, *Report on the Production of Beet Sugar as an Agricultural Enterprise in Massachusetts* (Boston, 1871). Charles A. Goessmann, *Report on*

Sugar Beets, Raised upon the Farm of the Massachusetts Agricultural College (New York: S. Angell, 1872). William Clark Smith, *Beet Sugar* (Amherst, MA: Amherst Record Print, 1872).

26. Henry F. Q. D'Aligny, *The Manufacture of Beet Sugar and Alcohol, and the Cultivation of Sugar-Beet* (Washington, DC: Government Printing Office, 1869). London. International Exhibition, 1862, *Reports of the Juries* (London: Printed for the Society of Arts, 1863), Class III, section B, 21–22. Philadelphia. International Exhibition, 1876, *Reports and Awards* (Washington, DC: Government Printing Office, 1880), vol. 4, Group III, "Production of Sugar."

27. William McMurtrie, *Report on the Cultivation of the Sugar Beets and the Manufacture of Sugar Therefrom in France and the United States* (Washington, DC: Government Printing Office, 1880), 163–64. *Congressional Record* 10 (1880): 2031.

28. *Sorghum and Beet Sugar: Letter from the Commissioner of Agriculture*, 49th Cong., 1st sess., 1886, H. Mis. Doc. 284.

29. Harvey W. Wiley, "The Sugar Problem," read to the Louisiana Sugar Planters' Association, December 11, 1884. Harvey W. Wiley, *The Sugar Industry of the United States* (Washington, DC: Government Printing Office, 1885). USDA, *Encouragement to the Sorghum and Beet Sugar Industry* (Washington, DC: Government Printing Office, 1883). E. H. Dyer to H. W. Wiley, in *Report of the Commissioner of Agriculture for the Year 1884* (Washington, DC: Government Printing Office, 1885), 21.

30. Wiley, *The Sugar Industry*.

31. USDA, *Progress of the Beet-Sugar Industry in the United States in 1904* (Washington, DC: Government Printing Office, 1904). C. O. Townsend, "The Beet-Sugar Industry in the United States," USDA, *Bulletin* 721 (1918).

32. Lewis S. Ware, *The Sugar Beet* (Philadelphia: H. C. Baird & Co., 1880). "Announcement," *The Sugar Beet* 1 (1880): 9–10. "Lewis Sharpe Ware," in *National Cyclopaedia of American Biography*, 12:475. The Lewis S. Ware collections are at the Franklin Institute and the American Philosophical Society.

33. "Rachel Lloyd: Early Nebraska Chemist," *Bulletin for the History of Chemistry* 17 (1995): 9–14. "Nebraska Sugar School," *Science* 19 (1892): 324. *Preamble and Resolution of Beet Sugar Convention in Nebraska in Favor of an Appropriation by Congress of $50,000 for School of Instructions*, 52nd Cong., 1st sess., 1892, S. Mis. Doc. 71. Mr. Manderson and Mr. Paddock in *Congressional Record* 23 (1892): 1127 and 1167.

34. August Buechler and Robert Barr, *History of Hall County, Nebraska* (Lincoln: Western Publishing and Engraving Co., 1920), 241. Information from *The Sugar Beet* 11 (1890): 14 (from [Omaha] *Bee*) and 48–49 (from the [Chicago] *Interocean*). Jack R. Preston, "Heyward G. Leavitt's Influence on Sugar Beets and Irrigation in Nebraska," *Agricultural History* 76 (2002): 381–92.

35. "Sugar Men Divided by Threat of Jail," *New York Times* (June 17, 1911), 2. "Made Profit of $500,000," *Boston Globe* (June 17, 1911), 2.

36. "Beet-Sugar Production in Nebraska," *Rural New Yorker* (February 15, 1890), 104. "Nebraska's Beet Sugar," *Chicago Tribune* (March 29, 1890), 12. "Grand Island's Beet-Sugar Factory," *Chicago Tribune* (October 4, 1890), 9. For a typical contract, see "Beet Sugar in America," *Willett & Gray's Weekly Statistical Sugar Trade Journal* (January 2, 1891), 4.

37. Henry T. Oxnard is quoted in *The Sugar Beet* 11 (1890): 14.

38. "Nebraska's Beet-Sugar Palace," *Chicago Tribune* (September 3, 1890), 9. "Mortimer at the World's Fair," *Louisiana Planter* 11 (1893): 249.

39. *Report [from] Mr. Paddock, from the Committee on Agriculture and Forestry*, 51st Cong., 1st sess., 1890, S. Rep. 509. *Congressional Record* 21 (1889): 126. "More Beet-Sugar Factories," *Washington Post* (October 30, 1890), 7. *Congressional Record* 21 (1889): 2142, 2225, 2285, and 3216. J. B. Hawes, *Sugar Beet Industry in Bohemia* (Washington, DC: Government Printing Office, 1890).

40. "Beet Sugar Culture," *Los Angeles Times* (June 5, 1891), 8. "A Great Industry," *Los Angeles Times* (June 6, 1891), 4. "Beet Sugar," *Los Angeles Times* (July 8, 1891), 2. "World's Greatest Sugar Factory at Oxnard, Ventura County," *Los Angeles Times* (September 4, 1898), 23.

41. "American Beet Sugar," *Wall Street Journal* (February 18, 1899), 5. "Was Not a Sale," *Los Angeles Times* (April 4, 1899), 9. "Oxnard Admits Big Sugar Expenditure," *New York Times* (June 15, 1913), 11. "Oxnard Plans Big Free Sugar Trust," *New York Times* (June 17, 1913), 3.

42. Jacob Adler, *Claus Spreckels: The Sugar King in Hawaii* (Honolulu: University of Hawaii Press, 1966), 26–27.

43. "Profit in Raising Sugar Beets," *Chicago Tribune* (December 28, 1896), 12. "The Sugar Beet in This Country," *Los Angeles Times* (July 12, 1902), A3. Herbert Myrick, *Sugar: A New and Profitable Industry in the United States* (New York and Chicago: Orange Judd, 1897).

44. F. A. Stilgenbauer, "The Michigan Sugar Beet Industry," *Economic Geography* 3 (1927): 486–506. Esther Anderson, "The Beet Sugar Industry of Nebraska as a Response to Geographic Environment," *Economic Geography* 1 (1925): 373–86. James E. Rowan, "Mechanization of the Sugar Beet Industry of Scottsbluff County, Nebraska," *Economic Geography* 24 (1946): 174–80. Western Beet Sugar Producers, *Beet Sugar Handbook* (San Francisco: Western Beet Sugar Producers, 1961), quoted in M. John Loeffler, "Beet-Sugar Production on the Colorado Piedmont," *Annals of the Association of American Geographers* 53 (1963): 364–90.

45. "H. T. Oxnard Dies of Heart Attack," *New York Times* (June 9, 1922), 1.

46. "No Sugar Bounty Wanted," *New York Times* (January 7, 1890), 2. "The Plea of Oxnard," *New York Times* (January 23, 1902), 8. "Sugar-Growing Industry," *Washington Post* (February 15, 1890), 2.

47. "Sugar Trust Strikes Again," *New York Times* (October 9, 1901), 13. "Sugar Men Divided."

48. "Nigger in Sugar Bin," *Los Angeles Times* (January 23, 1902), 1. "Hawaii and the Sugar Question," *Chicago Tribune* (February 6, 1893), 2.

49. "Sugar Lobbying Was Nation Wide," *New York Times* (June 19, 1913), 1.

50. Truman G. Palmer, *Sugar at a Glance* (Washington, DC: Government Printing Office, 1912). Frank Clifford Lowry and F. W. Taussig, *Sugar at a Second Glance* (Washington, DC: Government Printing Office, 1913). "Bryan to Hear Foreign Protest," *Los Angeles Times* (May 28, 1913), 14. "Sugar Pamphlet Franked," *New York Times* (June 13, 1913), 4. "Aided in Sugar Fight," *Washington Post* (June 13, 1913), 4.

51. Harry A. Austin, *History and Development of the Beet Sugar Industry* (Washington, DC: U.S. Beet Sugar Association, 1928). "Beet Sugar Group Expended $500,000 on Lobby in 8 Years," *New York Times* (October 17, 1929), 1. "H. A. Austin Rites to Be Held Today," *Washington Post* (July 20, 1931), 16.

52. Reed Smoot quoted in "A Lesson in Tariff Making," *New York Times* (August 29, 1922), 10. James B. Allen, "The Great Protectionist: Sen. Reed Smoot of Utah," *Utah*

238 NOTES

Historical Quarterly 45 (1977): 325–45. Matthew C. Godfrey, *Religion, Politics, and Sugar: The Mormon Church, the Federal Government, and the Utah-Idaho Sugar Company, 1907–1921* (Logan: Utah State University Press, 2007), 2.

53. "The Greenhouse Farm Flower," *Fortune* 7 (February 1933): 54–58, 111–16. A. Farrell Wood, "Beet Labor from Stoop to Skilled," *Sugar* 41 (February 1946): 30–33.

54. David Lee Child, *The Culture of the Beet and the Manufacture of Beet Sugar* (Boston, 1840), 139. Harvey W. Wiley, *Special Report on the Beet-Sugar Industry in the United States* (Washington, DC: Government Printing Office, 1898), 190.

55. Masakazu Iwataka, "The Japanese Immigrants in California Agriculture," *Agricultural History* 36 (1962): 25–37.

56. Paul S. Taylor, "Hand Laborers in the Western Sugar Beet Industry," *Agricultural History* 41 (1967): 19–26. Sidney Heitman, *Germans from Russia in Colorado* (Fort Collins, CO: Western Social Science Association, 1978). Leonard J. Arrington, *Beet Sugar in the West: A History of the Utah-Idaho Sugar Beet Company, 1891–1966* (Seattle: University of Washington Press, 1966). Leonard J. Arrington, "Science, Government and Enterprise in Economic Development of the Western Beet Sugar Industry," *Agricultural History* 41 (1967): 1–17.

57. Edward N. Clopper and Lewis W. Hine, "Child Labor in the Sugar Beet Fields of Colorado," *National Child Labor Bulletin* 4 (1916): 5–34. U.S. Children's Bureau, *Child Labor and the Work of Mothers in the Beet Fields of Colorado and Michigan* (Washington, DC: Government Printing Office, 1923). Walter Armentrout, Sara A. Brown, and Charles Gibbons, *Child Labor in the Sugar Beet Fields of Michigan* (New York: National Child Labor Committee, 1923). Sara A. Brown and Robie O. Sargent, *Children Working in the Sugar Beet Fields of the North Platte Valley of Nebraska* (New York: National Child Labor Committee, 1924). Sara A. Brown, *Children Working in the Sugar Beet Fields of Certain Districts of the South Platte Valley* (New York: National Child Labor Committee, 1925). "Toil Cripples Child Workers in Beet Fields," *Chicago Tribune* (February 28, 1924), 18. "Colorado Favors Child Labor Law," *Christian Science Monitor* (April 3, 1924), 5. "Finds Child Labor Raises Sugar Beets," *New York Times* (September 7, 1930), 28.

58. "Sugar by Quota," *Time* (April 30, 1934), 12.

59. "Mexican Labor on Farms of Southwest," *Christian Science Monitor* (May 24, 1917), 9. "Mexican Labor Sought," *Christian Science Monitor* (October 18, 1917), 11. "Mexican Labor Needed," *Christian Science Monitor* (October 31, 1917), 13. "Says Mexican Labor Is Indispensable to State," *Los Angeles Times* (November 4, 1917), II1. "Mexicans Are Ordered Home," *Los Angeles Times* (November 16, 1917), II2. "Open Doors to Mexican Labor," *Los Angeles Times* (November 20, 1917), II2.

60. "Will Help on Sugar Beets," *Washington Post* (January 6, 1919), 6. "Importation of Labor from Mexico Is Vital," *Los Angeles Times* (April 11, 1920), IX9.

61. Manual Garcia y Griego, "The Importation of Mexican Contract Laborers to the United States, 1942–1964," in *Between Two Worlds: Mexican Immigrants in the United States*, ed. David Gutierrez (Wilmington: Scholarly Resources, 1996), 45–85.

62. "The Oxnard Factory and Its Relation to Progressive Agriculture," ad in *Los Angeles Times* (January 1, 1920), V20.

63. For Indians, see "Queer Labor Protest," *Washington Post* (June 9, 1891), 7. For Russians, see "Beet-Sugar Business," *Los Angeles Times* (July 13, 1901), 1 "Labor and Economics in Clash in Midwest Sugar Beet Fields," *Christian Science Monitor* (June 13, 1938), 2.

64. Tomás Almaguer, *Racial Fault Lines: The Historical Origins of White Supremacy in California* (Berkeley: University of California Press, 1994), ch. 7, "In the Hands of People Whose Experience Has Been Only to Obey a Master Rather Than Think and Manage for Themselves." Richard S. Street, *Beasts of the Field: A Narrative History of California Farmworkers, 1769–1913* (Stanford, CA: Stanford University Press, 2004), ch. 18, "Blood Spots on the Moon: The 1903 Oxnard Sugar Beet Workers Strike."

65. "Beet Crop in U.S. Depends on Labor," *Sugar* 37 (July 1942): 14–15. "Japs to Harvest Sugar Beets," *Los Angeles Times* (September 22, 1942), 11. "Into the Sugar Beet Fields: Japanese American Laborers," *Life on the Home Front: Oregon Responds to World War II*, Oregon State Archives, 2008, www.sos.state.or.us/archives/exhibits/ww2/threat/labor.htm (accessed December 2, 2010).

66. "Labor Shortage Threatens California Beet Crop," *Wall Street Journal* (May 25, 1942), 8. "Beet Crop in U.S. Depends on Labor." "Mexican Workers for California," *Sugar* 37 (October 1942): 16. Garcia y Griego, "The Importation of Mexican Contract Laborers." "The Axis Can't Lay a Finger on This Sugar Grown *Inside* Our State," ad in *Los Angeles Times* (February 8, 1943), 6.

67. Max Hollrung, "Nebraska and the Beet Sugar Industry," *Bulletin of the Agricultural Experiment Station of Nebraska* 7 (1894): 97–126. "Would Bar Foreign Labor," *New York Times* (April 18, 1920), 6.

68. "Machines: They Can Solve Labor Problems of Sugar Beet Farmers," *Sugar* 40 (December 1942): 16–20.

69. A. Farrell Wood, "Beet Crop Mechanization Goes Ahead," *Sugar* 40 (June 1945): 32–33. A. Farrell Wood, "Beet Industry Sets Up Foundation," *Sugar* 40 (September 1945): 31. Wood, "Beet Labor from Stoop to Skilled." Wayne D. Rasmussen, "Technological Change in Western Sugar Beet Production," *Agricultural History* 41 (1967): 31–36. Rowan, "Mechanization of the Sugar Beet Industry of Scottsbluff County," "Stoop Is Obsolete in Beet Field Now," *New York Times* (December 5, 1954), F1.

70. G. R. Larke, "The Status of Pelleted Beet Seed," *Sugar* 40 (April 1945): 32–33. "Sugar Beet Industry Puts Hope in New Seed," *New York Times* (September 18, 1956), 41.

71. "Water: The Life Blood of the Sugar Beet," *Sugar* 42 (March 1947): 28–29. Loeffler, "Beet-Sugar Production on the Colorado Piedmont."

72. William John May Jr., *The Great Western Sugarlands* (New York: Garland, 1989), ch. 12, "Most Useless of Rivers."

73. Wiley, *Special Report on the Beet-Sugar Industry*, 199.

74. "Sugar Beet Seed Supply Cut Off," *New York Times* (August 30, 1914), X12. "Desperate Shortage in Sugar-Beet Seed," *Los Angeles Times* (August 13, 1914), I11.

75. "U.S. Ready for Trade Imports from Germany," *Christian Science Monitor* (October 1, 1914), 7. "Sugar Beet Seed Import to Be Aided," *Christian Science Monitor* (June 2, 1915), 11. "Americans Get Russian Sugar," *Christian Science Monitor* (December 1, 1917), 7. "Dutch Ship Brings 250 Americans," *Washington Post* (May 16, 1917), 3. "Sugar Seed from Germany," *New York Times* (January 27, 1915), 8.

76. "Selection of Mother Beets," *Los Angeles Times* (September 7, 1902), B7. "Production of Sugar Beet Seed Is New Industry," *Christian Science Monitor* (November 10, 1916), 16. "City's Mecca for Sugar Interests," *Los Angeles Times* (February 13, 1916), I11.

77. "Seek Perfect Sugar Beet," *Los Angeles Times* (March 22, 1926), 5. "Bug War Waged for Sugar Beet," *Los Angeles Times* (December 11, 1928), 4.

78. C. O. Townsend, "The Curly Top or Western Blight of the Sugar Beet," *Science* 23 (1906): 426–27. E. D. Ball, "The Leafhoppers of the Sugarbeet and Their Relation to the 'Curly-Leaf' Condition," U.S. Department of Agriculture, Bureau of Entomology, *Bulletin* 66 (1909): 33–52. H. B. Shaw, "The Curly-Top of Beets," U.S. Department of Agriculture, Bureau of Plant Industry, *Bulletin* 181 (1910). H. B. Shaw, "The Curly Top Disease of Sugar Beets," *Science* 31 (1910): 756.

79. "Appropriations for the Research Work of the Department of Agriculture," *Science* 69 (1929): 372–73. George H. Coons, "The Sugar Beet: Product of Science," *The Scientific Monthly* 68 (1949): 149–64. "This Bug Bandit Must Go!" *Los Angeles Times* (January 31, 1932), 24C. "Leaf Hopper Control in California," *Facts about Sugar* 27 (September 1932): 397–98. E. W. Brandes and G. H. Coons, "Beet Crop Problems: Science Helps Find the Answers," *Facts about Sugar* 29 (1934): 83–85 and 117–21. G. H. Coons, "The Sugar Beet: Product of Science," *The Scientific Monthly* 68 (1949): 149–64.

80. *Beet Sugar Marketing Study* (1948). "Industry-Wide Advertising Campaign Now," *Sugar* 43 (July 1948): 19. "Domestic and Cuban Industry Favors Nation-Wide Advertising Campaign," *Sugar* 43 (September 1948): 27–28.

81. "Beet Sugar Drive," *Wall Street Journal* (August 3, 1949), 1. "Dram Cake with Lemon Fillin Is a Favorite per Irene McCarthy," *Chicago Tribune* (June 21, 1950), A3. "Food Editors Say: . . . 'Your Best Buy Is Beet Sugar,'" ad in *Chicago Tribune* (January 6, 1950), A3. Ad in (Spokane) *Spokesman Review* (January 27, 1950), 22. For a different point of view, see Mani Niall, *Sweet!* (Cambridge, MA: Da Capo Lifelong, 2008), 9.

Chapter 6: Corn, Chemistry, and Capitalism

1. B. Hernstein, "The Centenary of Glucose and the Early History of Starch," *Journal of Industrial and Engineering Chemistry* 3 (1911): 158–68. *The Grangers' Glucose Co.: A Very Safe and Most Profitable Investment for Farmers* (New York, 1875).

2. S. Guthrie, "Notice of the Vaporization of Mercury in the Fumes of Nitric Ether; also, Notices of Various Chemical Products," *American Journal of Science* 21 (1832): 90–94. S. Guthrie, "On Sugar from Potato Starch," *American Journal of Science* 21 (1832): 284–88. Jesse Randolph Pawling, *Dr. Samuel Guthrie: Discoverer of Chloroform* (Watertown, NY: Brewster Press, 1947).

3. "Grape Sugar Manufacture," *Massachusetts Ploughman and New England Journal of Agriculture* 29 (1870): 1, from *American Artisan.* "Grape Sugar," *Scientific American* 24 (1871): 343. E. J. Nieuland, "Profitable Industries the Best," *Manufacturer and Builder* 7 (1875): 203. Harvey W. Wiley, "Glucose and Grape-Sugar," *Popular Science Monthly* 19 (June 1881): 251–57. This was widely cited in newspapers and other journals.

4. Brian W. Peckham, "Economics and Inventions: A Technological History of the Corn Refining Industry of the United States" (PhD diss., University of Wisconsin, Madison, 1979).

5. Fred'k Gossling (or Goessling), "Improved Manufacture of Sugar," U.S. Patent 42,727 (1864). Fred'k Gossling, "Improved Process of Treating Indian Corn and Beet-Roots to Produce Sugar and Sirup," U.S. Patent 42,728 (1864). Fred'k Gossling, "Improvement in the Manufacture of Sugar," U.S. Patent 45,561 (1864).

6. N/t, *American Agriculturist* 24 (February 1865): 44–45. "Making Molasses from Corn," *Scientific American* 11 (1864): 393.

7. "Sugar and Syrup from Indian Corn," *Transactions of the American Institute* (1864–1865): 154. "Corn Syrup," *Maine Farmer* (April 26, 1866), 1, from *New York Tribune*. Fred'k Goessling, "Improved Process for Making Sirup and Sugar from Indian Corn or Other Grain," U.S. Patent 49,749 (1865). Fred'k Goessling, "Improved Method of Making Sugar from Indian Corn or Other Grain," U.S. Patent 49,750 (1865). Fred'k Gossling, "Improved Process for Making Sirup from Indian Corn or Other Grain," U.S. Patent 49,751 (1865). The fullest and most reliable accounts of this business are in "Glucose," *Chicago Tribune* (December 25, 1880), 8, and "The Glucose Industry," *American Grocer* 25 (1881): 10. See also "The List of Premiums," *New York Times* (October 23, 1865), 8, and n/t, *New York Times* (October 28, 1865), 11. Caroline W. Goessling, widow of Frederick W., is listed in the *Jersey City Directory* for 1865–1866.

8. "New Publications," *Manufacturer & Builder* 13 (August 1881): 190.

9. A. W. Fox & Co. ads in Buffalo city directory for 1870, 199, and for 1873, 109. Nieuland, "Profitable Industries the Best."

10. "Cicero J. Hamlin Dead," *New York Times* (February 21, 1905), 7. "C. J. Hamlin Dead," *Boston Globe* (February 21, 1905), 3.

11. Henry Perry Smith, ed., *History of the City of Buffalo and Erie County* (Buffalo: D. Mason, 1884), 2: 255.

12. "Glucose or Grape Sugar in the United States: Its Manufacture and Uses," *American Grocer*, quoted in D. A. Wells, *The Sugar Industry of the United States and the Tariff* (New York, 1878), 17–19.

13. *New York Grape Sugar Co. v. American Grape Sugar Co. and others*, Circuit Court, N.D. New York, 1882. *New York Grape Sugar Co. v. Buffalo Grape Sugar Co. and others; Same v. American Grape Sugar Co. and others*, Circuit Court, N.D. New York, 1883. *New York Grape Sugar Co. v. Buffalo Grape Sugar Co. and others; Same v. American Grape Sugar Co. and others*, Circuit Court, N.D. New York, 1884. *New York Grape Sugar Co. v. Buffalo Grape Sugar Co. and others; Same v. American Grape Sugar Co. and others*, Circuit Court, N.D. New York, 1885. *New York Grape Sugar Co. v. American Grape Sugar Co. et al.; Same v. Buffalo Grape Sugar Co. et al.*, Circuit Court, N.D. New York, 1888. *New York Grape Sugar Co. v. American Grape Sugar Co. et al.* Circuit Court, N.D. New York, 1890. "Starch from 'Sweet Mash,'" *New York Times* (November 28, 1881), 1. "The Grape-Sugar Case," *New York Times* (March 4, 1882), 2. "The Profits of Starch Making," *New York Times* (January 14, 1886), 3.

14. "An Industry Leaving Buffalo," *New York Times* (May 24, 1881), 1.

15. *Susan B. Anthony et al. v. The American Glucose Company*, Court of Appeals of New York, 1894–1895.

16. "Grape-Sugar Works Closed," *New York Times* (April 10, 1883), 1. "Labor in Buffalo," *New York Times* (January 18, 1884), 1.

17. *Ferguson v. The Firmenich Manufacturing Company*, Supreme Court of Iowa, 1889. "Shuts Off Pending Litigation," *Chicago Tribune* (January 31, 1891), 6. *The State of Iowa, Appellee, v. W. S. Smith, Appellant*, Supreme Court of Iowa, 1891. *The State of Iowa, Appellee, v. The Glucose Sugar Refining Company, Appellant*, Supreme Court of Iowa, October, 1902. *Glucose Sugar Refining Co. v. City of Marshalltown et al.*, Circuit Court, S.D. Iowa, 1907.

18. *A. H. Beck et al., Administrators, Appellants, v. Firmenich Manufacturing Company, Appellee*, Supreme Court of Iowa, 1891.

19. "Glucose," *Scientific American Supplement* 10 (1880): 4126. This was based on a paper that A. P. Redfield read at the eleventh annual meeting of the Fire Underwriters Association of the Northwest.

20. "Twelve Men May Be in the Ruins," *New York Times* (April 13, 1894), 1. "Victims of the Buffalo Fire," *New York Times* (April 22, 1894), 2.

21. "F. O. Matthiessen Dies in His Paris Home," *New York Times* (March 10, 1901), 7. For William Wiechers, see "Obituary Notes," *New York Times* (December 16, 1888), 6. The relationship between Matthiessen and Havemeyer appears in the testimony of Ernst Mas before the U.S. Industrial Commission, *Preliminary Report on Trusts and Industrial Combinations* (Washington, DC: Government Printing Office, 1900), 81.

22. Charles F. Chandler's remarks on giving the Perkin Medal to Arno Behr, in Box 8, Folder "Perkin Medal," C. F. Chandler Papers, Rare Books and Manuscripts Division, Baker Library, Columbia University. T. B. Wagner, "Efficiency in Chemical Industries: The Corn Products Industry," *Journal of Industrial and Engineering Chemistry* 5 (1913): 677–78. "Dr. Behr Honored," *Los Angeles Times* (December 27, 1908), 15.

23. "A Glucose Manufactory," *Chicago Tribune* (February 12, 1880), 8. "Licensed to Organize," *Chicago Tribune* (March 7, 1880), 3. "Glucose," *Chicago Tribune* (March 7, 1880), 6. "Glucose," *Chicago Tribune* (March 13, 1880), 8. "Chicago Sugar Refinery," (Chicago) *Daily Inter Ocean* (December 25, 1880), 3. "A Gigantic Corn Sugar Factory," *Scientific American* 44 (1881): 393.

24. "Glucose," *Chicago Tribune* (March 7, 1880), 6. "A Glucose Manufactory."

25. "Starch Dust Explodes," *Chicago Tribune* (November 6, 1888), 8. "The Cause of the Explosion," *Chicago Tribune* (March 28, 1890), 1.

26. "Deadly Mill Dust," *Chicago Tribune* (March 28, 1890), 1. "The Cause of the Explosion." "Sugar Refinery Inquest," *Chicago Tribune* (April 25, 1890), 3.

27. "Jump to Death in Fire Swept Glucose Plant," *Chicago Tribune* (October 22, 1902), 1. "List Eight Dead in Glucose Fire," *Chicago Tribune* (October 23, 1902), 3. "Orders Changes in Glucose Plant," *Chicago Tribune* (October 24, 1902), 5. For payroll, see H. E. Horton, "Notes on the Manufacture of Glucose, Grape Sugars, and Starches," Crerar MS 209, 1:88, Special Collections Research Center, University of Chicago Library.

28. "Glucose," *Chicago Tribune* (March 13, 1880), 8. Ernst Mas described these men as chemists "of the highest rank" in U.S. Industrial Commission, *Preliminary Report*, 86.

29. H. E. Horton to Dr. Breyer, June 9, 1893, in Horton, "Notes," 2:178–79. H. E. Horton, "Fermentation of Glucose Syrups," *Journal of the American Chemical Society* 16 (1894): 808–9. H. E. Horton, "Acidity of Glucose Syrup and Grape Sugar," *Journal of the American Chemical Society* 17 (1895): 402–3. H. E. Horton, "Ash in Glucose and Corn Syrup," *Journal of the American Chemical Society* 17 (1895): 403–5.

30. Paris. Exposition Universelle, 1878, *Official Catalogue of the United States Exhibitors* (London: Chiswick Press, 1878), 232. *Buffalo City Directory for the Year 1878* (Buffalo: The Courier Company, 1878), 383.

31. "A Glucose Monopoly," *New York Times* (March 21, 1883), 1.

32. "Glucose and Grape Sugar-Makers," *New York Times* (January 12, 1882), 3. "Glucose Manufacturers," *Chicago Tribune* (February 16, 1882), 8. "The City," *Chicago Tribune* (June 8, 1882), 8.

33. "Here's Another Trust," *New York Times* (April 28, 1888), 1. "Glucose Trust in Danger," *New York Times* (December 4, 1888), 3. "Krauesmer Gets His Profits," *New York Times* (December 3, 1890), 6.

34. "To Revive the Pool," *Chicago Tribune* (January 28, 1894), 1.

35. "Glucose in a Pool," *Chicago Tribune* (June 19, 1897), 1. "Talk of a Glucose Trust," *New York Times* (July 21, 1897), 10. "Glucose Trust News Is Confirmed," *Chicago Tribune* (July 27, 1897), 2. N/t, *New York Tribune* (July 27, 1897), 10. "Candy Trade Affected," *Washington Post* (August 5, 1897), 4. N/t, *New York Tribune* (August 14, 1897), 4.

36. "To Fight Glucose Trust," *Chicago Tribune* (Dec 24, 1899), 4.

37. "Fix Details of Glucose Deal," *Chicago Tribune* (January 29, 1902), 5. "Chicago As Starch Focus," *Chicago Tribune* (May 11, 1902), 8. "To Close a Starch Factory," *New York Times* (August 2, 1902), 6. "Declines $50,000 Salary Voted Him," *Chicago Tribune* (August 15, 1905), 2.

38. "Standard Oil Absorbs the Glucose Industry," *New York Times* (January 9, 1906), 10. Alfred D. Chandler Jr., *The Visible Hand: The Managerial Revolution in American Business* (Cambridge, MA: Belknap Press, 1977), 335–36. Arthur S. Dewing, *Corporate Promotions and Reorganizations* (Cambridge, MA: Harvard University Press, 1914), ch. 3, "The Starch Consolidations," and ch. 4, "The Reorganizations of the Glucose Combination."

39. "Plenty of Work," *Chicago Defender* (National Division) (July 8, 1922), 3. "Mamie Bradley's Untold Story," *Chicago Defender* (National Division) (April 21, 1956), 8.

40. "In Equity, No. In the District Court of the United States for the Southern District of New York. *The United States of America, Petitioner, vs. Corn Products Refining Company and Others, Defendants*," Answer (filed November 1913). "Court Dissolves Corn Products Co.," *New York Times* (June 25, 1916), 12. "Trade Board Gets Corn Products Case," *New York Times* (November 14, 1916), 16. "Corn Products Co. Must Be Broken Up," *New York Times* (April 1, 1919), 20. "Corn Products Settlement Best for All Concerned," *Wall Street Journal* (April 26, 1919), 7. "Corn Products Firms Cited for Price Fixing," *New York Times* (April 7, 1932), 24.

41. House Committee on Ways and Means, Subcommittee on Changes in the Internal Revenue Laws, *Hearing in Relation to the Bill (H.R. 3170) to Tax and Regulate the Manufacture and Sale of Glucose*, 47th Cong., 2nd sess., 1882.

42. National Academy of Sciences, *Report on Glucose* (Washington, DC: Government Printing Office, 1884). For S. 2500, a bill to regulate the sale of grape sugar and glucose, as well as to prevent the adulteration of sugar, molasses, and sirup made from beets, sorghum, or sugar cane, see 47th Cong., 2nd sess., *Congressional Record* 14 (1882): 3031. See also Henry Morton, "Science and Sugar," *New York Tribune* (December 21, 1878), 2, and Henry Morton, "Poisoned Sugars," *Chicago Tribune* (December 21, 1878), 5.

43. *Report by the National Academy of Sciences on Grape Sugar and Glucose. Covering Especially Their Use in Articles of Food and Drink. Prepared by request of the Commissioner of Internal Revenue of the United States* (Buffalo: Printed for the American Glucose Co., 1884). There is a copy in Horton, "Notes," 3: 77.

44. John Darby, "Sugar," *American Grocer* 1 (1869–1870): 1–3. "Candy: Its Manufacture and Adulteration," *American Grocer* 1 (1869–1870): 12–13. "Adulterated Syrups," *American Grocer* 3 (1870): 573.

45. "The Adulteration of Sugar," *Scientific American* 6 (1850): 88. "New Process Sugar," *American Grocer* 23 (1880): 925.

46. "Sugar Adulteration!" *New York Times* (December 4, 1878), 8, and several subsequent issues. Havemeyers & Elder repeated this text in weekly, full-page advertisements in *American Grocer* 21 (March 27, 1879): 796, and many later issues.

47. "The Alleged Adulteration of Sugar," *New York Times* (January 7, 1879), 5.

48. "Alleged Sugar Frauds," *New York Tribune* (August 19, 1878), 2. "Heavy Losses to the Revenue," *New York Tribune* (September 23, 1878), 2. "Poisoned Sugar," *Chicago Tribune* (November 15, 1878), 2.

49. "Adulteration of Sugar," *New York Times* (April 2, 1880), 2. N/t, *New York Times* (April 11, 1880), 6. N/t, *New York Times* (June 15, 1880), 4.

50. "New York Legislature," *New York Times* (January 27, 1881), 1. "Discussing State Laws," *New York Times* (July 13, 1881), 5. "Food Adulteration," *New York Times* (July 20, 1881), 4. Lorine S. Goodwin, *The Pure Food and Drug Crusaders, 1879–1914* (Jefferson, NC: McFarland, 1999).

51. Ad for Crook's Hyper Sweet Corn Syrup in *Chicago Tribune* (November 16, 1873), 1.

52. F. W. Wagner, "The American Corn Products Industry," *Pure Products* 6 (1910): 101–3.

53. "*The United States v. Corn Products Refining Co.*," *Federal Reporter* 234 (1916): 964–1019, on 989 and 1003. "Corn Products Operates at About 75% of Capacity," *Wall Street Journal* (August 31, 1921), 4.

54. Ad in *Chicago Tribune* (August 3, 1903), 4; *Indiana Farmer's Guide* 31 (December 20, 1919): 21; *American Journal of the Medical Sciences* 168 (December 1924): 24. Ad in *Better Homes and Gardens* (October 10, 1930), 81. "Quintuplets Able to Take Nourishment," *New York Times* (May 30, 1934), 19. Ad in *Atlanta Constitution* (April 14, 1940), A18.

55. *Karo in the Kitchen* (ca. 1904), *Karo Recipes for Cooking and Candy Making* (1908), Emma C. Churchman, *Karo Cook Book* (New York: Corn Products Refining Co., 1910), Ida C. B. Allen, *The Modern Method of Preparing Delightful Foods* (New York: Corn Products Refining Co., 1926), *Proven Recipes Showing the Uses of the Three Great Products from Corn* (192?), *KaroKookery* (New York: Corn Products Sales Co., 1941), and *Newest Recipes for Better Meals* (New York: Corn Products Refining Co., 1952).

56. "Southern Pecan Pie Wins Recipe Award," *Washington Post* (April 8, 1933), 8.

57. Oscar E. Anderson Jr., "The Pure-Food Issue: A Republican Dilemma, 1906–1912," *American Historical Review* 61 (1956): 550–73, on 554–55. Harvey W. Wiley, *An Autobiography* (Indianapolis: Bobbs-Merrill, 1930), 268–72. "Corn Syrup vs. Glucose. Brief of H. W. Wiley, December 31, 1907," in RG 16, Records of the Secretary of Agriculture, General Correspondence, Box 7, NARA.

58. "Food Inspection Decision 87," *American Food Journal* 3 (March 1908): 24. Corn Products Refining Co., *Copies of Opinions of Chemists and Others in the Meaning of the Word "Syrup"* (New York: Corn Products Refining Co., 1907). "Dr. Wiley on Witness Stand," *Wall Street Journal* (August 18, 1911), 7. "Bribes Offered to the Chemists," *Atlanta Constitution* (August 18, 1911), 7.

59. "An Absurd Ruling on Labeling of Syrup," *American Grocer* 79 (February 19, 1908): 8. "The Administration Nullifies the Food Law," *American Grocer* 79 (February 26, 1908): 8. "Precedent, Politics, Paternalism," *American Grocer* 79 (March 4, 1908): 9.

60. "Dr. Wiley in Sugar Trust," *New York Times* (April 3, 1914), 1. "Wiley Called in Glucose Suit," *The Modern Grocer* (April 11, 1914), 7. "Dr. Wiley Goes to Sea," *Washington Post* (January 13, 1908), 3.

61. "*Corn Products Refining Co. v. Weigle*," *Federal Reporter* 221 (1915): 988–96. T. B. Wagner, "Uniformity of Pure Food Laws," *Pure Products* 11 (1915): 115–23.

62. *McDermott v. Wisconsin*, 228 U.S. 115 (1913). "Pure Food as between the State and the United States," *Virginia Law Register* 19 (1913): 148–51.

63. *Corn Products Refining Co. v. Eddy*, 249 U.S. 427 (1919). Thomas Reed Powell, "Constitutional Law in 1918–1919," *American Political Science Review* 13 (1919): 607–33, on 615.

64. RG 88, Food and Drug Administration, Records of the Board of Food and Drug Inspection, Hearings, 1907–1915, Box 2, NARA. Also RG 16, Records of the Secretary of Agriculture, General Correspondence, Box 32, Folder: Corn Syrup Corporation, NARA.

65. Louis Chiozza, "Improvement in Processes of Treating Maize," U.S. Patent 167,224 (1875; reissued in 1881 and assigned to Erhard Matthiessen). Arno Behr, "Process of Treating Corn in the Manufacture of Starch, Glucose, and Other Products Therefrom," U.S. Patent 247,152 (1881). Arno Behr, "Apparatus for Treating Corn in the Manufacture of Starch, Glucose, and Other Products Therefrom," U.S. Patent 247,153 (1881).

66. *Chicago Sugar-Refining Co. v. Charles Pope Glucose Co. et al.*, Circuit Court, N.D. Illinois, 1897. *Chicago Sugar-Refining Co. v. Charles Pope Glucose Co. et al.*, Circuit Court of Appeals, Seventh Circuit, 1898. "Glucose Case Near a Decision," *Chicago Tribune* (January 13, 1897), 7. "Has No Right to a Patent," *New York Times* (February 9, 1897), 2. "Scout a Sugar Victory," *Chicago Tribune* (February 5, 1898), 14. "Victory for Glucose Trust," *Washington Post* (February 5, 1898), 1. "Glucose Trust Wins," *Los Angeles Times* (February 5, 1898), 2.

67. N/t, *Chicago Tribune* (November 16, 1889), 4. "Oil from Corn," *Kansas Weekly Capital and Farm Journal* (September 29, 1892), 9. "Corn Oil" (September 1, 1893), in Horton, "Notes," 1:248.

68. "Corn Oil Mill Is in Operation," *Chicago Tribune* (January 4, 1897), 10. U.S. Industrial Commission, *Preliminary Report*, 77–78.

69. "Corn Oil versus Olive Oil," *Los Angeles Times* (June 14, 1901), 8. "Corn Products Refining," *Wall Street Journal* (February 1, 1910), 5.

70. "Mix in White Bread," *Chicago Tribune* (February 16, 1898), 12. "Flour Fight to End," *Chicago Tribune* (February 19, 1898), 13. "Against Glucose Trust," *Chicago Tribune* (June 14, 1899), 5. "The Industrial Commission," *New York Times* (June 14, 1899), 4. "Salary Was Cut Off," *Washington Post* (June 14, 1899), 9. "White Flour Is Impure," *Chicago Tribune* (November 12, 1899), 2. U.S. Industrial Commission, *Preliminary Report*, 77–78.

71. Arno Behr, "Method for Making Starch and Cattle Feed," U.S. Patent 491,234 (1893). Circular issued by the Chicago Sugar Refining Co. (December 1892), in Horton, "Notes," 1:250. *Buffalo Gluten Feed* (American Glucose Co., ca. 1894), in Horton, "Notes," 2:10. Form letter from Rockford Sugar Refinery Co., March 25, 1896, touting its new "first-class gluten feed," in Horton, "Notes," 2:8.

72. "To Displace Rubber," probably from (Chicago) *Evening Post*, n.d., in Horton, "Notes," vol. 2, CDEF. "A Rubber Substitute from Corn," *Scientific American* 79 (1898): 24. "Rubber from Corn Oil," *Chicago Tribune* (August 7, 1898), 10. "Will Push Corn Rubber," *Chicago Tribune* (October 13, 1898), 10. U.S. Industrial Commission, *Preliminary Report*, 77–78. William K. Leonard, "Process of Producing Rubber Substitutes," U.S. Patents 615,863 (1898) and 615,864 (1898).

Chapter 7: Cane Syrup and Corn Syrup

1. Ad in *Macon Weekly Telegraph* (March 24, 1863), 1. "Ribbon Cane Juice," *Atlanta Constitution* (December 1, 1882), 2.

2. Bennett Battle Ross, *Cane Syrup* (Auburn, AL: Agricultural Experiment Station of the Agricultural and Mechanical College, 1895). Bennett Battle Ross, *Experiments in Syrup Making* (Auburn, AL: Agricultural Experiment Station of the Agricultural and Mechanical College, 1899). Bennett Battle Ross, *The Manufacture of Cane Syrup* (Auburn: Agricultural Experiment Station of the Alabama Polytechnic Institute, 1905). Horace E. Stockbridge, *Cane, Syrup, Sugar* (Lake City, FL: Florida Agricultural Experiment Station, 1898). A. P. Spencer, *Sugar Cane and Syrup Making* (Gainesville, FL: University of Florida Agricultural Experiment Station, 1913). William Fremont Blackman, *Sugar and Cane Syrup in Florida* (Jacksonville, FL: s.n., 1921). Henry H. Harrington, *The Manufacture of Cane Syrup* (College Station, TX: Texas Agricultural Experiment Stations, 1903). Hamilton Pope Agee, *Cane Syrup Making* (New Orleans, LA: Sugar Experiment Station of the Louisiana State University and A&M College, 1911). E. B. Ferris, "Sugar Cane for Syrup Making," Mississippi Agricultural Experiment Station, *Bulletin* 129 (1909).

3. "Georgians Study Sugar Growing," *Atlanta Constitution* (December 8, 1899), 3. William Carter Stubbs's comments at a meeting of the Louisiana Sugar Planters Association are in *Louisiana Planter* 50 (1913): 191–92.

4. D. G. Purse, "Sugar-Cane Culture in Georgia," *Southern Cultivator* (October 15, 1899), 7. For a biographical sketch of Purse, see Charles C. Jones, *History of Savannah, Ga.* (Syracuse, NY: D. Mason & Co., 1890). "Atlantan Uses Knife on Wrists," *Atlanta Constitution* (August 8, 1908), 1. D. G. Purse, *Sugar Cane: Its History in Georgia, Florida, and South Carolina, 1767 to 1900* (Savannah, GA: Morning News Print, 1900).

5. "Cane Growers Come Wednesday," *Florida Times-Union* (May 2, 1904), 6. "Raising of Sugar Cane in Georgia Draws an Appropriation of $11,000," *Atlanta Constitution* (March 16, 1902), 19.

6. Harvey W. Wiley, *Manufacture of Sirups from Sugar Cane* (Washington, DC: Government Printing Office, 1902). Harvey W. Wiley, *Sugar Cane Culture in the Southeast for the Manufacture of Table Sirup* (Washington, DC: Government Printing Office, 1903). Harvey W. Wiley, *Experiments in the Culture of Sugar Cane and Its Manufacture into Table Sirup: A Report on the Investigations Conducted at Waycross and Cairo, Ga., in 1903 and 1904* (Washington, DC: Government Printing Office, 1903). Harvey W. Wiley, "Table Sirups," in *Yearbook of the United States Department of Agriculture for 1904* (Washington, DC: Government Printing Office, 1905), 241–48. Harvey W. Wiley, *Experimental Work in the Production of Table Sirup at Waycross, Ga., 1905* (Washington, DC: Government Printing Office, 1906).

7. *Memoirs of Georgia*, vol. 2 (Atlanta: Southern Historical Association, 1895). William Harden, *A History of Savannah and South Georgia*, vol. 2 (Chicago: Lewis Publishing Co., 1913). Ad in *Macon Daily Telegraph* (April 8, 1921), 11. Bill Baab, "The Georgia Mini Jug Story," *Bottles and Extras* 16 (summer 2005): 2–7.

8. "Those Who Grow the Sugar Cane," *Atlanta Constitution* (May 7, 1903), 3. "Sugar Growers Close Meeting," *Atlanta Constitution* (May 8, 1903), 3.

9. W. B. Roddenbery, "The Cultivation of Sugar Cane in Georgia," *Louisiana Planter* 30 (1903): 318. "Georgia's Opportunity in Sugar Cane Growing," *Macon Weekly Telegraph* (October 28, 1905), 4. "Mr. Roddenbery Gives His Plan of Cultivating Sugar Cane," *Columbus Enquirer Sun* (March 23, 1902), n.p.

10. "Cane Growers Come Wednesday." "Preparing for Convention," *Florida Times-Union* (May 3, 1904), 6. "Second Annual Convention Has Auspicious Opening," *Florida*

Times-Union (May 5, 1904), 6. "Cane Growers' Convention Ends Its Deliberations," *Florida Times-Union* (May 6, 1904), 5–7. Papers by Rose and by Stubbs, *Florida Times-Union* (May 6, 1904), 6–7. "Resolutions of the Cane Growers' Convention," *Sugar Planters' Journal* 34 (1903–1904): 492–93.

11. "Mr. Roddenbery Talks," *Florida Times-Union* (May 6, 1904), 6. D. G. Purse, *Georgia Cane Syrup Adulteration* (Savannah, GA, 1902). "Dr. J. M. McCandless Shows Need for Pure Food Law," *Atlanta Constitution* (May 11, 1903), 7. "By W. B. Broddenberry [sic]," *Florida Times-Union* (May 4, 1904), 6.

12. "Preparing for Convention." D. G. Purse, *Immigration . . . Address Delivered . . . before the Agricultural Society of Effingham County, Ga. . . . July 27, 1904* (Savannah, GA, 1904). "Captain Purse Talks of Cane," *Atlanta Constitution* (January 14, 1905), 6. Jean Ann Scarpaci, *Italian Immigrants in Louisiana's Sugar Parishes* (New York: Arno Press, 1980). Jeannie M. Whayne, *Shadows over Sunnyside: An Arkansas Plantation in Transition, 1830–1945* (Fayetteville: University of Arkansas Press, 1993).

13. "Dr. W. C. Stubbs' Lecture," *Southern Cultivator* 63 (March 1, 1905): 13. "Hundreds of Earnest Southern Men Gather to Discuss Cane Growing," *Montgomery Advertiser* (January 26, 1905), 1 and 8. "Cane Growers Complete Their Work and Adjourn," *Montgomery Advertiser* (January 27, 1905), 5. *Proceedings of the Fourth Annual Convention of the Inter-State Sugar Cane Growers Association* (Milwaukee, WI: S.E. Tate, 1906). USDA, Consumer and Marketing Service, *United States Standards for Grades of Sugarcane Sirup* (Washington, DC: U.S. Department of Agriculture, 1957).

14. *Proceedings of the Fourth Annual Convention of the Inter-State Sugar Cane Growers Association.*

15. Ads in *Atlanta Constitution* (November 30, 1911), SWG16 and SWG10.

16. "Roddenbery Wars on Red Rot," *Atlanta Constitution* (March 30, 1910), 5. "Boon to Farmers of South Georgia," *Atlanta Constitution* (April 23, 1916), 14.

17. Julian K. Dale, "Cooperative Cane-Sirup Canning: Producing Sirup of Uniform Quality," USDA Circular 149 (1920). "Syrup Association Started in Thomas," *Atlanta Constitution* (December 7, 1921), 3. "Wilder Is Honored by Cane Syrup," *Atlanta Constitution* (May 20, 1922), 7. "Thomas Sugar Cane Outlook Promising," *Atlanta Constitution* (July 23, 1922), E4.

18. "Woman Wins Prize," *Chicago Defender* (October 27, 1917), 11. "The Farmers' Column," *Chicago Defender* (November 28, 1931), 15.

19. Ads in *American Grocer* 103 (January 7, 1920): 2 and 6; 103 (February 4, 1920): 35. Ad in *Grocery World* 53 (January 29, 1912): 28.

20. Alaga ads in *Atlanta Constitution* (November 6, 1910), D5B; (November 8, 1910), 7; (July 9, 1912), 6. Penick & Ford ads in *Atlanta Constitution* (November 6, 1910), D5B. "Pure Gold Cane Syrup Wins Praise," *Chicago Defender* (June 10, 1933), 12.

21. Alaga ads in *Atlanta Constitution* (November 20, 1910), E8B; (October 25, 1911), 7. Cane Patch ads in *Atlanta Constitution* (December 23, 1920), 2, and *Macon Daily Telegraph* (May 6, 1921), 5. Marilyn Kern-Foxworth, *Aunt Jemima, Uncle Ben, and Rastus: Blacks in Advertising, Yesterday, Today, and Tomorrow* (Westport, CT: Greenwood Press, 1994).

22. "Pure Gold Cane Syrup Wins Praise."

23. Marcus Alexis, "Pathways to the Negro Market," *Journal of Negro Education* 28 (1959): 114–27. Robert Weems, *Desegregating the Dollar* (New York: New York University Press, 1998).

24. Alaga ads in *Ebony* 11 (November 1955): 167, and every month thereafter for many years. "Civil Rights Movement Was Nourished at Edna's," *Chicago Tribune* (January 15, 1996), 1. "Black Family Finds Suburban Neighbors Neighborly," *Chicago Tribune* (July 11, 1971), N3. "Pick 1st Target of Ala. Boycott," *Chicago Defender* (April 10, 1965), 1.

25. "Georgia Syrup Best, U.S. Officials Shown," *Macon Weekly Telegraph* (July 14, 1913), 3. Alaga ad in *Atlanta Constitution* (July 9, 1912), 6.

26. Peter A. Yoder, "Sugar-Cane Culture for Sirup Production in the United States," USDA, *Bulletin* 486 (1917). Julian K. Dale and C. S. Hudson, "Sugar-Cane Juice Clarification for Sirup Manufacture," USDA, *Bulletin* 921 (1920).

27. *Annual Report of the Department of Agriculture for the Year Ended June 30, 1920* (Washington, DC: Government Printing Office, 1921), 267. *Supplemental Estimates for Department of Agriculture*, 67th Cong., 2nd sess., Senate Doc. 175 (1922), 3–4. H. S. Paine to Hon. Ross Collins, July 27, 1921, in RG 88, Box 255, Folder 4455, NARA. Julian K. Dale and C. S. Hudson, *Manufacture of Sugar Cane Sirup by the Invertase Process* (Washington, DC: U.S. Department of Agriculture, 1922). Charles F. Walton, "Process for Preparing Sirup," U.S. Patent 1,465,459 (1923).

28. "Labeling of Corn Sirup," *Pure Products* 4 (1908), 120.

29. "Wilder & Buchanan and Their Great Success As Manufacturers," *Atlanta Constitution* (January 11, 1903), B9. "Government Moves against D. R. Wilder," *Atlanta Constitution* (March 17, 1909), 1. "Submitted His Label to the Department," *Atlanta Constitution* (March 19, 1909), 14. W. H. Thornton, *The Law of Pure Food and Drugs, National and State* (Cincinnati, OH: W. H. Anderson Co., 1912), 357–58.

30. "A Conviction under the Pure Food Law for Misbranding Syrup," *Modern Sugar Planter* 41 (July 22, 1911): 6.

31. Lannen & Hickey, *Food Laws Affecting Syrups and Syrup Mixtures* (Chicago: Barnard & Miller, 1915).

32. Rachel Maines, *Hedonizing Technologies: Paths to Pleasure in Hobbies and Leisure* (Baltimore: The Johns Hopkins University Press, 2009). "Cane Syrup Makes a Comeback, with Help from UF," *University of Florida News* (October 5, 2001).

33. C. S. Steen Syrup Mill, *The Story of Steen's Syrup and Its Famous Recipes* (Abbeville, LA[?]: s.n., 1966[?]).

34. Howard S. Paine and C. F. Walton, *Sugar-Cane Sirup Manufacture* (Washington, DC: Government Printing Office, 1925). C. F. Walton and I. K. Ventre, *How to Prevent Sugaring of Sugarcane Sirup* (Washington, DC, 1935). Edward W. Brandes, S. F. Sherwood, and B. A. Belcher, "Sugarcane for Sirup Production," *USDA Circular* 284 (1933). I. E. Stokes et al., *Culture of Sugarcane for Sirup Production* (Washington, DC: Agricultural Research Service, U.S. Department of Agriculture, 1961).

Chapter 8: Dextrose, High-Fructose Corn Syrup, and Specialty Sugars

1. Scientists used the words "glucose," "dextrose," and "d-glucose" interchangeably to mean a simple sugar (or monosaccharide) that had the chemical formula $C_6H_{12}O_6$ and that rotated the plane of polarized light to the right. In a similar vein, "fructose," "levulose," and "l-glucose" referred to a simple sugar that had the same chemical formula but rotated the plane of polarization to the left.

2. Corn Industries Research Foundation, *The Use of Corn in Industry* (New York: Corn Industries Research Foundation, 1936).

3. Arno Behr, "Process of Manufacturing Crystallized Anhydride of Grape-Sugar from a Watery Solution of Grape-Sugar," U.S. Patent 250,333 (1881). Arno Behr, "Method of Refining Grape-Sugar," U.S. Patent 250,334 (1881). Arno Behr, "Method of Manufacturing Crystallized Anhydride of Grape-Sugar from a Watery Solution of Grape-Sugar," U.S. Patent 256,622 (1882). Arno Behr, "Process of Manufacturing Crushed Anhydrous Grape-Sugar," U.S. Patent 256,623 (1882). Arno Behr, "Crystallized Anhydrous Grape-Sugar," U.S. Patent 259,794 (1882). Arno Behr, "On Crystallized Anhydrous Grape Sugar," *Journal of the American Chemical Society* 4 (1882): 11–15. T. B. Wagner, "Presentation of Behr Portrait Recalls His Pioneer Work on Chemistry of Corn," *Chemical and Engineering News* 2 (July 10, 1924): 3. "Anhydrous Sugar Inventor Passes," *Los Angeles Times* (June 26, 1921), I10.

4. H. E. Horton, "Notes on the Manufacture of Glucose, Grape Sugars, and Starches," University of Chicago Library, Special Collections, Crerar MS 209, 2:241. T. B. Wagner, "Efficiency in Chemical Industries: The Corn Products Industry," *Journal of Industrial and Engineering Chemistry* 5 (1913): 677–80.

5. T. B. Wagner, "Process of Manufacturing Anhydrous Grape-Sugar," U.S. Patent 835,145 (1906). "Theodore Brentano Wagner," in *National Cyclopaedia of American Biography* 27:231–32. Christian Porst and Nicholas Mumford, "Manufacture of Chemically Pure Dextrose," *Industrial and Engineering Chemistry* 14 (1922): 217–18.

6. Christian Porst, "Manufacture of High Purity Crystalline Anhydrous Dextrose," *Louisiana Planter* 67 (1921): 14–15. Christian Porst, "Making Crystalline Dextrose," *Sugar* 23 (1921): 380–81. Porst and Mumford, "Manufacture of Chemically Pure Dextrose." R. F. Jackson, "The Saccharimetric Normal Weight and Specific Rotation of Dextrose," *Scientific Paper of the Bureau of Standards* 293 (1916[?]).

7. W. B. Newkirk, "Method of Making Grape Sugar," U.S. Patent 1,471,347 (1923). Newkirk, "Manufacture and Uses of Refined Dextrose," *Industrial and Engineering Chemistry* 16 (1924): 1173–75. Brian W. Peckham, "Economics and Invention: A Technological History of the Corn Refining Industry of the United States" (PhD diss., University of Wisconsin, Madison, 1979).

8. "Sugar from Corn," *Wall Street Journal* (January 7, 1926), 17. "Farm Market Is Widened by Corn Discovery," *Chicago Tribune* (January 10, 1934), 9. Ad for dextrose in *Wall Street Journal* (January 2, 1936), 38. "Corn Products Giant Industry in Chicago Area," *Chicago Tribune* (January 23, 1938), A5. W. B. Newkirk, "Development and Production of Anhydrous Dextrose," *Industrial and Engineering Chemistry* 28 (1936): 760–66.

9. *Congressional Record* 67 (March 22, 1926): 6038.

10. "Sugar Substitutes Are Proving Popular," *Atlanta Constitution* (May 26, 1918), B9. "Corn Syrup as a Substitute for Sugar," *Atlanta Constitution* (June 16, 1918), C10. "Substitute Corn Syrup for Cane," *Los Angeles Times* (November 22, 1917), 17. Ads in *National Carbonator and Bottler* 61 (September 15, 1935): 39, and *American Carbonator and Bottler* 81 (September 1945): 73.

11. Ad for Dextrons in *Atlanta Constitution* (August 3, 1928), 12. Ad for Krystal Rock in *National Carbonator and Bottler* 67 (August 1938): 36. Ad for Kre-Mel in *Christian Science Monitor* (May 8, 1930), 5.

12. Carl Fellers, "Dextrose in the Food Industries and Its Health Status," *American Journal of Public Health* 29 (February 1939): 135–38. W. J. Corbett and P. H. Tracy, "Dextrose in Commercial Ice-Cream Manufacture," University of Illinois, Agricultural Experiment Station, *Bulletin* 452 (1939). Corn Products Refining Co., *Flavor and Color in Bread and Sweet Yeast Doughs* (New York: Corn Products Refining Co., 1935). Ad in *The Ice Cream Trade Journal* 21 (January 1925): 11.

13. "Corn Sugar of Dry Midwest Is Basis of 'Moon,'" *Chicago Tribune* (February 18, 1929), 6. "Corn Sugar Usage Hugely Increased by Rum Producers," *Washington Post* (March 16, 1930), 17. "5,000 Fine Is Imposed for Corn Sugar Sale," *Food Industries* 3 (1931): 363.

14. Ad for Pabst Blue Ribbon Syrup in *Los Angeles Times* (August 19, 1921), I13.

15. "Use of Corn Sugar," *Science* 72 (January 23, 1931): xii.

16. Ad in *Wall Street Journal* (January 4, 1943), 30. "Corn Refining Industry to Help Offset Lower Far East Imports of Starch and Sugar," *Wall Street Journal* (December 29, 1941), 1. "U.S. Cooks Testing Sugar Substitutes," *New York Times* (January 30, 1942), 15. Martha Ellyn, "Cater to Sweet Tooth with Sugar Substitutes," *Washington Post* (February 6, 1942), 13.

17. "Corn Product Refining," *Wall Street Journal* (October 11, 1923), 8. "Corn Products—Cerelose," *Wall Street Journal* (July 18, 1925), 2.

18. Harvey W. Wiley, *A History of the Crime against the Food Law* (Washington, DC: H. W. Wiley, 1929), ch. 9, "The Bureau of Standards."

19. Harvey W. Wiley, "Corn-Sugar Bill Is Up Again, and It Should Be Put Down," *Good Housekeeping* 90 (January 1930): 92.

20. "Court Construes Pure Food Act," *Christian Science Monitor* (April 15, 1919), 6.

21. Senate Committee on Manufactures, *Hearings on S. 481, a Bill to Amend the Food and Drugs Act*, 69th Cong., 2nd sess., 1926. House Committee on Interstate and Foreign Commerce, *Hearing . . . on H.R. 39, a Bill to Amend Section 8 of . . . "An Act for Preventing the Manufacture or Sale of Deleterious Foods, Drugs, Medicines, and Liquors, and for Regulating Traffic Therein"*, 69th Cong., 2nd sess., 1926. See also, for instance, *Congressional Record* 74 (1931): 2501; 65 (1924): 7492–99; 67 (1926): 6039–43.

22. "Corn Sugar Battle Rages for 3 Hours before Hyde," *Chicago Tribune* (July 26, 1930), 7.

23. "Curb on Corn Sugar in Foods Is Ended," *New York Times* (December 27, 1930), 3. "How Sugar Is Manufactured from the Humble Corn Grain," *New York Times* (January 4, 1931), XX4. "Glucose in Foods, Undeclared on Labels," *American Journal of Public Health* 20 (October 1930): 1121–22. "Company to Double Corn Sugar Output," *New York Times* (December 30, 1930), 38.

24. "Big Corn, Cane Fight," *Business Week* (April 29, 1939), 36. "What Kind of Sugar in Canned Fruit? Wallace to Decide if User Shall Know," *Christian Science Monitor* (September 29, 1939), 3.

25. "Food Act Ruling to Get Court Text," *New York Times* (May 31, 1940), 65. "Dextrose Victory," *Business Week* (January 17, 1942), 51–52. "Food Trade Approves New Labeling Rules," *New York Times* (March 10, 1942), 28.

26. *Fair Packaging and Labeling Act*, Public Law 89-755. "F.D.A. proposes Food Label Rules," *New York Times* (March 17, 1967), 43. "Sugar Producers and Label Laws," *Los Angeles Times* (August 17, 1969), D8.

27. A. E. Staley Manufacturing Co., *Corn Syrup in Jams, Jellies and Preserves* (Decatur, Ill: A. E. Staley Manufacturing Co., n.d.). "Improved Sweetening Agent Developed from Corn Starch," *Food Industries* 12 (January 1940): 62. "The Cornfield That Fed a Miracle," ad in *Barron's* (May 12, 1947), 4. Dan J. Forrestal, *The Kernel and the Bean: The 75-Year Story of the Staley Company* (New York: Simon and Schuster, 1982), 106–8, 205.

28. Richard O. Marshall and Earl R. Kooi, "Enzymatic Conversion of D-Glucose to D-Fructose," *Science* 125 (April 5, 1957): 648–49. R. O. Marshall, "Enzymatic Process," U.S. Patent 2,950,228 (1960).

29. "The Wet Millers of Corn," *Washington Post* (June 11, 1981), A1 and A8. E. K. Wardrip, "High Fructose Corn Syrup," *Food Technology* 25 (1971): 47–50. Yoshiyuki Takasaki and Asamu Tanabe, "Enzyma Method for Converting Glucose in Glucose Syrups to Fructose," U.S. Patent 3,616,221 (1971).

30. "Chemistry in 1972: Not a Leap, but an Inch," *Science* 179 (1973): 554–55. Norman E. Lloyd et al., "Process for Isomerizing Glucose to Fructose," U.S. Patent 817,832 (1974; assigned to Standard Brands Inc.).

31. J. W. Robinson, "Will High Fructose Corn Syrup Sweeten Your Future?" *Food Engineering* 27 (May 1975): 57–61. Forrestal, *The Kernel and the Bean.*

32. ADM three-page ad in *Food Engineering* 27 (December 1975): 35–37. "H. J. Heinz Co. to Buy Hubinger for $41.4 Million," *Wall Street Journal* (September 10, 1975), 10. "Heinz Unit Will Spend $30 Million to Produce High-Fructose Syrup," *Wall Street Journal* (March 26, 1976), 22. "Amstar to Increase Corn Syrup Output," *New York Times* (August 20, 1974), 54.

33. "When Competition against Sugar Turned Sour," *Business Week* (November 15, 1976), 136, 138. "Darkness before Dawn," *Forbes* (June 26, 1978), 67–68.

34. "Cargill's New $100 Million Corn Milling Plant," *Food Engineering* 55 (August 1983): 198. "CPC to Decide on High-Fructose Facility, Indicating Industry Recovery from Slide," *Wall Street Journal* (May 15, 1979), 38. "Tate & Lyle Sets Deal to Buy Amstar Sugar," *New York Times* (September 27, 1988), D2.

35. "Soft-Drink Prices Raised by Bottlers," *Washington Post* (March 31, 1974), M3. "Use of Corn Syrup as Sugar Substitute in Soft Drinks Set," *Wall Street Journal* (August 8, 1974), 20. "Heard on the Street," *Wall Street Journal* (August 9, 1974), 27.

36. "Coca-Cola OKs 50% Corn Sweetener Use," *Chicago Tribune* (January 29, 1980), C6. "Coca-Cola Increases Corn-Syrup Content to 100% at Fountains; Millers Helped," *Wall Street Journal* (May 9, 1984), 48.

37. "Why Did the 'Old Cola Drinkers of America' Turn Up Their Noses at Classic Coke?" *Los Angeles Times* (August 14, 1985), A17. "Tempests in a Pop Bottle," *Time* (August 26, 1985), 45. "Advertising; Coke Held Not to Be Real Thing," *New York Times* (August 15, 1985), D19.

38. "Pepsi Approves Syrup Change," *New York Times* (April 22, 1983), D3.

39. "New Sweetening Agent Takes on Sugar," *Chicago Tribune* (April 8, 1976), E21.

40. "Sugar under Fire," *Food Engineering* 59 (April 1987): 107–18.

41. Stroud Jordan, "Commercial Invert Sugar: Its Manufacture and Uses," *Industrial and Engineering Chemistry* 16 (1924): 307–10.

42. Noah W. Taussig, "Process of Making Inverted-Sugar Syrup," U.S. Patent 1,181,086 (1916). *"Proto-Fax" for Increased Sales* (New York: Nulomoline, 1936). Sucrest Corp., *"Proto-Fax" Cookies* (New York: Nulomoline, 1940). American Molasses

Co., *Basic Candy Formulas* (New York: Nulomoline Division, American Molasses Co., 1959).

43. "Washington Letter," *Journal of Industrial and Engineering Chemistry* 9 (1917): 815. Charles W. Taussig, *Some Notes on Sugar and Molasses* (New York, 1940), 116–20.

44. "American Sugar in Syrup," *Wall Street Journal* (May 5, 1919), 2. Ad in *Christian Science Monitor* (March 27, 1923), 3.

45. Ad in *Chicago Tribune* (June 23, 1929), 5.

46. "Bottlers Again Consider Use of Sugar Stretchers," *American Bottler* (March 1953): 18, refers to a 1946 paper, "The Use of Invert Syrup in Carbonated Beverage Manufacture."

47. U.S. Bureau of Chemistry and Soils, *Preparation of Invert Sugar by Inversion with Tartaric Acid* (Washington, DC, 1925). "Florida Scientists Tell How to Stretch Sugar," *New York Times* (May 8, 1942), 23.

48. American Molasses Co., *Basic Candy Formulas*. H. S. Paine, "Candy Makers Control Softening of Cream Centers," *Food Engineering* 1 (1928–1929): 200–2.

49. "Sees Liquid Sugar Expansion," *New York Times* (November 11, 1945), 59. B. W. Dyer & Co., *Liquid Sugar* (New York: Self-published, 1945).

50. *The Story of Liquid Sugar* (Yonkers, NY, 1944). *This Is Liquid Sugar* (Yonkers, NY, 1955, 1966). "Big Sugar Refinery to Open in Yonkers," *New York Times* (February 4, 1938), 6. "What the New York Seven-Up Bottling Co. Inc. Says about *Liquid Sugar*," full-page ad in *Food Industries* 22 (December 1950): 165. Ads in *Wall Street Journal* (April 24, 1952), 20, and (October 27, 1952), 7.

51. "Liquid Sugar Use by Industry Rise," *New York Times* (February 26, 1956), F1 and F12. "Frozen Food Business: How Sweet It Is for Sugar Industry," *Los Angeles Times* (February 3, 1971), H8 and H9.

Chapter 9: The Sorghum Rage of the Gilded Age

1. John J. Winberry, "The Sorghum Syrup Industry, 1854–1975," *Agricultural History* 54 (1980): 343–52. C. Wayne Smith and R. A. Frederikson, eds., *Sorghum: Origin, History, Technology and Production* (2000), says next to nothing about sugar or syrup.

2. D. R. B., "Researches on the Sorgho Sucré," in *Report of the Commissioner of Patents for the Year 1854* (Washington, DC: Government Printing Office, 1855), pt. 2, 219–23. "Chinese Sugar-Cane" in *Report of the Commissioner of Patents for the Year 1855* (Washington, DC: Government Printing Office, 1856), pt. 2, 279–81. John W. Chambers, "Production of Sugar at the North," *Transactions of the American Institute* (1863–1864): 395–97. C. R. B., "Daniel Jay Browne," in *Dictionary of American Biography* 3:164–65.

3. James F. C. Hyde, *The Chinese Sugar Cane* (Boston: J. P. Jewett & Co., 1857). Henry Steel Olcott, *Sorgho and Imphee, the Chinese and African Sugar Canes . . . with a Paper by Leonard Wray, Esq., of Caffraria, and His Patented Process for Crystallizing the Juice of the Imphee; to Which Are Added Copious Translations of Valuable French Patents* (New York: A. O. Moore, 1857). Charles F. Stansbury, *Chinese Sugar Cane and Sugar Making; Its History, Culture, and Adaptation to the Climate, Soil, and Economy of the United States, with an Account of Various Processes for Manufacturing Sugar* (New York: C. M. Saxton & Co., 1857). Hedges, Free & Co., *Experiments with Sorghum Sugar Cane; Including a Treatise on Sugar Making* (Cincinnati, OH: The Company, 1859).

4. Joseph S. Lovering, "A Detailed Account of Experiments and Observations upon the Sorghum Saccharatum or Chinese Sugar Cane, Made with the View of Determining Its Value as a Sugar Producing Plant . . . 1857 at Oakhill, Philadelphia County, Pennsylvania," *Journal of the Franklin Institute* 35 (1858): 125–36.

5. See, for instance, "Chinese Sugar Cane in Winnebago Co.," *Chicago Tribune* (March 8, 1861), 2; "Sorghum Sirup," *Scientific American* 6 (1862): 243; "Interesting from Chicago," *New York Times* (December 21, 1862), 2; and "The Sugar Growers of Michigan," *Chicago Tribune* (January 20, 1863), 3.

6. For "Dr. Emerson's" comments on sorghum, see *Proceedings of the American Philosophical Society* 9 (1862): 116–18. "The Challenge Cane Crusher," *Scientific American* 13 (1859): 208.

7. "The New Sugar Plant: Sorghum Saccharum; or, China Sugar Cane," *New York Times* (November 8, 1856), 6. C. T. Jackson, "Chemical Researches on the Sorgho Sucré," and D. J. B., "Crystallization of the Juice of the Sorgho Sucré," in *Report of the Commissioner of Patents for the Year 1856* (Washington, DC: Government Printing Office, 1857), pt. 2, 307–8 and 309–13. "Illinois Sorghum Molasses," *Scientific American* 6 (1862): 243.

8. "Sugar and Sorghum," *Scientific American* 21 (1857): 411. "Sorghum Sirups," *Scientific American* 24 (1858): 93.

9. I. A. Hedges to A. Lincoln, November 14, 1861, in Abraham Lincoln papers, Library of Congress. I. A. Hedges, "Sorghum Culture and Sugar Making," and D. M. Cook, "The Culture and Manufacture of Sugar from Sorghum," in *Report of the Commissioner of Patents for the Year 1861* (Washington, DC: Government Printing Office, 1862), 293–311 and 311–14. I. A. Hedges, *Sorgho or the Northern Sugar Plant, by Isaac A. Hedges, the Pioneer Investigator in the Northern Cane Enterprise, with an Introduction by William Clough, President, Ohio State Board Sorgo Culture* (Cincinnati, OH: Applegate & Co., 1863).

10. *Letter from the Commissioner of the Department of Agriculture, in Answer to Resolutions of the House of 3d and 5th December, in Regard to the Expenditure of the Agricultural Fund*, 37th Cong., 3rd sess., 1862, H. Ex. Doc. 14.

11. William Clough, "Sorghum, or Northern Sugar-Cane," in *Annual Report of the Commissioner of Agriculture for the Year 1864* (Washington, DC: Government Printing Office, 1865), 54–87.

12. "Chicago Sugar Refinery," *Chicago Tribune* (December 15, 1859), 1. "Refined Sorghum Syrup," *Chicago Tribune* (December 2, 1861), 4. "Cotton Packing and Sorghum Refining in Chicago," *Chicago Tribune* (January 29, 1863), 4. "A New National Product—Beet-Root Sugar," *New York Times* (January 23, 1863), 4. "A Sugar Refiner's Opinion of Sorghum," *Scientific American* 10 (1864): 360.

13. David W. Blymer, *The Sorghum Hand Book: A Treatise on the Sorgho and Imphee Sugar Canes* (Cincinnati, OH: Blymer Manufacturing Co., 1887), 4. *Clough's Refining & Deodorizing Process as Applied to the Manufacture of Syrup and Sugar from Sorghum* (Cincinnati, OH: Clough Refining Co., 1867). William Clough, "Clarifying of Sorghum Syrup," *Transactions of the American Institute* (1868–1869): 202–4.

14. *Report of the Department of Agriculture for the Year 1877* (Washington, DC: Government Printing Office, 1878), 23–47. John Alfred Heitmann, *The Modernization of the Louisiana Sugar Industry, 1830–1910* (Baton Rouge: Louisiana State University Press, 1987), 116–24, "USDA and Sorghum Sugar Fever: Early Post Reconstruction Attempts to Assist the Louisiana Sugar Industry."

15. Peter Collier, *Report of Analytical and Other Work Done on Sorghum and Corn-stalks* (Washington, DC: Government Printing Office, 1881). Harvey W. Wiley, "William McMurtrie," *Journal of Industrial and Engineering Chemistry* 5 (1913): 616–18, on 616, suggests that LeDuc fired McMurtrie because he "evidently had some political debts to pay."

16. Francis L. Stewart, "Improvement in Processes for Clarifying the Juices of Sorghum and Maize," U.S. Patent 203,507 (1878). Francis L. Stewart, "Improvement in Compounds for Defecating Saccharine Liquids," U.S. Patent 202,295 (1878). Francis L. Stewart, "Maize and Sorghum as Sugar Plants," in *Report of the Commissioner of Agriculture for 1877* (Washington, DC: Government Printing Office, 1878), 228–36. Francis L. Stewart, *Sugar Made from Maize and Sorghum. A New Discovery* (Washington, DC: The Republic Co., 1878). Francis L. Stewart, *Sorghum and Its Products: An Account of Recent Investigations Concerning the Value of Sorghum in Sugar Production, Together with a Description of a New Method of Making Sugar and Refined Syrup from the Plant* (Philadelphia: J. B. Lippincott & Co., 1867). "The Manufacture of Sugar from Cornstalks and Sorghum Cane," *Chicago Tribune* (September 18, 1878), 2.

17. William Gates LeDuc, *Circular Letter from the Commissioner of Agriculture Relative to the Manufacture of Maize and Sorghum Sugars* (Washington, DC: Government Printing Office, 1879).

18. "Immediate Needs of the Department," in *Report of the Commissioner of Agriculture for 1879* (Washington, DC: Government Printing Office, 1880), 28.

19. *Letter of the Commissioner of Agriculture to the Hon. Jno. W. Johnston, Chairman of the Committee on Agriculture, U.S. Senate, on Sorghum Sugar* (Washington, DC: Government Printing Office, 1880). *Congressional Record* 10 (1880): 2813.

20. *Preliminary Report of the Department of Agriculture for the Year 1880* (Washington, DC: Government Printing Office, 1881) consists almost entirely of reports about sorghum from around the country. "Sugar from Sorghum," *Washington Post* (October 18, 1880), 2.

21. "The Other Side of the Loring-Collier Difficulty," *Colman's Rural World* (May 3, 1883), 1. "Dr. Collier's Side," *Colman's Rural World* (May 10, 1883), 1.

22. "Success with Sorghum Sugar," *Washington Post* (October 31, 1881), 4. N/t, *Washington Post* (June 15, 1882), 2. "Manufacture of Sugar from Sorghum" in *Report of the Commissioner of Agriculture for 1881 and 1882* (Washington, DC: Government Printing Office, 1882), 19–20.

23. "The National Academy of Science [*sic*]," *Scientific American* 45 (1881): 353. National Academy of Sciences, *Investigation of the Scientific and Economic Relations of the Sorghum Sugar Industry* (Washington, DC: Government Printing Office, 1883). *Congressional Record* 13 (1882): 5609; 14 (1883): 1126.

24. "The Report on Sorghum," *Colman's Rural World* (July 19, 1883), 1. Heitmann, *The Modernization of the Louisiana Sugar Industry*, 122–23.

25. Charles Anthony Goessmann, *Contributions to the Knowledge of the Nature of the Chinese Sugar-Cane* (Albany: Printed by Charles van Benthuysen, 1862). Massachusetts Agricultural College, *Charles Anthony Goessmann* (Cambridge, MA: Printed at the Riverside Press, 1917), esp. 64 and 96–97.

26. "Success with Sorghum Sugar." "Sugar-Culture Experiments," *New York Times* (June 12, 1882), 1. Loring's *Circular* of June 6, 1882, is reprinted in USDA, *Encouragement*

to the *Sorghum and Beet Sugar Industry* (Washington, DC: Government Printing Office, 1883), and in *Sorghum and Beet Sugar: Letter from the Commissioner of Agriculture*, 49th Cong., 1st sess., 1886, H. Mis. Doc. 284.

27. Henry A. Weber and Melville A. Scovell, *Report on the Manufacture of Sugar, Syrup, and Glucose from Sorghum* (Champaign, IL: Gazette Steam, 1881). Henry A. Weber and Melville A. Scovell, "Process of Manufacturing Sugar and Sirup," U.S. Patent 250,118 (1881), and "Manufacture of Glucose," U.S. Patent 250,117 (1881). "Henry Adam Weber," in *National Cyclopaedia of American Biography* 19:277–78, notes that Weber had been "instrumental in establishing the new industry of sorghum-sugar manufacture on a large commercial scale throughout Illinois." "Melville Amasa Scovell," in *National Cyclopaedia of American Biography*, 15:242.

28. W. A. Henry, "Experiments in Making Sugar from Early Amber Cane," *American Agriculturist* 41 (1882): 153. "Agriculture," *Chicago Tribune* (December 15, 1883), 11. Olaf Hougen, "Magnus Swenson, Inventor and Chemical Engineer," *Norwegian-American Studies and Records* 10 (1948): 152–75. Madison, Wisconsin, Experimental Farm, *Experiments in Amber Cane and the Ensilage of Fodders* (Madison: David Atwood, 1882 and 1883).

29. Harold J. Abrahams, "The Sorghum Experiment at Rio Grande," *Proceedings of the New Jersey Historical Society* 83 (1965): 118–36. Jean Wilson Sidar, *George Hammell Cook: A Life in Agriculture and Geology, 1818–1889* (New Brunswick, NJ: Rutgers University Press, 1876).

30. For Wiley, see Oscar Anderson, *The Health of a Nation* (Chicago: University of Chicago Press, 1958). For the MVCGA, see notice in *Colman's Rural World* (May 17, 1883), 1. William M. Ledbetter, "Isaac Hedges' Vision of a Sorghum-Sugar Industry in Missouri," *Missouri Historical Review* 21 (1926): 361–69. See also "The Prospect of a National Fair," *Washington Post* (January 17, 1882), 4. Isaac A. Hedges, *Sugar Canes and Their Products: Culture and Manufacture* (St. Louis, MO: Author, 1879 and 1881). "Cane Growers," *Chicago Tribune* (December 4, 1879), 7.

31. "Mississippi Valley Cane Growers' Meeting," *Colman's Rural World* (March 1, 1883), 1. George B. Loring, *The Sorghum Sugar Industry: Address . . . before the Mississippi Valley Cane Growers' Association, Saint Louis, Mo., December 14, 1882* (Washington, DC: Government Printing Office, 1883). Harvey W. Wiley, *An Autobiography* (Indianapolis: Bobbs-Merrill, 1930), 153.

32. Wiley, *Autobiography*, 152–53 and 164–65. Harvey W. Wiley, *Sorghum, Its Success and Value* (Indianapolis: W. B. Burford, 1883). William Lloyd Fox, "Harvey W. Wiley's Search for American Sugar Self-Sufficiency," *Agricultural History* 54 (1980): 516–26.

33. "Sorghum as a Sugar-Producing Plant," *Manufacturer and Builder* 15 (1883): 274.

34. "Sorghum," *Chicago Tribune* (August 23, 1883), 8. Wiley, *Sorghum*. Noël Deerr, *The History of Sugar* (London: Chapman and Hall, 1950), 2:547–48. U.S. Centennial Exhibition, *Report of the United States Centennial Commission* (Washington, DC: Government Printing Office, 1880), Group III, 18–54.

35. "Sorghum Sugar Experiments," *New York Times* (November 24, 1883), 2. "Agriculture." Harvey W. Wiley, *Diffusion: Its Application to Sugar Cane, and Record of Experiments with Sorghum in 1883* (Washington, DC: Government Printing Office, 1884). "Sorghum." Harvey W. Wiley, "Experiments with Sorghum Cane, 1883," in *Report of the Commissioner of Agriculture for 1883* (Washington, DC: Government Printing Office, 1884), 423–44.

36. "Northern Cane Growers," *Colman's Rural World* (January 24, 1884), 1. "Sorghum Sugar Experiments," *New York Times* (January 5, 1884), 1. Harvey W. Wiley, *The Northern Sugar Industry: A Record of Its Progress During the Season of 1883* (Washington, DC: Government Printing Office, 1884).

37. William Elsey Connelley, *The Life of Preston B. Plumb* (Chicago: Browne & Howell Co., 1913), ch. 60, "Sugar." Harvey W. Wiley, "Report of the Chemical Division," in *Report of the Commissioner of Agriculture for 1884* (Washington, DC: Government Printing Office, 1885), 25.

38. "Loring's Costly Seeds," *Washington Post* (May 13, 1885), 1. "Ex-Commissioner Loring's 'Prizes,'" *Washington Post* (July 11, 1885), 1.

39. Harvey W. Wiley, *Experiments with Diffusion and Carbonatation at Ottawa, Kansas, Campaign of 1885* (Washington, DC: Government Printing Office, 1885); this is USDA, Chemical Division, *Bulletin* 6. Harvey W. Wiley, "Report O [*sic*] Then [*sic*] Ottawa Experiments," in *Report of the Commissioner of Agriculture for 1885* (Washington, DC: Government Printing Office, 1886), 123–38. "Sugar Made from Beets," *Washington Post* (November 4, 1885), 3. "The Extraction of Sugar from Sorghum and Sugar-Cane," *Science* 6 (1885): 524–25.

40. Harvey W. Wiley, *The Sugar Industry of the United States* (Washington, DC: Government Printing Office, 1885); this is USDA, Chemical Division, *Bulletin* 5. "The Sugar Industry at Hutchinson, Kan.," *Chicago Tribune* (December 15, 1883), 11. An illustration of the Hutchinson factory appears in R. R. Bowker, "A Lump of Sugar," *Harper's New Monthly Magazine* 73 (1886): 72–95, on 90–91.

41. Fox, "Harvey W. Wiley's Search for American Sugar Self-Sufficiency," 521.

42. *Debate on the Application of Diffusion in the Manufacture of Sugar from Sorghum and Tropical Sugar Cane*, 49th Cong., 1st sess., 1886. *Congressional Record* 17 (1886), 5495 and 6280.

43. "Sugar from Sorghum," *Los Angeles Times* (October 5, 1886), 1. Harvey W. Wiley, *Record of Experiments at Fort Scott, Kansas, in the Manufacture of Sugar from Sorghum and Sugar Canes, in 1886* (Washington, DC: Government Printing Office, 1887); this is USDA, Division of Chemistry, *Bulletin* 14. Harvey W. Wiley, "Experiments in the Manufacture of Sugar from Sorghum," in *Report of the Commissioner of Agriculture for 1886* (Washington, DC: Government Printing Office, 1887), 302–18. "Those Sugar Experiments," *New York Times* (November 26, 1886), 2.

44. "Making Beet Sugar in Nebraska," *Chicago Tribune* (October 23, 1890), 9. Harvey W. Wiley et al., *Record of Experiments with Sorghum in 1891* (Washington, DC: Government Printing Office, 1892); this is USDA, Division of Chemistry, *Bulletin* 34. "Experiments with Sorghum," *Report of the Secretary of Agriculture for the Year 1892* (Washington, DC: Government Printing Office, 1893), 136. "Experiments in the Improvement of Sorghum as a Sugar Producing Plant," *Report of the Secretary of Agriculture for the Year 1893* (Washington, DC: Government Printing Office, 1894), 186–87.

45. "Talking on a Sweet Subject," *Washington Post* (December 10, 1886), 1. "The Sugar Outlook," *Washington Post* (October 21, 1886), 2.

46. "Kansas a Sugar State," *Chicago Tribune* (September 14, 1887), 9. "No More Foreign Sugar," *Washington Post* (September 13, 1887), 1. Magnus Swenson, *Record of Experiments Conducted by the Commissioner of Agriculture in the Manufacture of Sugar from Sorghum and Sugar Canes at Fort Scott Kansas, Rio Grande New Jersey, and Lawrence, Louisiana, 1887–1888*; this is USDA, Division of Chemistry, *Bulletin* 17.

47. *Letter from the Commissioner of Agriculture, Transmitting . . . the Report of Professor Swenson on Sorghum Sugar*, 50th Cong., 1st sess., 1888, S. Ex. Doc. 68.

48. N/t, *Washington Post* (January 25, 1888), 4. "Big Money in Sorghum," *Chicago Tribune* (July 25, 1888), 10. "Sorghum Sugar," *American Grocer* 38 (October 19, 1887): 8–9. *Sorghum*, 50th Cong., 1st sess., 1888, H. Mis. Doc. 286.

49. Magnus Swenson, "Manufacturing of Sugar," U.S. Patent 371,528 (1887). Magnus Swenson, "Filter-Press," U.S. Patent 353,514 (1886); Magnus Swenson, "Concentrating Pan," U.S. Patent 353,515 (1886).

50. *Congressional Record* 19 (1887): 14 and 215. "Letters Patent Granted to M. Swenson," *Report of the Commissioner of Agriculture for 1887* (Washington, DC: Government Printing Office, 1888), 242–46. *Letter from the Commissioner of Agriculture in Response to Senate Resolution of December 7, 1887, Relative to Employees Obtaining Patents for the Process of Sugar Making*, U.S. Senate, 50th Cong., 1st sess., 1887, Ex. Doc. 24. "Prof. Swenson's Patent," *Washington Post* (December 16, 1887), 2. N/t, *New York Times* (December 27, 1887), 4. "An Interesting Patent Discussion in the Senate," *Scientific American* 57 (December 31, 1887), 417. "Sorghum Sugar Process," U.S. House of Representatives, 50th Cong., 1st sess., 1888, Mis. Doc. 71. "The Sorghum Patents," *Washington Post* (January 14, 1889), 2.

51. "To Control the Sorghum Crop," *New York Times* (March 3, 1888), 1. "Big Money in Sorghum." G. B. J., "Making Sorghum Sugar," *Chicago Tribune* (October 29, 1888), 7. Homer E. Socolofsky, "The Bittersweet Tale of Sorghum Sugar," *Kansas History* 16 (1993–1994): 276–89.

52. "Sugar Made from Sorghum," *New York Times* (July 31, 1894), 6.

53. "The Agricultural Appropriation Bill," *Washington Post* (July 12, 1888), 1. "Experiments in the Manufacture of Sugar" in *Report of the Commissioner of Agriculture for 1888* (Washington, DC: Government Printing Office, 1889), 23–28.

54. "Sugar from Sorghum Cane," *Washington Post* (September 28, 1889), 1. "Sorghum Manufacture," *Washington Post* (August 16, 1888), 7. "Production of Sorghum Sugar" in *Report of the Commissioner of Agriculture for 1889* (Washington, DC: Government Printing Office, 1890), 144–62, on 160–61. "Making Sugar from Sorghum," *Washington Post* (April 10, 1889), 1.

55. "Difficulty of Making Sorghum Sugar in Small Quantities" in *Report of the Secretary of Agriculture for the Year 1890* (Washington, DC: Government Printing Office, 1890), 148–49. "The Use of Alcohol in the Manufacture of Sugar from Sorghum," *Report of the Secretary of Agriculture for the Year 1891* (Washington, DC: Government Printing Office, 1892), 143–45. "Sugar from Sorghum," *Washington Post* (December 15, 1890), 2. N/t, *Washington Post* (March 9, 1891), 4. T. Berry Smith, "Sorghum Sugar," *Science* 23 (1894): 25–27.

56. "Domestic Sugar Production" and "Experiments in the Improvement of Sorghum as a Sugar Producing Plant" in *Report of the Secretary of Agriculture for the Year 1893* (Washington, DC: Government Printing Office, 1894), 33–34 and 186–90.

57. "Sorghum Sugar in Kansas," *New York Times* (February 2, 1898), 10. "End of a Government Experiment," *Washington Post* (February 2, 1898), 8. "Sugar Plant Sold," (Topeka) *Kansas Semi-Weekly Capital* (May 11, 1900), 7. "Syrup of Sorghum," (Topeka) *Kansas Semi-Weekly Capital* (February 8, 1901), 3. Dairy and Food Commissioner of Texas, *First Annual Report* (Austin, 1908), 13. Ad for Farmer Jones Sorghum Blend Syrup in *Dallas Morning News* (October 26, 1919), 9. Paul L. Vogt, *The Sugar Refining Industry in the United States* (Philadelphia: University of Pennsylvania, 1908), 3–4.

Chapter 10: Maple Sugar and Syrup

1. Robert Boyle, *Some Considerations Touching the Usefulnesse of Experimental Naturall Philosophy* (London, 1663), 2:112.

2. Dr. Robinson to Mr. Ray, March 10, 1684, and Mr. Ray to Dr. Robinson, March 13, 1684, in *The Correspondence of John Ray*, ed. Edwin Lankester (London, 1848), 162–64. "An Account of a Sort of Sugar Made of the Juice of the Maple, in Canada," *Philosophical Transactions of the Royal Society of London* 15 (1685): 988.

3. Carol Mason, "A Sweet Something: Maple Sugaring in the New World," in *The Invented Indian: Cultural Fictions and Government Politics*, ed. James Clifton (New Brunswick: Transaction Publishers, 1990), 91–106. Albert P. Sy, "History, Manufacture and Analysis of Maple Products," *Journal of the Franklin Institute* 166 (1908): 249–80, on 249–55.

4. André Thevet, *Les singularitez de la France antarctique, autremont nomée Amérique* (Paris, 1557). W. L. Grant, *History of New France* (Toronto: The Champlain Society, 1907–1914), 3:194; this is an English translation of Marc Lescarbot, *Histoire de la nouvelle France* (Paris, 1609). William F. Ganong, *The Description and Natural History of the Coasts of North America (Arcadia)* (Toronto: The Champlain Society, 1908), 380–81; this is an English translation of Nicolas Denys, *Description geographique et historique des costes de l'Amerique septentrionale* (Paris, 1672). William F. Ganong, *New Relation of Gaspesia by Father C. LeClercq* (Toronto: The Champlain Society, 1910), 122–23; this is an English translation of C. LeClercq, *Nouvelle relation de la Gaspesie* (Paris, 1691).

5. Thevet, *Les singularitez de la France antarctique, autremont nomée Amérique*, quoted in James J. Pendergast, *The Origin of Maple Sugar* (Ottawa: National Museums of Canada, 1982), 33.

6. Robert Beverley, *History and Present State of Virginia* (London, 1705), 118–19. Paul Dudley, "An Account of the Method of Making Sugar from the Juice of the Maple Tree in New England," *Philosophical Transactions of the Royal Society of London* 31 (1720): 27–28.

7. Susannah Carter, *The Frugal Housewife: Or, Complete Woman Cook* (New York, 1803), appendix.

8. Samuel Hopkins, *Historical Memoirs, Relating to the Housatonick Indians* (Boston, 1753), 26–27. S. W., "Review of *Historical Memoirs, relating to the Housatonick Indians* by the Rev. Mr. Hopkins," *Newport Mercury* (December 6, 1763), 3. Samuel Hopkins, *Dialogue Concerning the Slavery of the Africans* (Norwich: Printed and sold by J. P. Spooner, 1776). David S. Lovejoy, "Samuel Hopkins: Religion, Slavery and the Revolution," *New England Quarterly* 40 (1967): 227–43.

9. M. de Warville, "On Replacing the Sugar of the Cane by the Sugar of the Maple," *New York Magazine* 3 (1792): 484–86, cited in Wendy Woloson, *Refined Tastes* (Baltimore: The Johns Hopkins University Press, 2002), 29. The almanac is cited in Helen and Scott Nearing, *The Maple Sugar Book* (New York: J. Day Co., 1950), 19.

10. Benjamin Rush, *An Account of the Sugar Maple Tree of the United States and of the Methods of Obtaining Sugar from It* (London, 1792); the same text appeared in *Transactions of the American Philosophical Society* 3 (1793): 64–81. Agricola, "Advantages of the Culture of the Sugar Maple Tree," *American Museum* 4 (October 1788): 349–50. A Sugar Boiler, "Directions for Manufacturing Sugar from the Maple Tree," *American*

Museum 6 (August 1789): 100–1. A Friend of Manufactures, "American Maple Sugar and Melasses," *American Museum* 6 (September 1789): 209–10. *Remarks on the Manufacturing of Maple Sugar; with Directions for Its Further Improvement* (Philadelphia: James & Johnson, 1790). *Constitution of the Society for Promoting the Manufacture of Sugar from the Sugar Maple-Tree* (Philadelphia, 1793).

11. "Sugar Maple, from Thomas Jefferson Encyclopedia," Montecello.org, http://wiki .monticello.org/mediawiki/index.php/Sugar_Maple (accessed December 3, 2010).

12. Richard Saunders, *Poor Richard Improved* (Philadelphia, 1765).

13. "Portions of the Journal of André Michaux, Botanist, Written during His Travels in the United States and Canada, 1785–1796," *Proceedings of the American Philosophical Society* 26 (1889): 1–145. Daniel Jay Browne, *Sylva Americana; or a Description of the Forest Trees Indigenous to the United States Practically and Botanically Considered* (Boston, 1832). Daniel Jay Browne, "Sylva Americana; or a Description of the Forest Trees Indigenous to the United States Practically and Botanically Considered," *North American Review* 44 (1837): 334–61.

14. "Sugar," in *Report of the Commissioner of Patents for the Year 1853* (Washington, DC: Government Printing Office, 1854), pt. 2, 231–36.

15. William Freeman Fox and William F. Hubbard, *The Maple Sugar Industry* (Washington, DC: U.S. Department of Agriculture, 1905), 17. Federal Writers' Project of the Works Progress Administration of the State of Vermont, *Vermont: A Guide to the Green Mountain State* (Boston: Houghton, Mifflin, 1937), 42. "The Cigarettes We Smoke," *New York Times* (August 27, 1939), SM11.

16. See, for instance, Charles H. Jones and Arthur W. Edson, "The Maple Sap Flow," *Vermont Agricultural Extension Station Bulletin* 103 (1903): 45–184.

17. Orrin H. Jones, *Vermont Maple Sugar Industry* (Wilmington, VT: Deerfield Valley Times Print, 1889). Vermont Maple Sugar Makers' Association, *Vermont Maple Sugar and Syrup* (St. Albans, VT: Wallace Printing Co., 1894; rpt. Waitsfield, VT: The Association, 1993). Vermont Maple Sugar Makers' Association, *Pure Vermont Maple Sugar, Maple Syrup, Home of the Sugar Maple, History of the Maple Sugar Industry from the Indian Down to the Present Time* (St. Albans, VT: Vermont Maple Sugar Makers' Association, 1912).

18. Thomas E. Porter, *A Letter Concerning the Early Maple Sugar Industry of Paint Creek Valley* (Chillicothe, OH: Paint Creek Valley Folk Research Project, 1963). Gary William Graham, *Analysis of Production Practices and Demographic Characteristics of the Ohio Maple Syrup Industry* (PhD diss., Ohio State University, 2005). "Ohio Carries Off Most of the Medals for Maple Goods," *Chicago Tribune* (September 20, 1893), 4.

19. "Tin Plate and Sugar," *New York Times* (July 4, 1890), 3. "Will Receive a Bounty," *Boston Globe* (March 17, 1891), 5. "Maple Sugar Growers Protest," *New York Times* (September 22, 1894), 1. "More Patronage for Gabriel," *New York Times* (February 21, 1892), 2.

20. U.S. Internal Revenue Service, *Regulations Concerning Bounty on Maple Sugar* (Washington, DC: Government Printing Office, 1893). Fox and Hubbard, *The Maple Sugar Industry*, 17–18.

21. Albert Hugh Bryan, *Methods for the Analysis of Maple Products and the Detection of Adulterants* (Washington, DC: Government Printing Office, 1908). Albert Hugh Bryan, *Maple-Sap Sirup; Its Manufacture, Composition, and Effect of Environment Thereon* (Washington, DC: U.S. Department of Agriculture, 1910). Albert Hugh Bryan and W. F.

Hubbard, *Production of Maple Sugar and Syrup* (Washington, DC: Government Printing Office, 1912). Albert Hugh Bryan et al., *Maple Sugar: Composition, Methods of Analysis, Effect of Environment* (Washington, DC: U.S. Department of Agriculture, 1917). Julius Hortvet, "Chemical Composition of Maple Syrup," *Journal of the American Chemical Society* 26 (1904): 1523–46. Julius Hortvet, *Report on Saccharine Products (Maple Sugar and Syrup)* (Washington, DC: U.S. Department of Agriculture, 1906). H. W. Cowles Jr., "Suggested Standards for Maple Sugar and Syrup," *Journal of Industrial and Engineering Chemistry* 1 (1909): 773–75. R. T. Balch, "Maple Sirup Color Standards," *Journal of Industrial and Engineering Chemistry* 22 (1930): 255–57.

22. "Alleged Maple Sugar," *Chicago Tribune* (March 23, 1884), 11. N/t, *Chicago Tribune* (December 6, 1886), 4. Harvey W. Wiley, *The Sugar Industry of the United States* (Washington, DC: Government Printing Office, 1885), 189–215, on 195. Harvey W. Wiley, ed., *Sugar, Molasses and Sirup, Confections, Honey and Beeswax* (Washington, DC: Government Printing Office, 1892); this is pt. 6 of *Foods and Food Adulterants*. Fox and Hubbard, *The Maple Sugar Industry*, 9.

23. Ad in *The Cosmopolitan* 15 (October 1893): 29. Wiley, *Sugar, Molasses and Sirup.*

24. U.S. Department of Agriculture, Office of the Secretary, "Standards of Purity for Food Products" (June 26, 1906), reprinted in Arthur P. Greeley, *The Food and Drugs Act* (Washington, DC: J. Byrne & Co., 1907), 146–47. Food Inspection Decision 161 is quoted in Bryan et al., *Maple Sugar*, 2.

25. Food Inspection Decision 75, "The Labeling of Mixtures of Cane and Maple Sirups," in *American Food Journal* 2 (September 1907): 20. "Misbranding of Maple Sirup," *Pure Products* 5 (1909): 309, 446, and 616. Dairy and Food Commissioner of the State of Michigan, *Annual Report* 20 (1914): 38.

26. Ad in *Los Angeles Times* (September 2, 1900), III7.

27. Ads in *Ladies' Home Journal* 21 (August 1904): 25; 21 (April 1904): 58; 21 (December 1904): 22; 25 (November 1905): 29. Ad in *Lippincott's Monthly Magazine* 74 (December 1904): 95. "The National Pure Food Show," *American Food Journal* 1 (May 1906): 21–23; 1 (October 1906): 32.

28. Ads in *The Youth's Companion* 84 (November 10, 1910): 627; *Harper's Bazaar* 45 (November 1911): 527. Ads in *Good Housekeeping* 80 (January 1925): 146; 80 (July 1925): 157; 80 (November 1925): 284.

29. "A Sticky Situation: Maple Syrup Makers Can't Satisfy Demand," *Wall Street Journal* (June 23, 1967), 1. "How Much Maple Is in Maple Syrup?" *New York Times* (June 30, 1975), 36. "Maple Syrup Spigot Drips Slower," *Christian Science Monitor* (May 15, 1975), 2.

30. Helen Nearing and Scott Nearing, *The Maple Sugar Book* (New York: J. Day Co., 1950).

31. Josiah Daily, "Flavoring Extract for Sirup and Sugar," U.S. Patent 261,315 (1883). Ad for Mapleine in *Boston Globe* (December 5, 1891), 2. "Artificial Maple Sugar," *Chicago Tribune* (April 12, 1893), 16. Wiley, *Sugar, Molasses and Sirup*, 712–13.

32. "Pure Food Law Is Upheld," *New York Times* (May 2, 1909), 2. Ads for Mapleine in *International Confectioner* 19 (December 1910): 68; *American Food Journal* 7 (October 1911): 26. Archie Satterfield, *Crescent, 100 Years, 1893–1983* (Seattle: Crescent Manufacturing Co., 1983). *The Mapleine Cook Book* (Seattle: Crescent Manufacturing Co., 1910).

33. Ads in *International Confectioner* 26 (January 1917): 41; 19 (December 1910): 58.

34. John W. Sale and John B. Wilson, "Process for Manufacturing a True Maple Flavoring Product," U.S. Patent 1,642,789 (1927). "Maple Sirup—Yum Yum," *Washington Post* (April 20, 1928), 6. "New Product Gives Maple Taste to Sirup," *Chicago Tribune* (April 21, 1928), 12. "Maple Sugar and Chemistry," *Christian Science Monitor* (May 4, 1928), 20.

Chapter 11: Honey

1. Paul Dudley, "An Account of a Method Lately Found Out in New-England, for Discovering Where the Bees Hive in the Woods, in Order to Get Their Honey," *Philosophical Transactions of the Royal Society of London* 31 (1720): 148–51. Thomas Jefferson, *Notes on the State of Virginia* (Paris, 1785), query vi.

2. Tammy Horn, *Bees in America: How the Honeybee Shaped a Nation* (Lexington: University Press of Kentucky, 2005). Bee Wilson, *The Hive: The Story of the Honeybee and Us* (London: Thomas Dunne, 2004).

3. Richard Saunders, *Poor Richard's Almanac* (Philadelphia, 1735). Richard Saunders, *Poor Richard Improved . . . for the Year of our Lord 1765* (Philadelphia: B. Franklin and D. Hall, 1764).

4. "Statistics of Bee-Keeping," in *Report of the Commissioner of Agriculture for the Year 1868* (Washington, DC: Government Printing Office, 1869), 272–81, on 272. "The Honey Trade of the United States, Domestic and Foreign," *American Journal of Pharmacy* 43 (February 1871): 70–71.

5. Florence Naile, *America's Master of Bee Culture: The Life of Lorenzo Langstroth* (Ithaca, NY: Cornell University Press, 1976). Lorenzo Langstroth, "Bee Hive," U.S. Patent 9,300 (1852; reissued 1863).

6. L. L. Langstroth and S. Wagner, "Apparatus for Extracting Honey from the Comb," U.S. Patent 61,216 (1867). L. L. Langstroth & Son, "Honey Emptying Machine," *American Bee Journal* 3 (1868): 189.

7. M. Quinby and L. C. Root, *Quinby's New Bee-Keeping* (New York: Orange Judd Co., 1885), 90–92.

8. "Samuel Bowne Parsons," in *Dictionary of American Biography* 14:269–70.

9. Ernest R. Root, "A Son's Memories of A. I. Root," *Gleanings in Bee Culture* 51 (1923): 410–14. A. I. Root, *The ABC of Bee Culture* (Medina: A. I. Root, 1877). "Amos Ives Root," in *Dictionary of American Biography* 16:114 and in *National Cyclopaedia of American Biography* 25:253–54.

10. Frances Willard and Mary Livermore, ed., *A Woman of the Century* (Buffalo: Moulton, 1893), 726. Ellen S. Tupper, "Bee Keeping," *Report of the Commissioner of Agriculture for the Year 1865* (Washington, DC: Government Printing Office, 1866), 458–75. Ellen S. Tupper, "Chapters on the Honeybee," *Prairie Farmer* 16 (1865): 400, 433; 17 (1866), 10, 68, etc. "An Iowa Woman," *Chicago Tribune* (July 30, 1873), 2. "Forgeries by a Woman," *Chicago Tribune* (February 2, 1876), 3. "Mrs. Ellen S. Tupper," *The American Bee Journal* 24 (1888): 179.

11. Quinby is quoted in Eva Crane, *The World History of Beekeeping and Honey Hunting* (New York: Routledge, 1999), 495. Frank Pellett, *History of American Beekeeping* (Ames, IA: Collegiate Press, 1938), ch. 25, "Adulteration of Honey and the Pure Food Legislation."

12. Harvey W. Wiley, "Glucose and Grape Sugar," *Popular Science Monthly* 19 (June 1881): 251–57, on 254. Allen Pringle, "Artificial Honey and Manufactured Science," *Popular Science Monthly* 37 (May 1890): 70–75.

13. Harvey W. Wiley, "Honey and Its Adulterations," *American Apiculturist* 3 (1885), reprinted in Harvey W. Wiley, ed., *Foods and Food Adulterants* (Washington, DC: Government Printing Office, 1892), 801–8.

14. "Notice of Judgment Nos. 18–21, Food and Drugs Act," *Pure Products* 4 (1908): 510–16. "An Important Case of Honey Adulteration Won by the Government," *Gleanings in Bee Culture* 42 (1914): 43–44.

15. C. A. Browne, "Methods of Honey Testing for Bee Keepers," in *Miscellaneous Papers on Agriculture*, ed. U.S. Department of Agriculture, Bureau of Entomology (Washington, DC: Government Printing Office, 1911), 16–18. Browne, "Chemical Analysis and Composition of American Honeys," Bureau of Chemistry, *Bulletin* 110.

16. Notices in *American Bee Journal* 14 (1878): 6 and 106.

17. "Honey for Export," *New York Times* (October 8, 1877), 5. "Honey and Hives," *Massachusetts Ploughman* 37 (November 17, 1877): 1. "American Honey in England," *American Bee Journal* 15 (1879): 393. "Introducing Our Honey into England, and the Prejudice against the Yankees," *Gleaning in Bee Culture* 7 (1879): 474.

18. "The Fight against Commission Frauds and the Honey-Adulterers," *American Bee Journal* 37 (1897): 37. "Is Bee-Keeping Profitable?" *American Bee Journal* 18 (1882): 297.

19. "Nelson Wylie McLain," in *National Cyclopaedia of American Biography*, 17:338–39. Frank Benton, *The Honeybee* (Washington, DC: Government Printing Office, 1895). "Frank Benton," in *Who Was Who in America*, 4:78.

20. Mary G. Phillips, *The Bee Man: Life and Letters of Everett Franklin Phillips* (Ithaca, NY, 1967). E. F. Phillips, *The Rearing of Queen Bees* (Washington, DC: Government Printing Office, 1905). E. F. Phillips, *Miscellaneous Papers on Apiculture* (Washington, DC, 1907, 1909, and 1911). E. F. Phillips and G. F. White, *Historical Notes on the Causes of Bee Disease* (Washington, DC: U.S. Department of Agriculture, Bureau of Entomology, 1912). E. F. Phillips, *The Temperature of the Honeybee Cluster in Winter* (Washington, DC: Government Printing Office, 1914). E. F. Phillips, *Beekeeping* (New York: Macmillan, 1915). E. F. Phillips, *Beekeeping* (New York: Macmillan, 1928), 435–36.

21. E. F. Phillips, "A Wasted Sugar Supply," in *Yearbook of the United States Department of Agriculture for 1917* (Washington, DC: Government Printing Office, 1918), 395–400. U.S. Department of Agriculture, Office of the Secretary, "Honey," pt. 4 of *The Agricultural Situation for 1918* (Washington, DC, January 31, 1918).

22. "Color Grades for Honey," *Science* 59 (Supplement) (1924): xiv. E. L. Sechrist, *The Color Grading of Honey* (Washington, DC: Department of Agriculture, 1925). E. L. Sechrist, "Honey Grades Set Aim for Beekeepers," in *Yearbook of the United States Department of Agriculture for 1926* (Washington, DC: Government Printing Office, 1927), 435–36. "Beekeeping, Still in Commercial Infancy, Aided by U.S. Grades," in *Yearbook of the United States Department of Agriculture for 1927* (Washington, DC: Government Printing Office, 1928), 129. USDA, *United States Standards for Honey Recommended by the United States Department of Agriculture* (Washington, DC: Government Printing Office, May 1927; revised December 1927). "Pfund Color Grader," fact sheet of the Munsell Color Co. (Baltimore, n.d.); a copy is in the Trade Literature Collection,

Smithsonian Libraries. "The Root Honey Grader," *Root Quality Bee Supplies* (Medina, OH, 1922), 41.

23. Erik Skokstad, "Entomology: The Case of the Empty Hives," *Science* 316 (May 18, 2007): 970–72.

24. "The War Conference," *American Bee Journal* 57 (1917): 199–200. "Newspaper Specials," *Wall Street Journal* (May 1, 1917), 2. "The Agricultural Situation for 1918: Honey, More Honey Needed," *USDA Circular* 87 (1918). "Bee-Culture Investigations," *Annual Reports of the Department of Agriculture for 1918* (Washington, DC: Government Printing Office, 1919), 250. "Honey Market Reports Now Issued," in *Yearbook of the USDA for 1926*, 434.

25. "Senate Favors Busier Busy Bees, Adds $33,000 to Bee Culture Budget," *Christian Science Monitor* (April 21, 1941), 3. "Bees Working Overtime Trying to Keep Up with Defense," *Los Angeles Times* (May 23, 1941), 29.

26. Ads in *New York Times* (May 8, 1942), 25; (March 15, 1942), AG5.

27. "Uncle Gets Stung on a Purchase," *Washington Post* (February 15, 1948), B5. House Committee on Agriculture, Subcommittee No. 3, *Hearing on Price Support for Honey*, 81st Cong., 1st sess., 1949. Frederic L. Hoff and Jane K. Phillips, "Beekeeping and the Honey Program," *National Food Review* 13 (January–March 1990): 62–65. Mary Muth et al., "The Fable of the Bees Revisited: Causes and Consequences of the U.S. Honey Program," *Journal of Law and Economics* 46 (2003): 479–99.

28. "Beekeepers Abuzz As U.S. Mulls Ending Honey Price Supports," *Wall Street Journal* (September 20, 1982), 46. "Honey Imports Produce Sticky Situation for Beekeepers and Government," *Washington Post* (June 3, 1983), A19. "And Now the Problem Is Honey," *New York Times* (March 13, 1984), B6. "A USDA Policy for the Birds—and the Bees," *Wall Street Journal* (October 4, 1984), 32.

29. "Clinton May Have Thought It Safe Since They're Used to Being Stung," *Wall Street Journal* (June 29, 1992), B1. "Is U.S. Stuck with Honey Subsidies?" *Los Angeles Times* (March 21, 1993), A1. House Committee on Agriculture, Subcommittee on Specialty Crops and Natural Resources, *Review of Fiscal Year 1994 Budget Proposal to Eliminate the Honey Price Support Program*, 103rd Cong., 1st sess., 1993.

30. "Conference Report on H.R. 2493. Agriculture, Rural Development, Food and Drug Administration, and Related Agencies Appropriations Act," *Congressional Register*, 103rd Cong., 1st sess., 139 (1993), 6345–49.

31. House Committee on Ways and Means, Subcommittee on Trade, *Hearing on H. Con. Res. 80 to Disapprove the Determination of the President Denying Import Relief under the Trade Act of 1974 to the United States Honey Industry*, 95th Cong., 1st sess., 1977.

32. "Special Interests on the Attack. In this Case, They're Beekeepers," *New York Times* (February 3, 1994), D2. "Battered but Not Broken, the Honey Lobby Is Back and Winning," *New York Times* (April 13, 1995), D2.

33. "FDA Seizes More Than $32,000 Worth of Bulk Honey from Philadelphia Distribution Center," *FDA News Release* (June 10, 2010).

34. House Committee on Agriculture, Subcommittee on Research and Extension, *Hearing on H.R. 15936, H.R. 16455, H.R. 16617, and H.R. 16909*, 90th Cong., 2nd sess., 1968. House Committee on Agriculture, Subcommittee on Domestic Marketing and Consumer Relations, *Hearing on H.R. 9655, H.R. 9948, H.R. 11049, H.R. 11790*, 91st Cong., 1st sess., 1969, quote on 31–35.

35. House Committee on Agriculture, Subcommittee on Livestock, Dairy and Poultry, *Hearing on H.R. 5358*, 98th Cong., 2nd sess., 1984. Senate Committee on Agriculture, Nutrition and Forestry, Subcommittee on Agricultural Research and General Legislation, *Hearing on S. 2857*, 98th Cong., 2nd sess., 1984. "Honey Research, Promotion, and Consumer Information Act," 7 U.S.C. 4601.

36. Charles D. Michener, "The Brazilian Honeybee," *BioScience* 23 (1973): 523–27. Strom Thurmond's remarks in *Congressional Register*, 94th Cong., 1st sess., 121 (1975), 18946–47. P. L. 94-319; also S. Rpt. 94-193.

37. E. F. Phillips, "The Status of Apiculture in the United States" (1909), reprinted in "Miscellaneous Papers on Apiculture," USDA Bureau of Entomology, *Bulletin* 75 (1911), 59–80. "Bees Valuable to Growers of Fruit," *Los Angeles Times* (August 7, 1921), IX9. *Congressional Record* 78 (1934): 3365; 89 (1943): 3475; 90 (1944): A2986; and 95 (1949): 10630.

38. "Talk of Vicious Bees Stirs Research Group to Fly Team to Brazil," *Wall Street Journal* (September 13, 1971), 12. House Committee on Ways and Means, Subcommittee on Trade, *Hearing on H. Con. Res. 80*, 13. Roger A. Morse and Nicholas Calderone, *The Value of Honeybees as Pollinators of U.S. Crops in 2000* (Ithaca, NY: Cornell University, 2000).

39. Ad in *New York Times* (February 24, 1904), 6. D. Everett Lyon, *How to Keep Bees for Profit* (New York: Macmillan, 1910), 236–37. House Committee on Ways and Means, Subcommittee on Trade, *Hearing on H. Con. Res. 80*, 26–27.

40. "Uses of Honey Enters Extensively into Manufacturing," *Fort Worth Star Telegram* (September 25, 1910), 19. "Firm Uses Honey in Its Golf Balls," *Washington Post* (February 21, 1934), 16. "Added Use of Honey Promises Busier Times for Busy Bees," *Christian Science Monitor* (December 3, 1936), 4.

41. L. Langstroth, *The Hive and the Honeybee* (Northampton, MA: Hopkins, Bridgman & Co., 1853), 64. "Longevity Secrets Sought in Queen Bee 'Royal Jelly,'" *New York Times* 1 (November 29, 1928), 16. "Queen Bee's Food Gives Science a Clue to Secret of Longevity," *New York Times* (September 16, 1947), 25. Ads in *Los Angeles Times* (July 8, 1954), 5; *New York Times* (July 11, 1954), 71.

42. "Condemn Royal Jelly," *Science News Letter* 73 (March 29, 1958): 198.

Chapter 12: Saccharin

1. George B. Kauffman and Paul M. Priebe, "The Discovery of Saccharin: A Centennial Retrospect," *Ambix* 25 (1978): 191–207.

2. Henry E. Roscoe, "On Recent Progress in the Coal-Tar Industry," *Proceedings of the Royal Institution of Great Britain* 11 (1886): 450–66, on 462–66. William Crookes, "Chemistry To-Day and Its Problems," *The Forum* 1 (May 1891): 32–329, on 327.

3. Deborah J. Warner, "Ira Remsen, Saccharin, and the Linear Model," *Ambix* 55 (2008): 50–61.

4. Crystalose Heyden ad in *New Haven Register* (February 27, 1900), 5. Merck ad in *New York Journal* (January 7, 1898), 2.

5. *Lutz v. Magone*, 153 U.S. 105 (1894). Transcript of Record. File Date: 10/28/1890. Term Year 1893. *U.S. Supreme Court Records and Briefs, 1832–1978*. See Gale, Cengage Learning, Library of Congress (December 23, 2010).

6. "The Board of Customs," *New York Times* (April 6, 1898), 9. "The Board of Customs," *New York Times* (April 29, 1898), 10. *Comparison of the Tariff Act of August 28, 1894, with the Bill (H.R. 379) as Passed by the House, Amended and Passed by the Senate, and as It Became a Law, July 24, 1897* (Washington, DC: Government Printing Office, 1897), 48–49.

7. U.S. Department of Commerce and Labor, Bureau of Statistics, *The Foreign Commerce and Navigation of the United States Ending June 30, 1905* (Washington, DC: Government Printing Office, 1905), 283.

8. Dan Forrestal, *Faith, Hope and $5,000: The Story of Monsanto* (New York: Simon and Schuster, 1977). Monsanto ad, April 10, 1903, in Monsanto Papers, Products, Series 03, Box 03, folder "Saccharin History," Washington University Archives. "John Francis Queeny," in *National Cyclopaedia of American Biography*, 24:395. "Jules Bebie," in *National Cyclopaedia of American Biography*, 44:75–76. "Gaston DuBois," in *Who Was Who in America* 3:239. Jules Bebie, "Process for Producing Saccharin," U.S. Patent 1,366,349 (1921).

9. "Food News: Sweetening," *New York Times* (October 22, 1953), 37. Abbott Laboratories, *Annual Report* (1950–1953). Fred J. Helgren, assignor to Abbott Laboratories, "Sweetening Compositions and Method of Producing the Same," U.S. Patent 2,803,551 (1957).

10. "Saccharin in Packets," *New York Times* (August 18, 1954), 19. Rich Cohen, *Sweet and Low* (New York: Farrar, Strauss and Giroux, 2006). Benjamin Eisenstadt, "Sweetening Compound," U.S. Patent 3,259,506 (1966; assigned to Cumberland Packing Co.). Marvin Eisenstadt, "Cyclamate-Free Artificial Sweetener," U.S. Patent 3,625,711 (1971).

11. Constantin Fahlberg, "Manufacture of Saccharine Compounds," U.S. Patent 319,082 (June 2, 1885; reissued December 1, 1885). Fahlberg, List & Co., *Saccharine* (1893). Ad in Chicago Columbian Exposition, *Official Catalogue Exhibition of the German Empire* (Berlin: Imperial Commission, 1893), 23.

12. Nannie M. Tilley, *The R. J. Reynolds Tobacco Company* (Chapel Hill: University of North Carolina Press, 1985), 84–88.

13. Charles Herman Sulz, *A Treatise on Beverages* (New York: C. H. Sulz, 1888), 595–602. "The Case for Soda Water," *New York Times* (June 9, 1909), 6.

14. Forrestal, *Faith, Hope and $5,000*. Monsanto ad, April 10, 1903, in Monsanto Papers, Products, Series 03, Box 03, folder "Saccharin History," Washington University Archives.

15. Cash Book, 1908–1909, in Monsanto Papers, Series 10, subseries 02, flat box 23, Washington University Archives. Mark Pendergrast, *For God, Country and Coca-Cola* (New York: Basic Books, 2000), 109–10. "Soda and Iced Tea," *Biloxi Daily Herald* (August 24, 1903), 3. "Alarm over the Soft Drink Habit," (San Jose) *Evening News* (March 17, 1909), 4.

16. "Oppose Use of Saccharin," *Pure Products* 16 (1920): 59. "The Non-Alcoholic Drink Situation," *Pure Products* 17 (1921): 59. C. Alsberg, "Importance of the Bottling Industry," *Pure Products* 17 (1921): 9–14. Monsanto full-page advertisements in *American Bottlers' Journal* in each monthly issue in the early 1920s and occasional smaller ads in *National Carbonator and Bottler* in the mid-1930s.

17. "Self-Denial or Suicide," *National Carbonator and Bottler* 67 (August 1938): 61–63. "Saccharine Drinks Studies in Chicago," *National Carbonator and Bottler* 67

(January 1939): 20–21. "Wagner Urges Bottlers to Avoid Sugar Substitutes" *American Carbonator and Bottler* 81 (September 1945): 68. O. H. Lamborn, "The Industry's View on Sugar Problem," *American Carbonator and Bottler* 82 (December 1946): 78–79, 117–24, on 118.

18. "Proposals for Subsistence Supplies," *New York Times* (February 4, 1897), 5. "The Soldier's Emergency Ration," *Current Literature* 21 (March 1897): 245.

19. Harvey W. Wiley, "Saccharin in Food Products," *Pure Products* 1 (1905): 282–84.

20. Harvey W. Wiley in *Report of the Industrial Commission on Agriculture* (Washington, DC: Government Printing Office, 1901), 11:105. Harvey W. Wiley in *Adulteration of Food and Drink*, 56th Cong., 1st sess., 1901, S. Rep. 516, 44. Harvey W. Wiley, "Bitter-Sweet: The Story of the Fight against the Use of Saccharin in Foods," *Good Housekeeping* 57 (November 1913): 689–93. Harvey W. Wiley, *Foods and Their Adulterations* (Philadelphia: P. Blackiston's Son & Co., 1907), 311.

21. Wiley's testimony in U.S. Industrial Commission on Agriculture, *Report*, 53. Harvey W. Wiley to James Wilson, December 6, 1911, in RG 16, Records of the Office of the Secretary of Agriculture, General Correspondence, Box 37, Folder: Saccharin 2, NARA.

22. Arthur P. Greeley, *The Food and Drugs Act, June 30, 1906: A Study of the Text of the Act Annotated* (Washington, DC: J. C. Byrne & Co., 1907). "Food Inspection Decisions," *Pure Products* 3 (1907): 258–61. "Misbranding of Canned Corn (As to Presence of Saccharin)," *Pure Products* 5 (1909): 250. A. Klipstein & Co. ad for Heyden Sugar in *Journal of Proceedings of the Annual Convention of the National Association of State Dairy and Food Departments* (1903): 316.

23. Harvey W. Wiley, *An Autobiography* (Indianapolis: Bobbs-Merrill, 1930), 239–41.

24. "Special Food Commission," *Washington Post* (January 19, 1908), 8. "Referee Board Named," *Washington Post* (February 21, 1908), 16. U.S. Department of Agriculture, Referee Board of Consulting Scientific Experts, *Influence of Saccharin on the Nutrition and Health of Man* (Washington, DC: U.S. Department of Agriculture, 1911).

25. Ira Remsen to Russell Chittenden, February 6, 1911, in Box 6, Ira Remsen Papers, Special Collections, Milton S. Eisenhower Library, Johns Hopkins University.

26. Food Inspection Decisions 135 and 138. All the early food inspection decisions can be seen in Henry Sherman, *Food Products* (New York: Macmillan, 1915). "Saccharin Manufacturers Protest against Decision," *American Food Journal* 6 (June 1911): 24–25.

27. U.S. Department of Agriculture, Office of the Secretary, *Saccharin under the Food and Drugs Act of June 30, 1906* (Washington, DC: U.S. Department of Agriculture, 1911). George McCabe to James Wilson, October 23, 1911, in RG 16, Records of the Office of the Secretary of Agriculture, General Correspondence, Box 37, Folder: Saccharin 1, NARA.

28. "Conclusions of the Referee Board of Consulting Scientific Experts on Saccharin," *American Food Journal* 6 (September 1911), 51. "Hearing of Saccharin to Manufacturers, November 22, 1911," in USDA, Office of the Secretary, *Saccharin under the Food and Drugs Act*, 18–90. For an earlier request, see "Saccharine Men Petition," *New York Times* (September 30, 1911), 22. RG 88, Records of the Food and Drug Administration, Records of the Board of Food and Drug Inspection Hearings, 1907–1915, Box 25, NARA.

29. Food Inspection Decisions 142 and 146. "Saccharin in Foodstuffs," *Atlanta Constitution* (December 31, 1911), C3. "Clearing the Atmosphere in the Saccharin

Controversy," *American Food Journal* 7 (January 1912): 16–17. "Secretary MacVeagh's Views on Saccharin," *American Food Journal* 7 (March 1912): 9.

30. "Decision in Saccharin Question," *Wall Street Journal* (March 1, 1912), 6. All the early food inspection decisions can be seen in Sherman, *Food Products*.

31. Monsanto advertisement and article titled "Saccharin Not Prohibited," *American Food Journal* 7 (April 1912): 29 and 17–18.

32. Ad for Jireh in *Atlanta Constitution* (May 4, 1916), 3. Ad for Genesee in *American Journal of the Medical Sciences* 165 (January 1923): A21. Chicago Dietetic Supply House, *Foods and Equipment for Use in the Control of Sugar and Starch Restricted Diets* (Chicago, 1929).

33. Francis Hamilton to James Wilson, October 15, 1912, and "Memo of Action," October 18, 1912, in RG 16, Records of the Secretary of Agriculture, General Correspondence, Box 67, NARA. "The Sugar Question," full-page ad in *St. Louis Post-Dispatch* (October 27, 1919): 17.

34. Carl Alsberg to the Secretary of Agriculture, August 19, 1914, in RG 16, Records of the Secretary of Agriculture, General Correspondence, Box 174, NARA. C. Alsberg to Senator E. S. Johnson, December 5, 1919, in RG 88, Box 54, NARA. For a typical warning, see *Yearbook of the United States Department of Agriculture for 1926* (Washington, DC: Government Printing Office, 1927), 324.

35. "Saccharine Case to Be Tried," *American Food Journal* 12 (March 1917): 127. For Alsberg's testimony, see *Congressional Record* 58 (1919): 7410–16. Joseph S. Davis, ed., *Carl Alsberg, Scientist at Large* (Stanford: Stanford University Press, 1948). James Harvey Young, "Saccharin: A Bitter Regulatory Controversy," in *Research in the Administration of Public Policy*, ed. Frank B. Evans and H. T. Pinkett (Washington, DC: Howard University Press, 1975), 38–49.

36. Mira Wilkins, *The History of Foreign Investments in the United States to 1914* (Cambridge, MA: Harvard University Press, 1989), 399–400.

37. Senate Select Committee on Agriculture and Forestry, *Use of Saccharin for the Relief of the Present Sugar Shortage. Hearing Pursuant to S. Res. 209*, 66th Cong., 1st sess., 1919.

38. "Saccharin Proposed as Substitute for Sugar," *Wall Street Journal* (October 29, 1919), 2. Monsanto ad in *Washington Post* (October 22, 1919), 9. "Saccharin Still Forbidden," *Christian Science Monitor* (October 24, 1918), 6.

39. "Label Statements Relating to Nonnutritive Constituents," *Federal Register* 6 (1941): 5926. George Larrick to Lee Drug Sales Co., April 17, 1947, RG 88, Box 963, NARA.

40. R. F. Kneeland Jr. to Dr. Herman Sharlit, January 30, 1950, RG 88, Box 1295, NARA. L. M. Beacham Jr. to Mr. Kirk, September 10, 1954, RG 88, Box 1847, NARA.

41. Thomas Lannen, "Letter to the Members of the National Manufacturers of Soda Water Flavors," probably from *National Bottlers' Gazette* (October 5, 1918); copy in Monsanto Papers, Products, Series 03, Box 03, Saccharin, Folder 2, Washington University Archives.

42. "North Dakota First," *Grand Forks Daily Herald* (October 15, 1903), 4.

43. "*State v. Empire Bottling Co.*, Supreme Court of Missouri," in *Southwestern Reporter (July 29–August 26, 1914)* (St. Paul: West Publishing Co., 1914): 1176–78. Monsanto ad in *American Food Journal* 10 (1915): 25.

44. *"The People of New York, Respondent, v. Excelsior Bottling Works, Inc., Appellant,"* in *Reports of Cases Heard and Determined in the Appellate Division of the Supreme Court of the State of New York* 184 (1919): 45–52. "Saccharin Wins in New York," *American Food Journal* 13 (August 1918): 434–35. "New York Court Upholds Use of Saccharin," *Carbonator and Bottler* (June 1928), copy in Monsanto Papers, Products, Series 03, Box 03, Saccharin, Folder 1, Washington University Archives.

45. *"Cott Beverage Corporation v. Horst, Supreme Court of Pennsylvania"* in *West's Atlantic Reporter,* 2nd Series, 110:405. "Pennsylvania High Court Bars Ban on Dietetic Soft Drinks," *Wall Street Journal* (January 7, 1955), 4. "Law, Science Offer Dietetic Paradox," *American Bottler* (February 1955): 19–20. "Cola Producers Enter New Field," *New York Times* (May 19, 1963), 7.

46. Food Nutrition Board, *The Safety of Artificial Sweeteners for Use in Food* (Washington, DC: National Academy of Sciences, National Research Council, 1955), 2–8.

47. Harvey Levenstein, *Paradox of Plenty: A Social History of Eating in Modern America* (New York: Oxford University Press, 1993), 113.

48. *U.S. Statutes at Large* 72 (1958): 1784–89. For the complete GRAS list, see *Federal Register* 24 (1959): 9368–69.

49. "Congressman Says Actress's Speech Helped Bar Cyclamates," *New York Times* (October 22, 1969), 26.

50. Food Nutrition Board, *Policy Statement on Artificial Sweeteners* (Washington, DC: National Academy of Sciences, 1962).

51. Congressman Pogue to Secretary Richardson, September 25, 1970, in RG 88, Box 4374, NARA.

52. "Saccharin Study Planned by F.D.A." *New York Times* (March 20, 1970), 35. "No Health Hazard Found in Saccharin," *New York Times* (July 23, 1970), 1.

53. "F.D.A. Removes Saccharin from List of Safe Foods," *New York Times* (January 29, 1972), 27. *Congressional Record* 118 (1972): 6712. Art Buchwald, "The Diet Connection," *Washington Post* (March 22, 1977), B1. Art Buchwald, "Sneak Preview: The Bionic Rat," *Los Angeles Times* (June 12, 1977), J2.

54. National Research Council, *Safety of Saccharin and Sodium Saccharin in the Human Diet* (Washington, DC: National Academy of Sciences, 1974).

55. "Saccharin and Its Salts," *Federal Register* 42 (1977): 19996–20010.

56. *U.S. Statutes at Large* 91 (1977): 1451–54. Paul M. Priebe and George B. Kauffman, "Making Governmental Policy under Conditions of Scientific Uncertainty: A Century of Controversy about Saccharin in Congress and the Laboratory," *Minerva* 18 (1980): 556–74.

57. "Food-Pesticide Overhaul Wins House Approval," *New York Times* (July 24, 1996), A18.

58. "Congress Clears Saccharin," *Food Engineering* 73 (March 2001): 16.

Chapter 13: Cyclamates

1. "Michael Sveda, the Inventor of Cyclamates, Dies at 87," *New York Times* (August 21, 1999), A11. Michael Sveda and Ludwig Audrieth, "Sulphamic Acid and Salts," U.S. Patent 2,275,125 (1942). Edward Matson, "The Sucaryl Story," *American Bottler* (April 1953): 40–41. Herman Kogan, *The Long White Line* (New York: Random House, 1963), 234–36.

2. B. J. Vos, "Memorandum of Interview," May 5, 1949, in RG 88, Food and Drug Administration, General Subject Files, Box 1174, NARA.

3. House Select Committee to Investigate the Use of Chemicals in Food Products, *Chemicals in Food Products*, 81st Cong., 2nd sess., 1951, 1:385–86. The latter Lehman quote appears in James S. Turner, *The Chemical Feast* (Washington, DC: Center for Study of Responsive Law, 1970), 6.

4. "Sucaryl Joins Saccharin, Sugar as New Sweetener," *Wall Street Journal* (May 25, 1950), 5. "New Sweetener Special-Diet Aid," *New York Times* (May 25, 1950), 34. "Chemist Finds Synthetic Sweet for Cooking Use," *Los Angeles Times* (May 25, 1950), 27.

5. Abbott Laboratories, *Annual Report* (1950–1953).

6. Food Nutrition Board, *Summary Statement on Artificial Sweeteners* (Washington, DC, 1954); there is a copy in RG 88, Box 1849, NARA.

7. "DuPont Producing Calcium Cyclamate," *Journal of Agricultural Food Chemistry* 3 (1955): 798. "Merck Makes Calcium Cyclamate," *Journal of Agricultural Food Chemistry* 3 (1955): 873. "Non-Caloric Sweetener Added to Chemists' Line," *Food Engineering* 55 (September 1955): 155–56.

8. Memo of meeting between George Meeks of FDA and Howard J. Cannon, special representative of the Chemical Sales Division of Abbott Laboratories, October 2, 1951, and J. K. Kirk to Multiphase Inc., June 8, 1951, both in RG 88, Box 1444, NARA.

9. "Agricultural and Food Newsletter," *Journal of Agricultural Food Chemistry* 2 (1954): 591. Food Nutrition Board, *The Safety of Artificial Sweeteners for Use in Food* (Washington, DC: National Academy of Sciences, National Research Council, 1955).

10. House Committee on the Judiciary, Subcommittee No. 2, *Hearings*, 92nd Cong., 1st sess., 1971, 182–92, for testimony of Anita Johnson. For an example, see the label for "Diet Delight Low Calorie Artificially Sweetened Bartlett Pears," in RG 88, Box 4357, NARA. "Waistline Whittlers," *Wall Street Journal* (November 27, 1953), 1.

11. "B-1 Introduces 'Trim' for Dietetic Market," *American Bottler* (April 1953): 16–18. "Battle of the Bulge," *Time* (August 10, 1953), 82.

12. Fred J. Helgren, assignor to Abbott Laboratories, "Sweetening Compositions and Method of Producing the Same," U.S. Patent 2,803,551 (1957). Ads in *Food Engineering* 27 (September 1955): 34ff; *National Geographic* 112 (August 1957): n.p. Recipe for "Luscious—Low Calorie—Charlotte Russe," *New York Amsterdam News* (September 11, 1954), 13. Myra Waldo, *The Slenderella Cook Book* (New York: Putnam, 1957), 13.

13. Ad for Cheer Freeze in *Washington Post* (December 8, 1957), A4. Ad for Diet-Delight Dietetic Foods in *Woman's Day* (February 1951): 93. Ad for Lucky Pop in *Los Angeles Times* (July 10, 1955), 64E.

14. "Soda Pop Surge," *Wall Street Journal* (July 15, 1953), 1. "Sugar-Free No-Cal Makes Sales History," *Tobacco and Confectionery Guide*, reprinted in *Congressional Record* 99 (1953): A2957. "Bubbling Along," *Time* (August 7, 1964), 97. "Low-Calorie Sweeteners," *Time* (January 3, 1969), 41. "Crisis in the Diet Market," *Time* (October 24, 1969), 97.

15. Sample ad shown in American Sugar Refining Co., *Annual Report for 1957*, 13. Savannah Sugar Refining Corp., *Annual Report for 1966*, 6.

16. "Artificial Sweeteners as Presently Used Termed Safe by FDA," *Wall Street Journal* (May 20, 1965), 7.

17. P. O. Nees and P. H. Derse, "Calcium Cyclamate Feeding Study," *Nature* 208 (1965): 81–82. "Sugar Industry Group Again Questions Use of Artificial Sweeteners," *Wall Street Journal* (October 4, 1965), 18.

18. Memorandum of meeting between Goddard and Kirk of FDA and Derse and Nees of WARF, October 30, 1967, RG 88, Box 4354, NARA.

19. W. David Gardner, "Bitter News about Sweeteners," *New Republic* (September 14, 1968), 17–18. W. David Gardner, "Safety of Sweeteners," *New Republic* (January 4, 1969), 13–14.

20. "Abbott Hits FDA's Cyclamate Proposal," *Chicago Tribune* (April 13, 1969), C17. Douglas Cray, "Battle over Sweeteners Turns Bitter," *New York Times* (June 1, 1969), F12. Robert S. Goodhart, *Cyclamate Sweeteners in the Human Diet: A Scientific Evaluation* (Chicago: Published for the Calorie Control Council by Abbott Laboratories, 1968).

21. "Sugar Substitute Brings a Warning," *New York Times* (December 14, 1968), 1 and 32.

22. "Food Additives: Cyclamic Acid and Its Salts," *Federal Register* 34 (April 5, 1969): 6194.

23. *Consumer Reports Buying Guide* (New York: Consumers Union, 1965), 156. Jack Anderson, "FDA Questions Sweetener Safety," *Washington Post* (August 30, 1967), B11. Food Protection Committee, *Interim Report on Non-nutritive Sweeteners* (Washington, DC: National Academy of Sciences, 1968).

24. Gaylord Nelson to the Department of Health, Education and Welfare, May 2, 1969, RG 88, Box 4226, NARA.

25. House Committee on Interstate and Foreign Commerce, Subcommittee on Public Health and Welfare, *FDA Consumer Protection Activities*, 91st Cong., 1st and 2nd sess., 1970, 165–223. M. S. Legator et al., "Cytogenetic Studies in Rats of Cyclohexylamine, a Metabolite of Cyclamate," *Science* 165 (September 12, 1969): 1139–40.

26. Morton Mintz, "Rise and Fall of Cyclamates," *Washington Post* (October 26, 1969), 1 and 16. "Finch Raps FDA on Sweeteners," *Washington Post* (October 8, 1969), A6.

27. Mintz, "Rise and Fall of Cyclamates." J. M. Price et al., "Bladder Tumors in Rats Fed Cyclohexylamine or High Doses of a Mixture of Cyclamate and Saccharin," *Science* 167 (1970): 1131–32.

28. *U.S. Statutes at Large* 72 (1958): 1784–89.

29. "Food and Drugs," *Federal Register* 34 (October 21, 1969), 17063–64. Finch's statement can be found in House Committee on Government Operations, Subcommittee, *Cyclamate Sweeteners*, 91st Cong., 1st sess., 1970, 27.

30. "Sweetness and Light?" *Barron's* (November 17, 1969), 1.

31. *Message from the President of the United States Transmitting Recommendations Concerning the Protection of the Interest of Consumers*, 91st Cong., 1st sess., 1969, H. Doc. 91–188.

32. Quoted in Mintz, "Rise and Fall of Cyclamates," 1.

33. "Business Newsletter," *Canner/Packer* (January 1970): 9. "New Drugs," *Federal Register* 34 (1969): 20426–27. "New Drugs," *Federal Register* 35 (1970): 6574 and 5008.

34. House Committee on Government Operations, Subcommittee, *Hearing on Cyclamate Sweeteners*, 91st Cong., 2nd sess., 1970.

35. House Committee on Government Operations, *Regulation of Cyclamate Sweeteners*, 91st Cong., 2nd sess., 1970, 14–15.

36. "Food and Drugs," *Federal Register* 35 (1970): 13644–45.

37. Robert Dickinson, "Memorandum of Telephone Conversation," October 23 and 26, 1970, RG 88, Box 4357, NARA.

38. "FDA Chief Concedes Error on Sweetener Regulation," *Washington Post* (May 4, 1971), A3.

39. House Committee on the Judiciary, Subcommittee No. 2, *Hearings. Congressional Record* 118 (1972): 24865–90, on 24866.

40. Ronald G. Wiegand, "Cyclamate," in *Sweeteners: Issues and Uncertainties* (Washington, DC: National Academy of Sciences, 1975), 177–80; see also Michael Sveda's comments on 180–81. "FDA Asks Cancer Institute to Evaluate Cyclamates' Disease-Causing Potential," *Wall Street Journal* (March 20, 1975), 40.

41. "Cyclamates Safe, U.S. Panel Feels," *Washington Post* (December 11, 1975), A21. "Cylamate Warning Refused by Panel," *Washington Post* (January 14, 1976), B5. "F.D.A. Refuses to Lift '69 Cyclamate Ban Because of 'Unanswered Questions' about Safety," *New York Times* (May 12, 1976), 16. National Cancer Institute, *Report of the Temporary Committee for the Review of Data on Carcinogenicity of Cyclamate* (Bethesda: National Cancer Institute, 1976).

42. "Abbott Laboratories Sues FDA to Reinstate the Use of Cyclamates," *Wall Street Journal* (September 11, 1979), 17. "Success Is Sweet, Depending on Rulings," *Chicago Tribune* (February 10, 1980), E3. "Abbott Labs Gives Up Its Fight for Cyclamates," *Chicago Tribune* (September 21, 1980), W5.

43. National Research Council, *Evaluation of Cyclamate for Carcinogenicity* (Washington, DC: National Academy Press, 1985), 2.

44. "Abbott's Sweet on Its Sucaryl," *Chicago Tribune* (January 20, 1975), C8.

Chapter 14: Aspartame and Sucralose

1. Alan G. Robinson and Sam Stern, *Corporate Creativity: How Innovation and Improvement Actually Happens* (San Francisco: Berrett-Koehler Publishers, 1997), 34–38. "Annual Meeting Briefs," *Wall Street Journal* (April 28, 1969), 15.

2. James M. Schlatter, "Peptide Sweetening Agents," U.S. Patent 3,492,131 (1970). Robert H. Mazur, Arthur Goldkamp, and James M. Schlatter, "Aspartic Acid Containing Dipeptide Lower Alkyl Esters," U.S. Patent 3,475,403 (1969). "Deerfield Scientist Discovers Possible Artificial Sweetener," *Chicago Tribune* (November 6, 1969), N13. "Searle Says Research on Artificial Sweetener Progressing Smoothly," *Wall Street Journal* (December 22, 1971), 24. "Canners Say Ban on All Cyclamates Is 'Severe Blow,'" *Wall Street Journal* (August 17, 1970), 3.

3. Jean Mayer, "Sweet but Sad," *Chicago Tribune* (November 16, 1972), A14. "G. D. Searle & Co. Is Boosting Outlays to Develop Sweetener," *Wall Street Journal* (September 6, 1972), 8. "Royal Crown Cola Says '73 Earnings Will Top Record Set Last Year," *Wall Street Journal* (October 22, 1973), 19. "Searle Plans Plant for New Sweetener," *Chicago Tribune* (April 30, 1973), C7.

4. "Searle Awaits Go-Ahead for Low-Calorie Sweetener," *Food Engineering* 46 (July 1974): 30. "Sour Response to New Sweetener," *Washington Post* (July 27, 1974), B3. "Dipeptide Sweetener Cleared for Limited Use," *Food Engineering* 46 (September 1974): 33. "Sweetener under Fire," *Washington Post* (September 24, 1974), B3.

5. "Aspartame Sweetener and Liver Damage," *Washington Post* (May 14, 1975), B5. "Aspartame Suspended," *Chicago Tribune* (December 5, 1975), C11. "FDA Unexpectedly Stays Searle's Right to Market Its New Artificial Sweetener," *Wall Street Journal*

(December 5, 1975), 8. "F.D.A. Bars Sales of a Sweetener," *New York Times* (December 5, 1975), 56.

6. Andrew Cockburn, *Rumsfeld: His Rise, Fall, and Catastrophic Legacy* (New York: Scribner, 2007), ch. 4, "How $weet It Is."

7. "Scientists Seek New Sweetener," *New York Times* (April 9, 1977), 41. "Report Says Rumsfeld to Get Top Post at Searle," *Chicago Tribune* (April 16, 1977), 7. "Rumsfeld Seeking a Turnaround for Searle," *New York Times* (January 16, 1978), D1–D2. "A Politician-Turned Executive Surveys Both Worlds," Donald Rumsfeld interview in *Fortune* (September 10, 1979), 88–94. "Searle Aiming High, Seeks Equal Chance," *Chicago Tribune* (April 30, 1981), C10.

8. "Searle Prods the FDA to Lift Suspension of Sweetener Product," *Wall Street Journal* (April 5, 1979), 16. "FDA Rejects Searle's Plea for Withdrawal of Stay on Aspartame," *Wall Street Journal* (May 8, 1979), 21. Cockburn, *Rumsfeld*. "Aspartame: Ruling on Objections and Notice of Hearing before a Public Board of Inquire," *Federal Register* 44 (1979): 31716–18.

9. "Searle's Battle to Introduce Aspartame, Artificial Sweetener, Enters Key Round," *Wall Street Journal* (January 28, 1980), 10. "Aspartame Wins a Nod," *Barron's* (February 4, 1980), 26. "Aspartame Opposed by FDA," *Chicago Tribune* (October 2, 1980), C8. "FDA Should Postpone Approval for Searle's Aspartame, Panel Says," *Wall Street Journal* (October 2, 1980), 22. Lewis D. Steginik and L. J. Filer Jr., *Aspartame: Physiology and Biochemistry* (New York: M. Dekker, 1984), preface.

10. "Searle Aiming High, Seeks Equal Chance." "Artificial Sweetener Wins FDA Approval," *Washington Post* (July 16, 1981), A10. "Searle Has FDA Approval to Sell Aspartame in U.S.," *Chicago Tribune* (July 16, 1981), C7. "Aspartame; Commissioner's Final Decision," *Federal Register* 46 (July 24, 1981): 38283–308. "Aspartame Approved Despite Risks," *Science* 213 (1981): 986–87.

11. "Food Additives Permitted for Direct Addition to Food for Human Consumption: Aspartame," *Federal Register* 48 (July 8, 1983): 31376–82. "Monsanto Says Its Aspartame Sweetener Is Cleared by FDA for Much Wider Uses," *Wall Street Journal* (November 26, 1986), 9. "NutraSweet Says Sweetener Cleared for Expanded Use," *Wall Street Journal* (June 8, 1988), 22.

12. "Monsanto's NutraSweet Unit Is Seeking Approval to Use Baked-Goods Sweetener," *Wall Street Journal* (October 23, 1987), 16. "Good News for Sweet Teeth," *Newsweek* (May 11, 1992), 69.

13. "Coke, Searle Sign Aspartame Pact," *Washington Post* (August 3, 1983), F1. "Metzenbaum: NutraSweet Inquiry Needed," *Washington Post* (February 7, 1986), B9.

14. Cockburn, *Rumsfeld*, ch. 4, "How $weet It Is," esp. 68–69. See also, "Monsanto's NutraSweet Co. Loses Aspartame Monopoly," *Wall Street Journal* (July 8, 1987), 28.

15. "Monsanto to Buy Searle for $2.7 Billion," *Chicago Tribune* (July 19, 1985), 1. Joseph E. McCann, *Sweet Success: How NutraSweet Created a Billion Dollar Business* (Homewood, IL: Business One Irwin, 1990).

16. *Aspartame Safety Issues: A Scientific Update* (Washington, DC: International Food Information Council, 1986).

17. "Metzenbaum: NutraSweet Inquiry Needed." Senate Committee on Labor and Human Resources, *"NutraSweeet"—Health and Safety Concerns*, 100th Cong., 2nd sess., 1988.

18. "Monsanto Selling Sweetener Ingredient Business," *New York Times* (March 28, 2000), C4. "Equal's Maker, Merisant Worldwide, Files for Chapter 11 Bankruptcy," *Chicago Tribune* (January 10, 2009), 17.

19. "Point of Purchase," *New York Times* (November 19, 2000), SM146.

20. Leslie Hough et al., "Sweeteners," U.S. Patent 4,435,440 (1984).

21. Sylvia Wilson, n/t, *New York Times* (February 12, 1987), D4.

22. "Food Additives Permitted for Direct Addition to Food for Human Consumption; Sucralose," *Federal Register* 63 (1998): 16417–33; 64 (1999): 43908–9.

23. "Sticky Success," *New York Times* (March 13, 2005), E30.

24. "Advertising," *New York Times* (October 26, 2000), C10.

25. "A Something among the Sweet Nothings," *New York Times* (December 22, 2004), C1. "Sweet Stand-ins," *Time* (December 30, 2004), 161. "Splenda, the Artificial Sweetener, Adds a Brown Sugar Blend," *New York Times* (April 4, 2005), C6.

26. "Splenda Settles Lawsuit over 'Sugar' Claim," MSNBC, May 11, 2007, www.msnbc.msn.com/id/18618557/ns/business-us_business (accessed December 3, 2010).

SELECTED BIBLIOGRAPHY

Conrad, Glenn R., and Ray F. Lucas. *White Gold: A Brief History of the Louisiana Sugar Industry, 1795–1995.* Lafayette: University of Southern Louisiana, 1995.

Deerr, Noël. *The History of Sugar.* London: Chapman and Hall, 1950.

Eichner, Alfred S. *The Emergence of Oligopoly: Sugar Refining as a Case Study.* Baltimore: Johns Hopkins University Press, 1969.

Heitmann, John A. *The Modernization of the Louisiana Sugar Industry, 1830–1910.* Baton Rouge: Louisiana State University Press, 1987.

Hollander, Gail M. *Raising Cane in the "Glades": The Global Sugar Trade and the Transformation of Florida.* Chicago: University of Chicago Press, 2008.

Horn, Tammy. *Bees in America: How the Honeybee Shaped a Nation.* Lexington: University Press of Kentucky, 2005.

Levenstein, Harvey. *Paradox of Plenty: A Social History of Eating in Modern America.* New York: Oxford University Press, 1993.

May, William John, Jr. *The Great Western Sugarlands.* New York: Garland, 1989.

McCann, Joseph E. *Sweet Success: How NutraSweet Created a Billion Dollar Business.* Homewood, IL: Business One Irwin, 1990.

Mintz, Sidney. *Sweetness and Power: The Place of Sugar in Modern History.* New York: Viking, 1985.

Peña, Carolyn de la. *Empty Pleasures: The Story of Artificial Sweeteners from Saccharin to Splenda.* Chapel Hill: University of North Carolina Press, 2010.

Sitterson, J. Carlyle. *Sugar Country: The Cane Sugar Industry in the South, 1753–1950.* Lexington: University of Kentucky Press, 1953.

INDEX

Abbott, Joseph F., 26
Abbott Laboratories, 195, *196*, 197, 198, 200, 202, 204, 206–7
Acarus sacchari, 20, *21*
Achard, Karl Franz, 85
Adams, John, 32, 35
additives, 41–42
adulteration: of cane syrup, 129, 131–32; Havemeyers & Elder and, 118; of honey, 173–74; of maple sugar and syrup, 142, 165–66; of molasses, 43; of sugar, 20, 24, 118–19
advertising, truth in, 138–39
A. E. Staley Manufacturing Company, 139
AFL. *See* American Federation of Labor
African American farmers, of cane syrup, 128
Agricultural Adjustment Act, 73–74
Alabama-Georgia Syrup Company, 128, 129, 132; Alaga, 128–29, *130*, 131–32
alcohol, industrial, 46. *See also* rum
alcoholism, 33–34
Allyn, L. B., 25
Alsberg, Carl, 188
American Beet Sugar Association, 96
American Beet Sugar Company, 95
American Bottlers of Carbonated Beverages, 184
American Chemical Society, 3, 19
American Crystal Sugar Company, *106*
American Federation of Labor (AFL), 62

American Glucose Company, 115–16, 118, 122
American Grape Sugar Company, 112, 115
American Maize Company, 136
American Molasses Company, 40, 44
American Revolution. *See* Revolution, American
American Sugar Cane League, 57, 63, 72
American Sugar Company, 29
American Sugar Growers' Society, 59
American Sugar Refining Company (ASRC): Crystal Domino Sugar and, 25; Cuba and, 28–29; employees of, 18, *18*; founding of, 12; law and, 14–17; Louisiana cane sugar *v.*, 59–60; as Sugar Trust, 11–12, 14, 59, 94, 97, 116; tariffs and, 13–14
Ames, Fisher, 32
Amolco, 39
Amstar, 19, 29–30, 141; Amstar-Holdings, 29
Anaheim Sugar Company, 103
Anderson, Jack, 202
Anderson, John A., 154–55
anhydrous dextrose, 134, 136
antislavery movement, 160–63
Army Corps of Engineers, 80–81, 83
artificial flavorings, maple syrup and, 165–66, 168
artificial rubber, 122
artificial sweeteners: consumption of, 1; side effects of, 3, 27; in soft drinks,

183–84, 190, 192, 197–98, 200, *201*,
211. *See also specific sweeteners*
aspartame: brands, 210–11; discovery
of, 3, 209; FDA and, 209–12; market
share of, 212
ASRC. *See* American Sugar Refining
Company
Aunt Dinah molasses, 40

Babbitt, Bruce, 82
Babst, Earl, 16, 25
Bache, Alexander Dallas, 53
bagasse, 47–48, 69
baked beans, 32, 41
Bantino, F. G., 180
Barbados, 31, 35–36, 85
Barker, George F., 117
Barlow, Joel, 85
Bayard, Nicholas, 6
Bebie, Jules, 182–83
Becnel, Thomas, 63
bees: Italian, 171–72; killer, 178–79;
pollination by, 179; symbolism of, 169
Bees in America (Horn), 169
beet sugar: as big business, 93–96, *95*;
in California, 94–96, 101–2; curly
top and, 104; in France, 85, 87–89;
in Germany, 90; immigrants and, 4,
99–102; industry, origin of, 2, 85–90;
irrigation and, 102–3; labor and, 4,
98–102, *100–101*; in late nineteenth
century, 91–93; leafhoppers and, 104;
in Massachusetts, 88–90; mechanized
cultivation of, 102; Mormons and, 90;
in Nebraska, 93–94, *95*, 96; seeds and,
103–4; selling of, 104–7, *105–6*; slaves
and, 4; tariff protection and, 96–98;
USDA and, 91–92, 102, 104; Wiley
and, 92, 96–97, 98, 103
Beet Sugar Convention, 93
Beet Sugar Society of Philadelphia, 87
Behr, Arno, 114, 121, 133–34
bellows smoker, 170
Benjamin, Judah P., 53
Benton, Frank, 175
Berle, Adoph A., Jr., 44

Berthelot, Marcellin, 109
betabeleros (sugar beet workers of
Mexican descent), 99–109, 102
Beverley, Robert, 160
big business: beet sugar as, 93–96, *95*;
molasses and, 44, *45*
Big Sugar (Wilkinson), 78
Biot, Jean-Baptiste, 22, 142
Bitting, Clarence, 72–74
blackstrap molasses, 44, 46
bladder cancer, 192–93, 204, 207
Blanchette, L.-J., 88
Blymer, David, 147
Board of Food and Drug Inspection, 41,
166, 186
boiler, sorghum, *156*
bone black, 8
Booke of Sweetmeats, 5
Boré, Étienne, 47
Bostick, J. E., 123
Bouchereau, Louis, 50, 55
Boudousquie, Antoine, 49, 50
Boudousquie, Charles, 50
Bourne, Benjamin, 70–71, 73
Boyle, Robert, 1, 159
Bracero Program. *See* Mexican Farm
Labor Program
Bradley, Mamie, 117
Brandes, E. W., 69
brands: aspartame, 210–11; cane syrup,
128–29, 132; cyclamate, 195–200, *196*,
199; dextrose, 134, *135*, 136; honey,
175; invert sugar, 142–43; liquid sugar,
144; maple sugar and syrup, 166–68;
molasses, 40; saccharin, 181–83;
sugar, 24–25, 30
Breaux, John, 79
Brer Rabbit molasses, 40
Brewer, John, 58
Broward, Napoleon Bonaparte, 68–69
Browne, Charles A., 20
Browne, Daniel Jay, 145, 164
Buchwald, Art, 192
Buel, Jesse, 87
Buffalo, glucose produced in, 111–14
Buffalo Grape Sugar Company, 112

Bufo marinus, 82
Bush's Maple Flavor No. 1617, 168

Cain, George, 202
Cairo, Georgia, 125, 127
California, beet sugar in, 94–96, 101–2
Camp Sorghum, 147
Canderel, 210
cane sugar: origin of, 2; USDA and, 51, 62, 68, 69–70. *See also* Florida, cane sugar in; Louisiana, cane sugar in
Cane Sugar Refiners' Association, 139
cane syrup: adulterated, 129, 131–32; African American farmers of, 128; brands, 128–29, 132; grading standard, 126–27; growth of, 127–28; as hobby, 132; ISCGA and, 125–27; labor and, 126, 128; origin of, 2, 123–24; scientists and, 129, 131; selling of, 128–29, *130*; USDA and, 124–25, 127, 131
carbonatation, 152, 154
Carey Act, 103
Cargill Inc., 141
Carlton, Doyle, 70
Carrington, M. L., 128
Carter, Susannah, 160
Cartier, Jacques, 159
Casamajor, Paul, 19
Casoid biscuits, 187
Castro, Fidel, 29
Catts, Sidney, 69
CCC. *See* Commodity Credit Corporation
Celotex, 69
centrifugal machines, 9, 170
Cerelose, 134, *135*, 136, 138
Chadbourne Plan of 1931, 73
Champaign Sugar Company, 149
Champion Molasses Gate, 39
Chandler, Alfred, 116
Chandler, Charles F., 19, 117
Chavez, Cesar, 77
The Chemical Feast (Turner), 206
chemistry: cane sugar in Louisiana and, 53–54; glucose and, 114–15

Chicago, glucose produced in, 114–15
Chicago Sugar Refining Company (CSRC), 114–15, 116, 121–22, 134
Child, David Lee, 89–90, 98
child labor, 99, *100*
Chino beet-sugar factory, 94–95
Chiozza, Luigi, 121–22
Church, Edward, 88–89
CIO. *See* Congress of Industrial Organizations
Clark, William Smith, 91
class, sugar consumption correlated with, 1–2, 32
Clay, Henry, 87
clayed sugar, 6
Cleveland, Grover, 14, 68, 152, 157
Clewiston Sugar House, 70, 73
Clinton, Bill, 82–83, 177
Clinton Corn Processing Company, 140
Clinton Corn Syrup Refining Company, 136
Clopper, Edward, 99
Clough, William, 146
Coca-Cola, 141, 184, 211
Cockburn, Andrew, 211–12
Cole, Cyrenus, 138
Collier, Peter, 147, 148–49, 150
Collinses, 28
Colman, Norman, 152, 154, 155
Colonial Sugar Company, 63
colony collapse disorder, 176
Columbus, Christopher, 5
Colwell Iron Company, 148
comb foundation, 171, *172*
commercially processed foods, 2. *See also* packaged foods
commercial refiners, 5–6
Commodity Credit Corporation (CCC), 177
Congress of Industrial Organizations (CIO), 62
conservation, in Florida, 80–83
Continental Congress, 37
Convention of the Representatives of the Louisiana Protected Industries, 59
Convertit, 143

Cook, Nevis, 206
Coolidge, Calvin, 138
Cooper, Rachel, 128
cooperatives, honey, 174–75
coopers, 17
corn: by-products, 121–22; glucose from, 3, 110–11, 117–21, 133; lobby, 121; truth in advertising and, 138–39. *See also* corn syrup; dextrose
Corn Products Company, 116
Corn Products Refining Company (CPRC), 116–17, 119–20, 121, 134, 136, 138–39, 140, 141
Corn Refiners Association, 142
Corn Sweeteners Inc., 140
corn syrup: consumption, 133, 142; high-fructose, 3, 139–42; Karo, 119–21; origin of, 3, 111; sorghum *v.*, 121; Wiley and, 120
Cott Beverage Corporation of Connecticut, 190
CPRC. *See* Corn Products Refining Company
Crevecoeur, Jean, 169
Crookes, William, 181
Crystal Domino Sugar, 24–25
CSRC. *See* Chicago Sugar Refining Company
Cuba: overproduction and, 73; reciprocity treaty with, 97; refining and, 28–29. *See also* Revolution, Cuban
Cummins, Albert, 138
curly top (disease), 104
cutters, West Indian, 76, 77–78
cyclamates: banned, 192, 205; brands of, 195–200, *196*, *199*; Delaney Clause and, 204; discovery of, 3, 195; FDA and, 195, 197, 200–207; GRAS, 191; health effects of, 200–207
Cyclamate Sweeteners in the Human Diet (Abbott Laboratories), 202
Cylan, 197, 198

Dahlberg, Bror E., 69–72
Dahlberg Sugar Cane Industries, 71
Dailey, Josiah, 168

Dairy and Food Commission of Ohio, 165
Daniels, Edward, 127
Delaney, James, 191
Delaney Clause: cyclamates and, 204; saccharin and, 191–94, *193*
demerara sugar, 9, 11
Denys, Nicolas, 160
Derse, Philip, 200
Deseret Manufacturing Company, 90
dextrose: anhydrous, 134, 136; brands, 134, *135*, 136; development of, 133–36, *135*; HFCS and, 139–42; meaning of, 248n1; packaged foods and, 136–38, *137*; in Splenda, 213; taste of, 136; truth in advertising of, 138–39; uses of, 133
d-glucose, 248n1
Diet Delight Foods, 198, *199*
diffusion process, 92, 151, 152
Dingley Tariff, 59, 97, 182
Disston, Hamilton, 67–68
Dole, Bob, 80
Domino brand, 24–25, 30
Dorsey, George Washington Emery, 94
double-refined sugar, 6, 11
Douglas, Marjorie Stoneman, 81
Downing, Emmanuel, 34
Drawbaugh, Daniel, 39
DuBois, Gaston, 182
Dubrunfaut, Augustin, 142
Dudley, Paul, 160, 169
Dumas, J. B., 109
DuPont, 195, 197, 198
Dutch Color Standard, 22–24, 126
Dyer, E. H., 92

Eagle Brand granulated sugar, 24
Ebert, Charles, 134
Edwards, Jonathan, 160
Eisenhower, Dwight D., 29, 63, 191
Elder, George, 7, 30
Elder, Louisine, 30
Elkins Act, 15
Elm Hall Plantation, 50–51, 54, 56, 57, 58, 60

Empire Bottling Company, 190
Enterprise Manufacturing Company, 39
The Everglades (Douglas), 81
Everglades, draining of, 67–69, 80–83
Everglades National Park, 81–82
Excelsior Bottling Works, 190
Experiments in the Culture of the Sugar Beet in Nebraska (Nicholson and Lloyd), 93

factory towns, 60–61, *61*
Fahlberg, Constantine, 181–82, 183
Fahlberg, List & Company, 181
Fair Packaging and Labeling Act, 139
Falkiner, Ralph S., 71
Fanjul, Alfonso, 30, 77
Fanjul, Alfonso, Jr., 77, 80, 82
Fanjul, J. Pepe, 30, 77, 80
Farmers Co-operative Cane Syrup Association of Georgia, 128
FDA. *See* Food and Drug Administration
Finch, Robert, 203–5
Firmenich, Joseph, 113, 115
Fleischmann, Charles Lewis, 86, 90
Florida, cane sugar in: conservation *v.*, 80–83; Dahlberg and, 69–72; Disston and, 67–68; Everglades drained for, 67–69, 80–83; government programs supporting, 79–80; immigrants and, 76–78; labor conditions in, 77–78; origin of, 67; tariffs and, 72; unions and, 77; U.S. Sugar Corporation and, 72–76, *75*, 79, 82
Florida Sugar and Food Products Company, 69
Florida Sugar Cane League, 76
Flo-Sun, 77, 82
Flo Sweet, 144
fluorine, 122
Foley, Mark, 80
Food, Drug, and Cosmetic Act of 1938, 189, 191
Food Additives Amendment, 191
Food and Drug Administration (FDA): aspartame and, 209–12; corn lobby and, 121; cyclamates and, 195, 197,
200–207; origin of, 41; saccharin and, 180, 189, 191–93; sucralose and, 213
Food Protection Committee, 191
Fort Scott, 154–55
Fox, Arthur W., 111–12
France, beet sugar in, 85, 87–89
Frankfurter, Felix, 15
Franklin, Benjamin, 163–64, 169–70
Franklin Sugar Company, 152
Franklin Sugar Refinery, 24
fraud, tariffs and, 13
French and Indian Wars, 36
fructose: glucose as isomer of, 140; meaning of, 248n1; vacuum pan and, 8
The Frugal Housewife (Carter), 160
Fulton Sugar Refining Company, 94

G. D. Searle & Company, 209–12
generally recognized as safe (GRAS), 191–92, 200, 204
Germany, beet sugar in, 90
Gerry, Elbridge, 32
Gilchrist, Albert, 69
Gilmore, J. Y., 50
glucose: in Buffalo, 111–14; chemistry and, 114–15; in Chicago, 114–15; consumption, 142; from corn, 3, 110–11, 117–21, 133; dangers of producing, 113–14, 115; fructose as isomer of, 140; from grapes, 111–12; health and, 117–18; horizontal and vertical integration of, 115–17; immigrants and, 115; meaning of, 248n1; origin of, 109; patents and, 110–11, 121; sucrose mixed with, 118; taste of, 109; vacuum pan and, 8. *See also* corn syrup
Glucose Sugar Refining Company, 116, 122
Glucose Trust. *See* Glucose Sugar Refining Company
Godchaux, Charles, 49, 58
Godchaux family, 48–53, *54*, 55–59, 60, 65
Godchaux, Jules, 49, 59
Godchaux, Leon, 48–53, 57, 59, 61

Godchaux, Leon, II, 63
Godchaux Sugars Inc., 60–65, *61, 64*
Goessling, Frederick W., 110–11, 112
Goessmann, Charles A., 91, 149
Goldberger, Joseph, 33
Gomez-Mena, Lillian, 77
Goodhue, Benjamin, 32, 38
Gore, Al, 82
Grand Island, Nebraska, 93–94, *95*
Grandma's Old Fashioned Molasses, 40
granulated sugar: Eagle Brand, 24; introduction of, 11
granulators, 11
grapes, glucose from, 111–12
GRAS. *See* generally recognized as safe
Gulf-to-Atlantic Canal, 69
Guthrie, Samuel, 110

Hall, Edward, 62
Hamilton, Alexander, 38
Hamlin, Cicero, 112–14, 115, 122
Hand, Learned, 117
Harding, Warren G., 138
Hardwick, Thomas, 16
Harkness, William, 151
harvesters, 71, *72*, 78–79
Hassall, Arthur Hill, 20
Hatch Act of 1887, 91, 123
Hatch, William, 155
Hauser, Gayelord, 46
Havemeyer, Frederick, 7
Havemeyer, Frederick C., Jr., 7
Havemeyer, George, 17
Havemeyer Hall, 19
Havemeyer, Henry Osborne, 8, 16, 28, 30, 114
Havemeyer, Theodore Augustus, 8–9
Havemeyer, William, 7, 23
Havemeyers & Elder: adulteration and, 118; brands, 24–25; in New York, 7–11, *9–10*, 15, 17, 19, 20
Hayes, Arthur Hull, 211
Hayes, Augustus Allen, 146
Hayes, Rutherford B., 23
Hedges, Isaac, 146, 152
Heflin, Howell, 211

Heitmann, J. A., 149
Hersey, Charles, 11
Hersey Manufacturing Company, 11
Heyden Sugar, Crystallose, and Garantose, 182, 183, 185
HFCS. *See* high-fructose corn syrup
high-fructose corn syrup (HFCS): dextrose and, 139–42; origin of, 3; in soft drinks, 141
Hine, Lewis, 99
The Hive and the Honeybee (Langstroth), 173
H. K. & F. B. Thurber, 174
honey: adulterated, 173–74; brands, 175; cooperatives, 174–75; federal scientific and technical support for, 175–76; federal support for business of, 176–79; in industry, 179–80; industry, origin of, 2; in New England colonies, 1, 169; pollination and, 179; selling of, 174–75; technology for, 170–71, *171–72*; USDA and, 175–78, 179; Wiley and, 174
Hoover, Herbert, 80
Hopkins, Samuel, 160, 162
Hopkins, Stephen, 37
horizontal integration, of glucose, 115–17
Horn, Tammy, 169
Hough, Leslie, 213
Howard, Edward Charles, 8
Howells, William Dean, 41
Hruschka, F. E., 170
Hunt Brothers, 65
Hyde, Arthur, 138

immigrants, labor by: beet sugar and, 4, 99–102; *betabeleros*, 99–100, 102; in Florida, 76–78; glucose and, 115; Italian immigrants, 126; Japanese immigrants, 101; refinery work by, 4, 17–18; Russian immigrants of German descent, 99
industrial alcohol, 46
Internal Improvement Fund, 67
Interstate Sugar Cane Growers Association (ISCGA), 125–27

invertase, 131, 143
invert sugar, 142–43, 174
irrigation, beet sugar and, 102–3
ISCGA. *See* Interstate Sugar Cane Growers Association
Islam, sugar cane spread by, 5
Isnard, Maximin, 88
Isoclear, 141
Isomerose, 140
Isosweet, 140
Italian bee, 171–72
Italian immigrants, 126

Jackson, Richard F., 134
Japanese immigrants, 101
Jarrett, Jacqueline, 203–4
Java, 69, 73
J. C. Hubinger Brothers, 140
Jebb, Thomas, 112–13
Jebb, William, 112–13
Jefferson, Thomas, 33, 162–63
Jennings, William, 68
Jones-Costigan amendment, 73–74, 79, 99
Joselyn, John, 34
Joy, Charles A., 111
J. S. Lovering & Company, 22
Judd, Orange, 111

Kansas, sorghum in, 152–56
Karo Corn Syrup, 119–21
killer bees, 178–79
King, Martin Luther, Jr., 129
Kirchhoff, Gottlieb, 109
Kissimmee River, 81, 82
Klein, G. L., 54
Klock, Joe, 79
Koenig, Harry, 93
Kolischer, Theodore, 148
Kooi, Earl, 140

labor: beet sugar and, 4, 98–102, *100–101*; cane sugar in Louisiana and, 54–57, *55–56*; cane syrup and, 126, 128; child, 99, *100*; conditions, in Florida, 77–78; by slaves, 3–4, 54–55, *55*; sugar

and, 3–4; temporary, 75–78. *See also* immigrants, labor by
Ladd, Edwin Freemont, 190
Lafayette Sorghum Sugar Refinery, 150
Lake Okeechobee, 68, 69, 74, 76, 80–81
Lamm, Justine, 49
Langstroth, Lorenzo, 170, 173
leafhopper, 104
LeDuc, William, 147–48
Lehman, Arnold, 195
Leon Godchaux Company Ltd., 58–59
Le Ray de Chaumont, James, 87–88
Lescarbot, Marc, 160
Letters from an American Farmer (Crevecoeur), 169
Levenstein, Harvey, 191
levulose, 248n1
Ley, Herbert, Jr., 203, 204–5
l-glucose, 248n1
Liebig, Justus, 149
Lincoln, Abraham, 146, 166
Linnaeus, Carl, 164
Liquid Carbonic Company, 184
liquid sugar, 143–44
Lloyd, Rachel, 93
Lobo, Julio, 65
Lo-Cal Dietetic Beverages, 189
Lodge, Henry Cabot, 97
Loeb, William, 15, 16
Log Cabin Maple Syrup, 166–68
Look Younger, Live Longer (Hauser), 46
Loring, George B., 148, 149, 150, 152
Louisiana, cane sugar in: ASRC *v.*, 59–60; chemistry and, 53–54; Godchaux, Leon, and, 48–53, 57, 59, 61; Godchaux Sugars Inc. and, 60–65, *61, 64*; labor and, 54–57, *55–56*; Leon Godchaux Company Ltd. and, 58–59; origin of, 47–48; quality control and, 53–54; slaves and, 54–55, *55*; sugar centrals in, 57–58; tariffs/trusts and, 59–60
Louisiana Planter, 51, 52, 56–57, 58, 59–60, 61, 126
Louisiana Purchase, 2, 47
Louisiana Sugar Chemists' Association, 54

Louisiana Sugar Experiment Station, 124
Louisiana Sugar Planters' Association
(LSPA), 50–51, 59
Loxahatchee Wildlife Refuge, 82
LSPA. *See* Louisiana Sugar Planters'
Association
Lucky Pop, 198
Lutz & Movius, 182

Madewood Plantation, 52
Madison, Dolley, 85
Madison, James, 35
Maines, Rachel, 132
maltodextrin, in Splenda, 213
Mapleine, 168
maple sugar and syrup: adulterated,
165–66; antislavery movement
and, 160–63; artificial flavorings
and, 165–66, 168; blends, 165–66,
167; brands, 166–68; in colonies, 1;
industry, origin of, 2; major producers
of, 164–65; money and, 163–65;
Native Americans and, 159–60, *161*;
sap gathered for, *161*, *163*; taste of,
164–65; USDA and, 165, 168
Mapletone, 168
Margraff, Andreas, 85
Marie Earle skin cream, 180
marketing, of molasses, 39–40
Marshall, Richard, 140
Marshalltown, Iowa, 113
Martin, John, 70
Mas, Ernest, 122
Massachusetts, beet sugar in, 88–90
Massachusetts Institute of Technology,
26, 44
Massachusetts Society for Promoting
Agriculture, 88, 90
Mather, Cotton, 33–34
Mather, Increase, 33
Matthiessen, Conrad H., 116
Matthiessen, Erhard, 121
Matthiessen, Franz O., 114, 116, 133–34
Mayer, Jean, 205
Mays, Willy, 129, *130*
McCabe, George, 187, 188

McCarthy, Colman, 78
McCarthy, Irene, 107
McCosh, James, 30
McCulloh, Richard, 53, 90
McCusker, John, 33
McDermott v. Wisconsin, 121
McKinley Tariff of 1890, 13, 51–52, 156
McKinley, William, 92
McLane, Louis, 13
McMurtrie, William, 24, 147
McNeil Specialty Products Company,
213–14
Merck, 182, 197
Merisant, 212, 214
Metzenbaum, Howard, 212
Mexican Farm Labor Program, 102
Miller, William Allen, 109
minimum-wage law, 62
Mintz, Sidney, 2
Mississippi Valley Cane Growers'
Association (MVCGA), 150, 151, 152
molasses: adulterated, 43; big business
and, 44, *45*; blackstrap, 44, 46; brands,
40; class and, 2, 32; early uses of,
31–33; federal government and,
40–43; marketing of, 39–40; natural
maple extract, 168; New Orleans,
43, 44; packaged foods and, 40–43;
production, 5; rum and, 31, 33–38;
slaves eating, 33; tariff and, 36–37,
38–39; transportation of, 39–40;
USDA and, 33, 43
Molasses Act, 36
molds. *See* sugar, molds
Möller, Peter, 9
Möller, William, 24
Monsanto Chemical Works, 182–84,
188–89, 190, 197, 211–12
moonshine, 136
Mormons, beet sugar and, 90
Morrill Land Grant Act of 1862, 91
Morton, J. Sterling, 157
Mott, Charles Stewart, 72, 79
movable frame hive, 170, *171*
Muench, Carl, 71
Müller, Hermann, 179

muscovado. *See* raw sugar
MVCGA. *See* Mississippi Valley Cane
Growers' Association
Mysteries of Bee-Keeping Explained
(Quinby), 173

Nabisco, 179–80
Nagel, Charles, 187
Napoleon Bonaparte, 85
National Academy of Sciences, 117, 148,
191–93, 202, 204
National Agricultural Workers Union,
62–63
National Bureau of Standards, 24
National Cancer Institute, 204, 207
National Glucose and Grape Sugar
Association, 116, 117
National Honey Board, 178, 180
National Research Council, 178–79, 207
National Sugar Refinery of New York, 65
Native Americans: alcoholism and, 34;
maple sugar and, 159–60, *161*
Nebraska, beet sugar in, 93–94, 96, *95*
Nees, P. H., 200
Negro Labor Committee, 63
Nelson, Gaylord, 203
New Deal programs, 73
New England colonies: honey in, 1, 169;
maple sugar and syrup in, 1; rum
produced in, 34–35, 38–39; trade with
West Indies, 1, 31–32
Newkirk, William, 134
Newlands Act of 1902, 103
New Orleans molasses, 43, 44
New York, maple sugar and syrup in,
164–65
New York, sugar refining in: end of era of,
29–30; by Havemeyers & Elder, 7–11,
9–10, 15, 17, 19, 20; philanthropy and,
30; preeminence of, 7; refinery work
and, 17–19, *18*; science and, 19–20,
21; sugar brands and, 24–25, 30; Sugar
Institute and, 25–27; Sugar Research
Foundation and, 25–27, *27*; tariffs
and, 12–14, 22–24. *See also* American
Sugar Refining Company

New York Central Railroad, 15
New York Steam Sugar Refinery, 19
New York Sugar Trade Laboratory, 20
Nichol, Robert, 20
Nicholson, Hudson, 93
Nixon, Richard M., 205
No-Cal, 198, 200, *201*
Nolen, John, 70
Northampton Beet Sugar Company,
88–89
Notice on the Beet Sugar (Church), 88
Novelty, 39
Nulomoline, 142–43
NutraSweet, 211–12, 214

obesity, 136, 142, 189, 198
Ockershausen, Adolphus, 111
Oglethorpe, James, 34
Ohio State Board of Sorgo Culture, 146
Old Colony Distillery, 46
Olney, John, 210
On the Culture of the Sugar Beet (Child),
90
Orphan Drug Act, 211
Otis, James, 36
Oxnard, Henry T., 93–96, 100

packaged foods: dextrose and, 136–38,
137; molasses and, 40–43; saccharin
and, 183, 190
Paddock, Algernon, 94
Palma, Tomás Estrada, 28
Palmer, Truman, 97
Park, Frank, 127
Parkinson, William, 153, 154, 155
Parkinson Sugar Works, 153–54, 157
Parr, Richard, 15
patents, glucose and, 110–11, 121
Patout, 65
pecan pie, 120
Pedder, James, 87, 88
pellagra, 33
Penick & Ford, 40, 41, 43, 44
Pennsylvania Sugar Company, 69
Penn, William, 34
Pepper, Claude, 74

PepsiCo, 141
Pfund, A. H., 176; Pfund color grader, 176
Phadnis, Shashikant, 213
philanthropy, by refiners, 30
Phillips, Everett Franklin, 175, 176, 177, 179
Pike, Benjamin, Jr., 22
Pitt, William, 36
plantations: Elm Hall, 50, 51, 54, 56, 57, 58, 60; Madewood, 52; motorized, 71; Raceland, 50, 51, 53, 54, 55, 57–58, 60, 65; Reserve, 49–50, 52, 57, 59, 60, 61, 65; Runnymede, 68; slaves on, 3–4; Souvenir, 50; St. Cloud, 68; in West Indies, 3, 5, 48
plantation white sugar, 9, 11
plantation yellow sugar, 9, 11
Plumb, Preston, 152, 154–55, 156
Poage, William R., 192
"poison squad," 41–42
polariscopes, 22–24
pollination, bees and, 179
Poor Richard Improved (Franklin), 163–64
Porst, Christian, 134
Postum, 40–41
powdered sugar, 7
Practical Treatise on the Manufacture of Starch, Glucose, Starch Sugar and Dextrine, 111
Prince, John, 88
Prohibition, 39, 136
Proofstation Oost Java, 69
Proxmire, William, 210
Pugh, Thomas, 52
Pure Food and Drugs Act, 41, 120, 167, 174, 185
Purse, Daniel Gurgel, 124, 127

quality control, cane sugar in Louisiana and, 53–54
Queeny, John F., 182
Quinby, Moses, 170, 173

Raceland Plantation, 50, 51, 53, 54, 55, 57–58, 60, 65
Randolph, John Fitz, 86–87

rationing, sugar, 26, 27, 177
raw sugar: class and, 2; instructions for refining, 5
Ray, John, 159
Reagan, Ronald, 79, 177, 211
Reconstruction Finance Corporation, 73
Referee Board of Consulting Scientific Experts, 43, 186
Refined Sugar & Syrups, 144
refinery work: dangers of, 17; by immigrants, 4, 17–18; in New York, 17–19, 18
refining. See sugar refining
Refining Sugar from Beets (Blanchette), 88
Remonstrance against the Sugar Act, 37
Remsen, Ira, 43, 117, 181–82, 186
Report . . . on the Culture, in France, of the Beet Root (Pedder), 87
Report on Manufactures (Hamilton), 38
Report on Methods of Sugar Analysis (Louisiana Sugar Chemists' Association), 54
Reserve Plantation, 49–50, 52, 57, 59, 60, 61, 65
Revere, Paul, 35
Revolution, American: rum and, 35–38; sugar tariff after, 12
Revolution, Cuban: of 1898, 28; of 1959, 76
Reynolds, R. J., 183
Richardson, Elliot, 192
Rillieux, Norbert, 58; Rillieux multiple-effect evaporation system, 58
Rivers and Harbors Act, 80
Roddenbery family, 125–26, 127, 128, 129, 132
Roddenbery, Julian, 127
Roddenbery, Seaborn A., Jr., 127
Roddenbery, Walter B., 125–26, 129
Rodrigue, John, 54, 55
Ronaldson, James, 87
Roosevelt, Theodore, 15, 16, 28, 43, 92, 120, 185–86
Rose, Rufus, 126
Ross, Bennett Battell, 123
royal jelly, 180
rubber, artificial, 122

rum: American Revolution and, 35–38; line, 39; molasses and, 31, 33–38; production, in northern colonies, 34–35, 38–39; slaves and, 34
Rumsfeld, Donald, 210, 211–12
Runnymede Plantation, 68
Rush, Benjamin, 162, 164
Rusk, Jeremiah, 155
Russian immigrants, of German descent, 99
Ryan, Thomas, 155

saccharimeter, 22, 53
saccharin: analysis of, 22; brands, 181–83; Delaney Clause and, 191–94, *193*; discovery of, 3; FDA and, 189, 191–93; federal government and, 185–89; GRAS, 191–92; health effects of, 185–89, 190–94; industrial uses of, 183–85; market share of, 212; origin of, 181–83; packaged foods and, 183, 190; proposed ban of, *193*, 193–94; at state level, 190–91; USDA and, 185; Wiley and, 184, 185–86, 188
Saccharin Study and Labeling Act, 195–94
Saccharin Warning Elimination via Environmental Testing Employing Science and Technology (SWEETEST) Act, 194
Sacrinpak, 183
samples, sugar, 15, 16
Schlatter, James, 209
Schmitz, John, 192
science: cane syrup and, 129, 131; refining in New York and, 19–20, *21*; sugar refining benefiting from, 2–3
Scovell, Melville Amasa, 149, 152
Searle, Daniel, 209
seeds, sugar beet, 103–4
Serres, Olivier de, 85
Sherman Antitrust Act, 15, 16–17, 26
Sherman, James S., 185
Sherman, John, 23
Silliman, Benjamin, 13, 47–48, 53, 54, 110
Silliman, Benjamin, Jr., 148–49
Silvester, Giles, 33

simple sugar, 248n1
Sioux Honey Association, 175
slaves: antislavery movement and, 160–63; beet sugar and, 4; cane sugar in Louisiana and, 54–55, *55*; emancipation of, 48–49; labor by, 3–4, 54–55, *55*; molasses eaten by, 33; rum and, 34; on sugar plantations, 3–4
Slenderella, 198
Smithson, James, 86
Smoot-Hawley Tariff of 1930, 72, 98
Smoot, Reed, 97–98
Snider, Jacob, Jr., 87
Society for Promoting the Manufacture of Sugar from the Sugar Maple Tree, 162
Society for Promoting Trade and Commerce within the Province, 36
soda. *See* soft drinks
sodium benzoate, 41–42, 186
soft drinks: artificial sweeteners in, 183–84, 190, 192, 197–98, 200, *201*, 211; HFCS in, 141
Soleil, Jean Baptist François, 22
Soleil-Scheibler polariscopes, 23
sorghum: boiler, *156*; Collier and, 147–50; corn syrup *v.*, 121; in Kansas, 152–56; LeDuc and, 147–48; origin of, 145–46; public-private partnerships bringing, 2; syrup, 145, 146; USDA and, 146–52, 156; Wiley and, 150–54, 155–56
The Sorghum Hand Book (Blymer), 147
Southern Industries, 65
Southern Sugar Company, 69–71
Souvenir Plantation, 50
Spanish-American War, 28, 185
Spencer, Guilford, 92
Splenda: ingredients of, 213–14; market share of, 212
Spreckels, Claus, 28, 96
Spreckels Sugar Company, 29, 104
starch sugar, 109–15, 118. *See also* glucose
Starke, Reinhold, 107
St. Cloud Sugar Plantation, 68
Stewart, Francis L., 147
Stimson, Henry L., 15–16
Stockbridge, Horace, 123–24

Stow, J. M. E., 54
Stowe, Harriet Beecher, 40
strikes, 63, *64*, 65
Stuart, Alexander, 8, 19, 30
Stuart, Robert, 8, 19, 30
Stubbs, William Carter, 124
Studniczka, Henry, 54
Sucaryl, 183, 195, *196*, 197–98, *199*
sucralose: discovery of, 3; FDA and, 213; in Splenda, 213–14
sucrose: consumption, 142; glucose mixed with, 118; taste of, 109, 136
sugar: adulterated, 20, 24, 118–19; analysis, tariff and, 22–24; beetle, 20, *21*; brands, 24–25, 30; centrals, 57–58; in commercially processed foods, 2; consumption, 1–2, 25–26, 32, 133, 142; cubes, 24–25; introduction of, 1–2; as investment, 4; labor and, 3–4; molds, 8–9; overproduction of, 73; rationing, 26, *27*, 177; samples, 15, 16. *See also specific types of sugar*
Sugar Act of 1764, 12, 37
Sugar Act of 1937, 62, 74
Sugar Association, 27, 141, 200, 214
Sugar at a Glance, 97
sugar bakers. *See* commercial refiners
The Sugar Beet (Ware), 92
The Sugar Beet Industry (Wiley), 92
sugar beets. *See* beet sugar
sugar cane, cultivation of, 5. *See also* cane sugar
Sugar Cane Growers Cooperative of Florida, 76
Sugar Information Inc., 27, 200
Sugar Institute, 25–27
Sugar Palace, 94, *95*
Sugar Refineries Company, 11–12
Sugar Refinery Workers Union, 18
sugar refining: commercial, 5–6; Cuba and, 28–29; double, 6, 11; of raw sugar, instructions for, 5; scientific research and analysis benefiting, 2–3; USDA and, 20. *See also* New York, sugar refining in; refinery work
Sugar Research Foundation, 25–27, *27*, 192, 200

Sugar Trust. *See* American Sugar Refining Company
sulfur, 42–43, 129, 131
Sullivan, Leonor, 207
Sulz, Charles Herman, 183
Sveda, Michael, 195
Swamp Lands Act of 1850, 67
Swanson, Gloria, 191
Sweeta, 183
SWEETEST Act. *See* Saccharin Warning Elimination via Environmental Testing Employing Science and Technology Act
Sweet'N Low, 183, 212
Sweetose, 139
Swenson, Magnus, 150, 153, 154–55, 157

Tadman, Michael, 54
Taft, William Howard, 58–59, 92
Takasaki, Yoshiyuki, 140
Talisman Sugar Corporation, 76
Tamiami Trail, 69
Tanabe, Osamu, 140
Tariff of 1870, 13
tariffs: of 1870, 13; after American Revolution, 12; ASRC and, 13, 14; beet sugar and, 96–98; Dingley Tariff, 59, 97, 182; Florida cane sugar and, 72; fraud and, 13; Louisiana cane sugar and, 59–60; LSPA and, 51; McKinley Tariff of 1890, 13, 51–52, 156; molasses, 36–39; origin of, 1–2, 12; quality control and, 53; refining in New York and, 12–14, 22–24; Smoot-Hawley Tariff of 1930, 72, 98; sugar analysis and, 22–24
Tate & Lyle, 18–19, 30, 141, 213, 214
Taussig, Charles William, 44
Taussig, Noah, 44, 142
Taylor, Fred, 90
Taylor, John, 90
temporary workers, 75–78
Thevet, André, 159
Thibodaux Massacre, 55
Thomas, Norman, 63
Thurmond, Strom, 179

Till, Emmett, 117
Tilley, Nannie, 183
Towle Maple Syrup Company, 166–67
Towle, Patrick J., 166–68
transportation, of molasses, 39–40
triangle trade, 34
Trubek, Moses, 54
trusts, cane sugar in Louisiana and, 59–60
Tugwell, Rexford, 44
Tupper, Ellen S., 173
turbinado sugar, 9, 11
Turner, James S., 206, 210

Uncle Tom's Cabin (Stowe), 40
unions, 18, 62–63, *64*, 65, 77
United Farm Workers, 77
United Packinghouse Workers Union, 63
United States v. E. C. Knight, 15
Universal Exposition (Paris, 1867), 91
U.S. Beet Sugar Association, 97, 102, 104, 139
U.S. Department of Agriculture (USDA): beet sugar and, 91–92, 102, 104; cane sugar and, 51, 62, 68, 69–70; cane syrup and, 124–25, 127, 131; honey and, 175–78, 179; invert sugar and, 143; maple sugar, syrup and, 165, 168; molasses and, 33, 43; refining and, 20; saccharin and, 185; sorghum and, 146–52, 156; Wiley at, 43, 68, 92, 120, 121, 124, 138, 165, 174, 184
U.S. Industrial Alcohol Company (USIAC), 46
U.S. Sugar Corporation, 72–76, *75*, 79, 82
USDA. *See* U.S. Department of Agriculture
USIAC. *See* U.S. Industrial Alcohol Company

vacuum pan, 8, 48
Vaughan, Benjamin, 162
Vaughan, John, 87
Veillon, Louis, 182
Velva Breakfast Syrup, 128
Vermont, maple sugar and syrup in, 164–65

vertical integration, of glucose, 115–17
Vilmorin, Louis, 145
vinegar, 40
Vines, Richard, 31

Wagner, Theodore B., 134
Wallace, Henry C., 138
Walton, Charles F., 131
Ward, Ned, 34
Ware, Lewis Sharp, 92–93
Washington, George, 31–32, 35, 38, 163
Webb, Electra Havemeyer, 30
Weber, Henry Adam, 149
Webster, Daniel, 86
Webster, George, *6*
Weinstein, I. Bernard, 207
Western Beet Sugar Company, 96
West Indies: cutters from, 76, 77–78; New England colonies' trade with, 1, 31–32; plantations in, 3, 5, 48
West Side Sorghum Refinery, 146
Wheelock, Eleazer, 34
White, Joseph M., 86
Wiechers, William, 114, 133–34
Wiley, Harvey: beet sugar and, 92, 96–97, 98, 103; corn syrup and, 120; honey and, 174; as pure-food crusader, 41–43, 120, 124, 126, 129, 138, 185; saccharin and, 184, 185–86, 188; sorghum and, 150–54, 155–56; at USDA, 43, 68, 92, 120, 121, 124, 138, 165, 174, 184
Wilkinson, Alec, 78
Williams, Horace, 111–12
Wilson, James, 41, 42, 92, 124, 187
Wilson, Woodrow, 16, 97
Winthrop, John, 31, 34, 159
Woolsey, Edward, 9
Woolsey, George, 9
World War I, 103, 131, 136, 182, 189
World War II, 136, *137*, 184
Wray, Leonard, 145

Young, Brigham, 90

Zeckendorf, William, 65
Zerban, Frederick W., 20

ABOUT THE AUTHOR

Deborah Warner is a historian of science and technology, and has written about scientific instruments, celestial cartography, and women in science. As a curator of the Physical Sciences Collection in the Smithsonian's National Museum of American History, she appreciates the extent to which objects can inspire historical analysis. As a foodie with a sweet tooth, she is especially interested in objects that pertain to sweet stuff. The historic mills on St. John in the U.S. Virgin Islands are monuments to the connections between sugar, slavery, and the American past. Saccharimeters, optical instruments designed for sugar analysis that had a remarkably difficult time gaining congressional acceptance for tax purposes, indicate the importance of sugar to the nation's economy. The discussion of saccharin in the museum's "Science in American Life" exhibit strengthened her interest in artificial sweeteners, practical science, and the long-standing debates over the relative merits of pure and applied science.